**McGRAW-HILL
BOOK COMPANY**

New York
St. Louis
San Francisco
Auckland
Düsseldorf
Johannesburg
Kuala Lumpur
London
Mexico
Montreal
New Delhi
Panama
Paris
São Paulo
Singapore
Sydney
Tokyo
Toronto

WILLIAM L. JOLLY
*Professor of Chemistry
University of California, Berkeley*

The Principles
of Inorganic
Chemistry

This book was set in Times New Roman.
The editors were Robert H. Summersgill and Anne T. Vinnicombe;
the cover was designed by Scott Chelius;
the production supervisor was Leroy A. Young.
The drawings were done by Oxford Illustrators Limited.
Kingsport Press, Inc., was printer and binder.

Library of Congress Cataloging in Publication Data

Jolly, William L.
    The principles of inorganic chemistry.

    Bibliography: p.
    1. Chemistry, Inorganic. I. Title.
QD151.2.J64   546   75-9808
ISBN 0-07-032758-0

**THE PRINCIPLES
OF INORGANIC
CHEMISTRY**

1234567890KPKP798765

CONTENTS

PREFACE

In spite of the enormous variety of compounds and reactions encompassed by inorganic chemistry, chemists continually aim to rationalize the known information regarding inorganic compounds with a few simple principles. This book attempts to outline these principles and to describe rules which are valuable for correlation and prediction. I have tried to give due emphasis to the thermodynamic and kinetic (as well as the structural) aspects of chemistry. I have included topics because of their importance, not because of the thoroughness with which they are understood or the ease with which they are taught. For example, the book includes a section on metals and alloys even though this topic has not been systematized as well as many other topics in inorganic chemistry.

The book should be intelligible to students who have completed a one-year course in general chemistry and an introductory course in organic chemistry. Some elementary topics, such as atomic quantum theory and Lewis structures, are included for review and emphasis. Data from advanced experimental techniques such as photoelectron spectroscopy and Mössbauer spectroscopy are occasionally included, after very brief introductions to these techniques. In such cases, literature references for further background are provided. The elementary concepts of symmetry are employed in several chapters, but in such a general way that the student need not have any previous formal training in the subject. Appendix B briefly covers the formal aspects of symmetry and group theory.

Some of the textbooks available for use in an advanced lecture course in inorganic chemistry contain extensive descriptive material covering the chemistry of the elements. When teaching such a course, I have seldom assigned students to read this descriptive material; practically all of my reading assignments have been to the introductory or "theoretical" parts of textbooks. I believe the same sort of reading assignments are made by most other teachers. Of course, chemical principles cannot be taught without illustrative facts. But the factual material must be limited to a digestible amount and must be carefully selected to show exceptions, as well as conformity, to theoretical concepts.

I wish to thank my colleagues and students for their help, both direct and indirect, in the preparation of this book. I hope readers will not hesitate to send me their criticisms and suggestions for improvement.

WILLIAM L. JOLLY

1

ELECTRONIC CONFIGURATIONS OF ATOMS AND THE PERIODIC TABLE

Most of the chemical properties of a compound can be correlated with the electronic configurations of the constituent atoms. Thus it is important to know and to understand the significance of the electronic structure of atoms. In this chapter we shall discuss the basis for predicting electronic structures, and we shall show how the periodic table may be used to correlate the properties of atoms in terms of their electronic structures.

Although some of the material in the chapter is found in elementary chemistry books, it should not be assumed that the material is easily understood. It is included here, in very brief form, because of its fundamental importance in the study of inorganic chemistry and to permit the reader the opportunity to review.

QUANTUM NUMBERS

In 1926 Schrödinger[1] proposed a differential equation (which now bears his name) for relating the energy of a system to the space coordinates of its

[1] E. Schrödinger, *Ann. Phys.*, **79**, 361, 489; **81**, 109 (1926).

constituent particles. For a particle in three dimensions, the equation may be written

$$\frac{\partial^2 \psi}{\partial x^2} + \frac{\partial^2 \psi}{\partial y^2} + \frac{\partial^2 \psi}{\partial z^2} + \frac{8\pi^2 m}{h^2}(E - V)\psi = 0$$

where ψ is the "wave function," x, y, and z are the cartesian coordinates of the particle, m is its mass, E is the total energy, and V is the potential energy. This equation incorporates both the wave character of the particle and the probability character of the measurements. The wave function ψ has properties analogous to the amplitude of a wave; its square, ψ^2, is proportional to the probability of finding a particle at the coordinates x, y, z.

When the Schrödinger equation is applied to the hydrogen atom or to any system with one electron and one nucleus, the solution includes three "integration constants." These are the familiar quantum numbers n, l, and m_l. The "principal" quantum number n may have any integral value from 1 to infinity:

$$n = 1, 2, 3, \ldots$$

The "azimuthal" or "orbital angular momentum" quantum number l may have any integral value from zero to $n - 1$:

$$l = 0, 1, 2, \ldots (n - 1)$$

However, for historical reasons, l is usually not specified by these integers, but rather by the letters s, p, d, f, g, ... (continuing alphabetically), which correspond to $l = 0$, 1, 2, 3, 4, ..., respectively. The n and l values of an electron are often designated by the notation nl, in which the value of l is indicated

Table 1.1 SOME ALLOWED VALUES OF THE HYDROGEN ATOM QUANTUM NUMBERS

n	l	m_l	m_s	No. of combinations
1	0	0	$\pm\frac{1}{2}$	2
2	0	0	$\pm\frac{1}{2}$	2
2	1	$-1, 0, +1$	$\pm\frac{1}{2}$	6
3	0	0	$\pm\frac{1}{2}$	2
3	1	$-1, 0, +1$	$\pm\frac{1}{2}$	6
3	2	$-2, -1, 0, +1, +2$	$\pm\frac{1}{2}$	10
4	0	0	$\pm\frac{1}{2}$	2
4	1	$-1, 0, +1$	$\pm\frac{1}{2}$	6
4	2	$-2, -1, 0, +1, +2$	$\pm\frac{1}{2}$	10
4	3	$-3, -2, -1, 0, +1, +2, +3$	$\pm\frac{1}{2}$	14

by the appropriate letter. Thus a $2p$ electron has $n = 2$ and $l = 1$. The "magnetic" quantum number m_l may have any integral value from $-l$ to $+l$:

$$m_l = -l, -(l-1), \ldots, -1, 0, +1, +2, \ldots, l-1, +l$$

Because an electron has spin, and consequently a magnetic moment which can be oriented either up or down, yet a fourth quantum number must be specified, the "spin" quantum number m_s. The permissible values of m_s are $\pm\frac{1}{2}$.

As a consequence of the restrictions on the quantum numbers, the electron of a hydrogen atom may be assigned only certain combinations of quantum numbers. These permissible combinations for $n = 1, 2, 3$, and 4 are indicated in Table 1.1. Each allowed combination of n, l, and m_l corresponds to an atomic *orbital*. We say that the electron may be "put into" or "assigned to" a particular orbital. Of course, in any orbital, the m_s quantum number may be either $+\frac{1}{2}$ or $-\frac{1}{2}$.

ORBITAL SHAPES AND ENERGIES FOR THE HYDROGEN ATOM

It is convenient to express the wave function of the hydrogen atom in terms of the polar coordinates r, θ, and ϕ and to factor the function into three separate parts, each of which is a function of only one coordinate:

$$\psi(r, \theta, \phi) = R(r) \cdot \Theta(\theta) \cdot \Phi(\phi)$$

The values of the three functions, and hence also the spatial distribution of the electron of a hydrogen atom, are markedly affected by the values of n and l. The spatial distributions can be indicated graphically in several ways. Let us first consider the case of $l = 0$, that is, s electrons. In Fig. 1.1, the radial wave function R is plotted as a function of r, the distance from the nucleus, for $n = 1, 2$, and 3. Three facts should be noted. First, in each case, the magnitude of the wave function has its maximum value at the nucleus. Second, for $n > 1$, the wave function is zero in certain regions called nodes. (As a general rule, there are $n - 1$ nodes in an atomic wave function.) Third, the sign of R changes as it passes through a node.

Although R is a function which is not directly related to any experimentally measurable quantity, the function R^2 is proportional to electron density and therefore has considerable physical significance. In Fig. 1.2, R^2 is plotted versus r for $1s$, $2s$, and $3s$ electrons. Note that R^2, like R, has its maximum value at the nucleus and that nodes appear for $n > 1$. Of course, R^2 is never negative; a negative electron density is physically meaningless. One important feature of

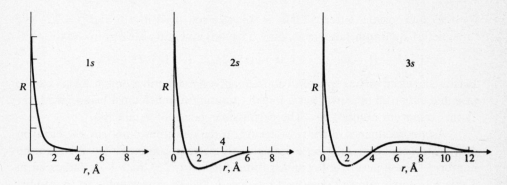

FIGURE 1.1
Plots of R versus r for 1s, 2s, and 3s orbitals of the hydrogen atom. The radius
scale is the same throughout, but the scale for R is changed for the various
orbitals.

any s-electron distribution that Figs. 1.1 and 1.2 do not make obvious is that
the distribution is independent of θ and ϕ; that is, it is spherically symmetric.
To show this feature in a graph or picture would be exceedingly difficult;
however, we have done the next best thing in Fig. 1.3, where we have plotted

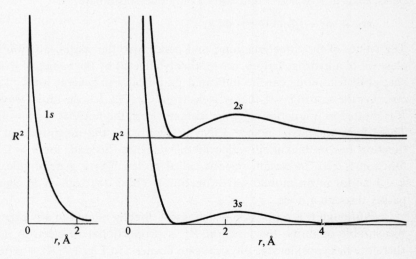

FIGURE 1.2
Plots of R^2 versus r for 1s, 2s, and 3s orbitals of the hydrogen atom. The radius
scale is the same throughout, but the density scale is changed for the various
orbitals.

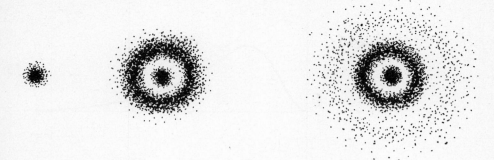

FIGURE 1.3
Electron density in a plane passing through the hydrogen atom for 1s, 2s, and 3s electrons.

electron density as a function of position on a plane passing through the nucleus for 1s, 2s, and 3s electrons. The electron density is indicated by the black dots; dark regions correspond to high electron density, and light regions correspond to low electron density. Figure 1.3 makes it easier to understand that the nodes in 2s and 3s wave functions are actually spherical *surfaces*.

Analogous plots can be drawn for 2p and 3p electrons. In Fig. 1.4, R is plotted versus r, and in Fig. 1.5, R^2 is plotted versus r. The angular distribution of a p electron is very different from that of an s electron. A nodal plane passes

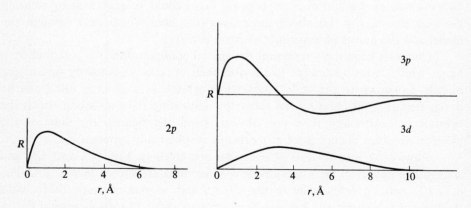

FIGURE 1.4
Plots of R versus r for 2p, 3p, and 3d orbitals of the hydrogen atom. The radius scale is the same throughout, but the scale for R is changed for the various orbitals.

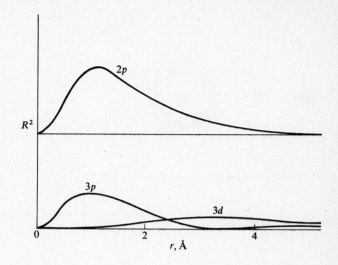

FIGURE 1.5
Plots of R^2 versus r for $2p$, $3p$, and $3d$ orbitals of the hydrogen atom. The density scale is different for the two plots.

through the nucleus, and the electron density is concentrated in lobes on both sides of the nodal plane. Electron density contour maps for $2p$ and $3p$ electrons are shown in Fig. 1.6. These maps show the distribution of electron density in planes which pass through the regions of maximum electron density and the nucleus. The p-electron distributions are cylindrically symmetric; a three-dimensional representation of the contour maps could be generated by rotating the diagrams in Fig. 1.6 about the z axes (the lines which pass through the nuclei and the points of maximum electron density).

Chemists commonly represent the spatial configuration of a p orbital by a figure which looks somewhat like a dumbbell or a sausage tightly tied at the middle, as shown in Fig. 1.7. The three p orbitals of a given n value can be designated as p_x, p_y, and p_z, the subscripts indicating the axes along which the orbitals lie. Although it is not obvious from the figures, the sum of the electron densities of such a set of p orbitals is spherically symmetric.

Plots of R and R^2 for $3d$ electrons are given in Figs. 1.4 and 1.5, and sausagelike figures of the five equivalent $3d$ orbitals are shown in Fig. 1.8. The d_{xy}, d_{xz}, and d_{yz} orbitals lie in the xy, xz, and yz planes, respectively, with the lobes directed midway between the cartesian axes. The $d_{x^2-y^2}$ orbital lies in the xy plane, with the lobes directed along the axes. The d_{z^2} orbital is a hybrid, or combination, of a $d_{z^2-x^2}$ and a $d_{z^2-y^2}$ orbital; it consists of two main lobes

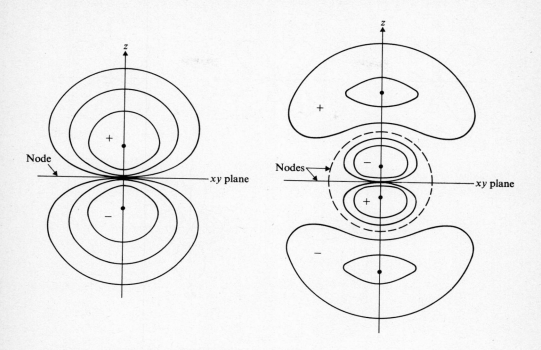

FIGURE 1.6
Electron density contours for 2p and 3p orbitals of the hydrogen atom.

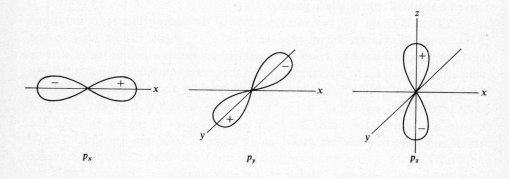

FIGURE 1.7
"Sausage" representations of p orbitals.

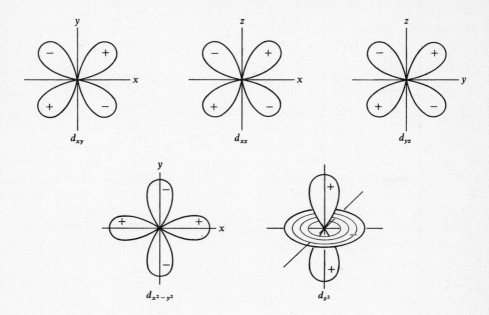

FIGURE 1.8
"Sausage" representations of five equivalent d orbitals.

directed along the z axis and a kind of torus or band lying in the xy plane. Note that each $3d$ orbital has two nodal surfaces which pass through the nucleus. The sum of the electron density distributions of the five d orbitals of a given n value is spherically symmetric.

In Figs. 1.7 and 1.8, the various lobes of the sausagelike representations of the p and d orbitals are marked with $+$ or $-$ signs. These refer to the sign of the original wave function ψ. Orbitals may be classified in terms of their symmetry with respect to inversion through the center (nucleus). When inversion (which simply involves changing the signs of the x, y, and z coordinates) causes no change in the sign of ψ, the orbital and its wave function are described as *gerade* (German for even); when inversion causes a change in sign, the orbital and wave function are described as *ungerade* (German for odd). Thus s and d orbitals are gerade, and p orbitals are ungerade. Later, when we consider the formation of molecular orbitals by the combination of atomic orbitals, we shall appreciate the importance of these symmetry designations and the signs of the wave functions.

If one solves the Schrödinger equation for the energy levels of the hydrogen atom, or of a one-electron ion, one obtains[1]

$$E = -\frac{2\pi^2 m Z^2 e^4}{n^2 h^2} = -13.60 \frac{Z^2}{n^2} \quad \text{eV}$$

where E is the energy relative to the infinitely separated electron and nucleus, m is the mass of the electron, Z is the charge on the nucleus, e is the electronic charge, and h is Planck's constant. It can be seen that the energy is independent of the values of the l and m_l quantum numbers but is dependent on the principal quantum number n. The absolute energy of the atom increases with increasing n, and the atom is in its lowest energy state (the ground state) when $n = 1$, that is, when the electron is in the $1s$ orbital. Because electrostatic interaction energy decreases with increasing distance, it is clear that the average distance of the electron from the nucleus increases with increasing n.

ELECTRONIC CONFIGURATIONS OF ATOMS AND IONS

The Schrödinger equation has never been solved in closed form except for the hydrogen atom and one-electron ions such as He^+, Li^{2+}, etc. However, spectroscopic data and accurate calculations based on successive approximations (such as the Hartree-Fock self-consistent field method) have shown that poly-electronic atoms possess atomic orbitals completely analogous to those of the hydrogen atom.[2] Thus we may speak of $1s$, $2p$, $4f$, etc., orbitals for any of the known elements. The Pauli exclusion principle states that in a given atom no two electrons can have the same values for all four quantum numbers. Consequently, each atomic orbital can contain no more than two electrons, of opposite spin. We can describe the building of a set of elements by a sequence of imaginary steps in each of which we simultaneously increase the nuclear charge by one unit and add one electron to an atomic orbital. If the electrons are added to the atomic orbitals in the proper order, the elements are generated in their ground states. In Table 1.2 we have listed, in order of atomic number, the elements and their electron configurations. By examination of Table 1.2, the reader can verify that, to a fair approximation, the order of filling atomic orbitals is the order of increasing $n + l$, and, in cases of ambiguity, the order of increasing n. This empirical

---

[1] In 1913 Niels Bohr derived this relation by combining classic physics with some of the ideas of quantum theory. See G. Herzberg, "Atomic Spectra and Atomic Structure," 2d ed., pp. 13–21, Dover, New York, 1944.

[2] H. F. Schaefer, "The Electronic Structure of Atoms and Molecules," Addison-Wesley, Reading, Mass., 1972.

Table 1.2 ELECTRON CONFIGURATIONS OF THE ELEMENTS

Z	Element	Electron configuration†	Z	Element	Electron configuration
1	H	$1s$	53	I	$[Kr]4d^{10}5s^25p^5$
2	He	$1s^2$	54	Xe	$[Kr]4d^{10}5s^25p^6$
3	Li	$[He]2s$	55	Cs	$[Xe]6s$
4	Be	$[He]2s^2$	56	Ba	$[Xe]6s^2$
5	B	$[He]2s^22p$	57	La	$*[Xe]5d6s^2$
6	C	$[He]2s^22p^2$	58	Ce	$*[Xe]4f5d6s^2$
7	N	$[He]2s^22p^3$	59	Pr	$[Xe]4f^36s^2$
8	O	$[He]2s^22p^4$	60	Nd	$[Xe]4f^46s^2$
9	F	$[He]2s^22p^5$	61	Pm	$[Xe]4f^56s^2$
10	Ne	$[He]2s^22p^6$	62	Sm	$[Xe]4f^66s^2$
11	Na	$[Ne]3s$	63	Eu	$[Xe]4f^76s^2$
12	Mg	$[Ne]3s^2$	64	Gd	$*[Xe]4f^75d6s^2$
13	Al	$[Ne]3s^23p$	65	Tb	$[Xe]4f^96s^2$
14	Si	$[Ne]3s^23p^2$	66	Dy	$[Xe]4f^{10}6s^2$
15	P	$[Ne]3s^23p^3$	67	Ho	$[Xe]4f^{11}6s^2$
16	S	$[Ne]3s^23p^4$	68	Er	$[Xe]4f^{12}6s^2$
17	Cl	$[Ne]3s^23p^5$	69	Tm	$[Xe]4f^{13}6s^2$
18	Ar	$[Ne]3s^23p^6$	70	Yb	$[Xe]4f^{14}6s^2$
19	K	$[Ar]4s$	71	Lu	$[Xe]4f^{14}5d6s^2$
20	Ca	$[Ar]4s^2$	72	Hf	$[Xe]4f^{14}5d^26s^2$
21	Sc	$[Ar]3d4s^2$	73	Ta	$[Xe]4f^{14}5d^36s^2$
22	Ti	$[Ar]3d^24s^2$	74	W	$[Xe]4f^{14}5d^46s^2$
23	V	$[Ar]3d^34s^2$	75	Re	$[Xe]4f^{14}5d^56s^2$
24	Cr	$*[Ar]3d^54s$	76	Os	$[Xe]4f^{14}5d^66s^2$
25	Mn	$[Ar]3d^54s^2$	77	Ir	$[Xe]4f^{14}5d^76s^2$
26	Fe	$[Ar]3d^64s^2$	78	Pt	$*[Xe]4f^{14}5d^96s$
27	Co	$[Ar]3d^74s^2$	79	Au	$*[Xe]4f^{14}5d^{10}6s$
28	Ni	$[Ar]3d^84s^2$	80	Hg	$[Xe]4f^{14}5d^{10}6s^2$
29	Cu	$*[Ar]3d^{10}4s$	81	Tl	$[Xe]4f^{14}5d^{10}6s^26p$
30	Zn	$[Ar]3d^{10}4s^2$	82	Pb	$[Xe]4f^{14}5d^{10}6s^26p^2$
31	Ga	$[Ar]3d^{10}4s^24p$	83	Bi	$[Xe]4f^{14}5d^{10}6s^26p^3$
32	Ge	$[Ar]3d^{10}4s^24p^2$	84	Po	$[Xe]4f^{14}5d^{10}6s^26p^4$
33	As	$[Ar]3d^{10}4s^24p^3$	85	At	$[Xe]4f^{14}5d^{10}6s^26p^5$
34	Se	$[Ar]3d^{10}4s^24p^4$	86	Rn	$[Xe]4f^{14}5d^{10}6s^26p^6$
35	Br	$[Ar]3d^{10}4s^24p^5$	87	Fr	$[Rn]7s$
36	Kr	$[Ar]3d^{10}4s^24p^6$	88	Ra	$[Rn]7s^2$
37	Rb	$[Kr]5s$	89	Ac	$*[Rn]6d7s^2$
38	Sr	$[Kr]5s^2$	90	Th	$*[Rn]6d^27s^2$
39	Y	$[Kr]4d5s^2$	91	Pa	$*[Rn]5f^26d7s^2$
40	Zr	$[Kr]4s^25s^2$	92	U	$*[Rn]5f^36d7s^2$
41	Nb	$*[Kr]4d^45s$	93	Np	$*[Rn]5f^46d7s^2$
42	Mo	$*[Kr]4d^55s$	94	Pu	$[Rn]5f^67s^2$
43	Tc	$[Kr]4d^55s^2$	95	Am	$[Rn]5f^77s^2$
44	Ru	$*[Kr]4d^75s$	96	Cm	$*[Rn]5f^76d7s^2$
45	Rh	$*[Kr]4d^85s$	97	Bk	$[Rn]5f^97s^2$
46	Pd	$*[Kr]4d^{10}$	98	Cf	$[Rn]5f^{10}7s^2$
47	Ag	$*[Kr]4d^{10}5s$	99	Es	$[Rn]5f^{11}7s^2$
48	Cd	$[Kr]4d^{10}5s^2$	100	Fm	$[Rn]5f^{12}7s^2$
49	In	$[Kr]4d^{10}5s^25p$	101	Md	$[Rn]5f^{13}7s^2$
50	Sn	$[Kr]4d^{10}5s^25p^2$	102	No	$[Rn]5f^{14}7s^2$
51	Sb	$[Kr]4d^{10}5s^25p^3$	103	Lr	$[Rn]5f^{14}6d7s^2$
52	Te	$[Kr]4d^{10}5s^25p^4$			

† Asterisks indicate exceptions to the rule for adding electrons.

rule is a useful mnemonic for the electron confiigurations of the elements. There are exceptions to this simple rule for adding electrons; these nonconforming atoms are indicated by asterisks in Table 1.2. (For example, Cu has a $3d^{10}4s$ configuration instead of the predicted $3d^94s^2$.) However, the exceptions correspond to minor perturbations and are relatively unimportant in correlating atomic properties and in understanding the periodic table.

We have already noted that the energy of a hydrogen atom is essentially independent of the l quantum number of the single electron. Therefore we might have expected, in building up the elements, that it would make no difference whether an electron was added to an empty $3s$ orbital, $3p$ orbital, or $3d$ orbital. The fact that it does make a difference and that the energy of an electron in a polyelectronic atom increases with increasing l as well as increasing n is a consequence of the fact that the shapes of the various orbitals are different. The lower the l value for a given n value, the more effectively the electron shields the nuclear charge for other electrons of the same n value. Thus an s electron (which has a high electron density in the region of the nucleus) more effectively shields the nucleus from a p electron (which has essentially no electron density at the nucleus) than vice versa. Consequently, both the effective nuclear charge felt by an electron and its binding energy decrease in the order $s, p, d, f, \ldots$.

The symbolic representations of electronic configurations in Table 1.2 give no details regarding the filling of orbitals in partly filled shells of p, d, and f electrons. For example, consider the carbon atom, which has two $2p$ electrons. In this case, if we do not distinguish between the three $2p$ orbitals, the orbitals can be occupied in three different ways:

According to a rule propounded by Hund, the ground state of a set of electrons of given n and l values corresponds to an orbital occupancy such that the multiplicity is maximized, i.e., the electrons are unpaired as far as possible. Application of this rule tells us that the first of the three indicated configurations is the ground state and that the other configurations are excited states. The fact that the third configuration is not the ground state is easy to rationalize: When two electrons are in the same orbital, the electrostatic repulsion between them is greater than when they are in separate orbitals. The fact that the first configuration is more stable than the second is more difficult to rationalize; it is a consequence of so-called exchange energy[1] which is lost when two electrons of the same spin are forced to have their spins opposed.

[1] See, for discussion, L. E. Orgel, "An Introduction to Transition-Metal Chemistry; Ligand Field Theory," 2d ed., pp. 42–45, Wiley, New York, 1966.

One might suppose that, in the successive removal of electrons from an atom to form a series of stable positive ions, the electrons would leave the atomic orbitals in the reverse order in which they were filled in the hypothetical buildup of the elements. To some extent this is true; however, there are many exceptions. For example, the Ti^{2+} ion has the configuration $1s^2 2s^2 2p^6 3s^2 3p^6 3d^2$, whereas the Ca atom (which has the same number of electrons) has the configuration $1s^2 2s^2 2p^6 3s^2 3p^6 4s^2$. Exceptions of this type are not surprising when one recognizes that the nuclear charge of titanium is two units more positive than that of calcium and that the process of removing electrons from an atom is not the same as removing electrons while simultaneously removing an equal number of protons from the nucleus. In the case of Ti^{2+}, all the atomic orbitals are smaller and closer to the nucleus than the corresponding orbitals of Ca, but the difference is more marked for d orbitals than for s and p orbitals. This effect is so great that in Ti^{2+} the $3d$ electrons are not shielded from the nucleus as well as $4s$ electrons would be.

As a general rule, ionization of an atom causes the relative energies of the orbitals to approach those of a hydrogenlike atom; that is, the orbital energies become relatively independent of l and mainly dependent on n. Usually the easiest electrons to remove from an atom or an ion are those with the maximum value of n; of this set of electrons, the easiest to remove are those with the maximum value of l.

The valence electrons of an atom are those which are relatively easily removed. Generally, removal of all the valence electrons of an atom yields an ion with an outer electronic configuration of the type $ns^2 np^6$ or $ns^2 np^6 nd^{10}$. Such ions are called atomic cores or kernels.

THE PERIODIC TABLE

A periodic table is a listing of the elements in an array such that atomic number increases from left to right in any row and the elements in any column have outer electronic structures which are similar except for principal quantum number. A common form of the periodic table is shown in Fig. 1.9. Such a table is a very useful device for recalling groups of related atoms and for correlating properties with systematic changes in electronic structure. Every inorganic chemist should have a periodic table displayed near his or her desk or workbench. Even a chemist who has memorized the periodic table will find that it often helps to look at the table when thinking about inorganic problems.

The elements may be classified in terms of their valence electrons. Those whose atoms contain d electrons in the valence shell are called "transition

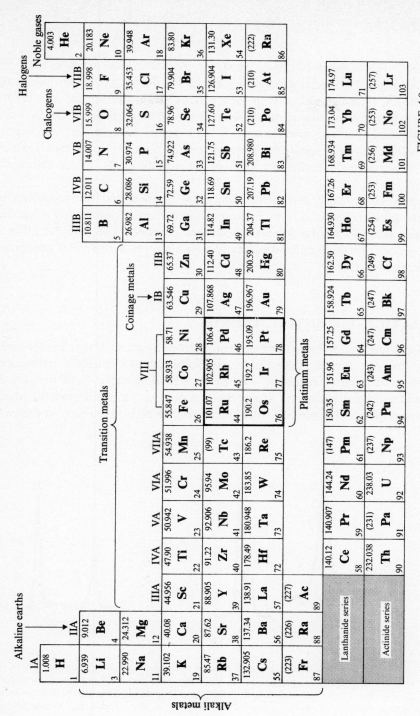

FIGURE 1.9
Periodic table of the elements.

elements." These elements appear near the middle of the periodic table, as shown in Fig. 1.9. Special groups of transition elements, the lanthanides and the actinides, contain f electrons in their valence shells and appear in separate rows below the main body of the table. All the other elements of the periodic table are called "nontransition elements," "main-group elements," or "representative elements." Special names are given to certain "families" of elements which appear in particular columns or blocks of the periodic table. These names are indicated in Fig. 1.9.

The periodic table derives its name from the fact that the properties of the elements, when listed in order of atomic number, are periodic; that is, certain properties recur at regular intervals of atomic number. The periodic table is simply a device for listing the elements so that elements with similar properties are grouped together.

IONIZATION-POTENTIAL TRENDS

The energy required to remove an electron from an atom, such that the removed electron has zero kinetic energy and is infinitely distant from the resulting ion, is called an "ionization potential" or "electron binding energy." In the case of a neutral hydrogen or helium atom, there is only one kind of electron which can be removed, and hence only one ionization potential. However, in the case

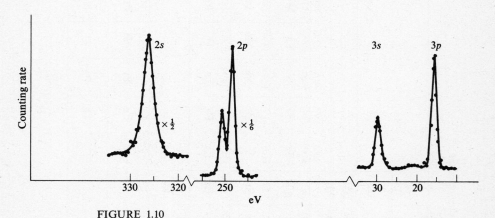

FIGURE 1.10
X-ray photoelectron spectrum of argon excited by Mg Kα X radiation. The X ray has insufficient energy to ionize the $1s$ electrons and is not sufficiently monochromatic to permit resolution of the $3p_{1/2}$ and $3p_{3/2}$ peaks. (*Reproduced with permission from K. Siegbahn et al., "ESCA Applied to Free Molecules," North-Holland, Amsterdam, 1969.*)

of a more complicated atom such as argon, there are separate ionization potentials corresponding to each occupied set of atomic orbitals. For example, in the case of argon, one can separately ionize a $1s$, $2s$, $2p$, $3s$, or $3p$ electron. In fact, in the case of the $2p$ or $3p$ electrons, there are two kinds of ionization processes (and corresponding ionization potentials). One can remove a p electron such that the spin magnetic moment of the remaining odd electron and the orbital magnetic moment of the set of five remaining p electrons are either aligned (yielding a $^2P_{3/2}$ ion) or opposed (yielding a $^2P_{1/2}$ ion). (See Appendix C for a discussion of the significance of term symbols such as $^2P_{3/2}$.) One method for measuring ionization potentials is photoelectron spectroscopy.[1] In this technique one irradiates a sample with monochromatic photons and measures the kinetic energy spectrum of the ejected photoelectrons. Electrons are observed at kinetic energies corresponding to the various electronic levels of the sample. It is then a simple matter to calculate the corresponding ionization potentials by using the Einstein relation

$$hv = I + E_k$$

where hv is the photon energy, I is the ionization potential of a given level, and E_k is the electron kinetic energy. An X-ray photoelectron spectrum of argon is shown in Fig. 1.10. The various ionization potentials of argon are listed in Table 1.3.

The lowest ionization potential of an atom (corresponding to ionization of the highest-energy occupied level and formation of the ground-state ion) is

---

[1] For a discussion of *X-ray* photoelectron spectroscopy, see J. M. Hollander and W. L. Jolly, *Acc. Chem. Res.*, **3**, 193 (1970). For a discussion of *ultraviolet* photoelectron spectroscopy, see A. D. Baker, *Acc. Chem. Res.*, **3**, 17 (1970).

Table 1.3 IONIZATION POTENTIALS OF ARGON†

Level	I, eV
$3p_{3/2}$	15.759
$3p_{1/2}$	15.937
$3s$	29.24
$2p_{3/2}$	248.52
$2p_{1/2}$	250.55
$2s$	326.3
$1s$	3205.9

† K. Siegbahn et al., "ESCA Applied to Free Molecules," North-Holland, Amsterdam, 1969.

often referred to simply as "the ionization potential" of the atom. The corresponding process may be represented by the following chemical equation:

$$M(g) \rightarrow M^+(g) + e^-(g)$$

To be precise, the energy of this process should be called the "first ionization potential" of the atom. The "second," "third," etc., ionization potentials correspond to the processes

$$M^+(g) \rightarrow M^{2+}(g) + e^-(g)$$
$$M^{2+}(g) \rightarrow M^{3+}(g) + e^-(g)$$
$$\text{(etc.)}$$

Such ionization potentials, corresponding to the formation of ground-state ions, are listed for the elements in Table 1.4. The ionization potentials vary in a fairly systematic way throughout the periodic table. From left to right in a given row (corresponding to the filling of a particular shell of electrons), the first ionization potential generally increases. This increase is due to the fact that the valence electrons do not efficiently shield one another from the nucleus. The ionization potential of a valence electron may be taken to be proportional to the expression $(Z - S)^2/n^2$, where Z is the atomic number, S is the screening constant, and n is the principal quantum number of the valence shell. The quantity $Z - S$ is often referred to as the effective nuclear charge. As a rough approximation,[1] we may assume that each ns or np electron in a particular valence shell contributes 0.35 to S, that each electron of principal quantum number $n - 1$ contributes 0.85 to S, and that all electrons of still lower principal quantum number contribute 1.00 to S. Thus, for a valence electron in aluminum $(1s^2 2s^2 2p^6 3s^2 3p)$, $S = 2 \times 1.00 + 8 \times 0.85 + 2 \times 0.35 = 9.5$, and $Z - S = 13 - 9.5 = 3.5$. As we move to the right, element by element, the quantity $Z - S$ gradually increases until it reaches a maximum value of $18 - 11.25 = 6.75$ at argon. By increasing the atomic number one more unit, we come to potassium $(1s^2 2s^2 2p^6 3s^2 3p^6 4s)$. The value of S for the valence electron in potassium is $2 \times 1.00 + 8 \times 1.00 + 8 \times 0.85 = 16.8$, and the effective nuclear charge is $19 - 16.8 = 2.2$. Thus this simple model correctly predicts that the effective nuclear charge (and hence the ionization potential) reaches a maximum at the noble gas argon. In the same way, we can rationalize the maxima in first ionization potentials observed for all the noble gases, which, except for helium, are characterized by valence shells containing complete sets of p electrons. Less pronounced maxima are observed for other elements whose valence electrons consist of complete shells. For example, the maxima at Be and Mg correspond to complete s shells, and the maxima at Zn and Cd correspond to

[1] J. C. Slater, *Phys. Rev.*, **36**, 57 (1930).

Table 1.4 IONIZATION POTENTIALS OF THE ELEMENTS (in electronvolts)‡‡

Z	Element	I	II	III	IV	V	VI	VII	VIII
1	H	13.598							
2	He	24.587	54.416						
3	Li	5.392	75.638	122.451					
4	Be	9.322	18.211	153.893	217.713				
5	B	8.298	25.154	37.930	259.368	340.217			
6	C	11.260	24.383	47.887	64.492	392.077	489.981		
7	N	14.534	29.601	47.448	77.472	97.888	552.057	667.029	
8	O	13.618	35.116	54.934	77.412	113.896	138.116	739.315	871.387
9	F	17.422	34.970	62.707	87.138	114.240	157.161	185.182	953.886
10	Ne	21.564	40.962	63.45	97.11	126.21	157.93	207.27	239.09
11	Na	5.139	47.286	71.64	98.91	138.39	172.15	208.47	264.18
12	Mg	7.646	15.035	80.143	109.24	141.26	186.50	224.94	265.90
13	Al	5.986	18.828	28.447	119.99	153.71	190.47	241.43	284.59
14	Si	8.151	16.345	33.492	45.141	166.77	205.05	246.52	303.17
15	P	10.486	19.725	30.18	51.37	65.023	220.43	263.22	309.41
16	S	10.360	23.33	34.83	47.30	72.68	88.049	280.93	328.23
17	Cl	12.967	23.81	39.61	53.46	67.8	97.03	114.193	348.28
18	Ar	15.759	27.629	40.74	59.81	75.02	91.007	124.319	143.456
19	K	4.341	31.625	45.72	60.91	82.66	100.0	117.56	154.86
20	Ca	6.113	11.871	50.908	67.10	84.41	108.78	127.7	147.24
21	Sc	6.54	12.80	24.76	73.47	91.66	111.1	138.0	158.7
22	Ti	6.82	13.58	27.491	43.266	99.22	119.36	140.8	168.5
23	V	6.74	14.65	29.310	46.707	65.23	128.12	150.17	173.7
24	Cr	6.766	16.50	30.96	49.1	69.3	90.56	161.1	184.7
25	Mn	7.435	15.640	33.667	51.2	72.4	95	119.27	196.46
26	Fe	7.870	16.18	30.651	54.8	75.0	99	125	151.06
27	Co	7.86	17.06	33.50	51.3	79.5	102	129	157
28	Ni	7.635	18.168	35.17	54.9	75.5	108	133	162
29	Cu	7.726	20.292	36.83	55.2	79.9	103	139	166
30	Zn	9.394	17.964	39.722	59.4	82.6	108	134	174
31	Ga	5.999	20.51	30.71	64				

(continued)

Table 1.4 (*continued*)

Z	Element	I	II	III	IV	V	VI	VII	VIII
32	Ge	7.899	15.934	34.22	45.71	93.5	127.6		
33	As	9.81	18.633	28.351	50.13	62.63	81.70		
34	Se	9.752	21.19	30.820	42.944	68.3	88.6	155.4	
35	Br	11.814	21.8	36	47.3	59.7	78.5	103.0	192.8
36	Kr	13.999	24.359	36.95	52.5	64.7	84.4	111.0	126
37	Rb	4.177	27.28	40	52.6	71.0	90.8	99.2	136
38	Sr	5.695	11.030	43.6	57	71.6	93.0	106	122.3
39	Y	6.38	12.24	20.52	61.8	77.0		116	129
40	Zr	6.84	13.13	22.99	34.34	81.5			
41	Nb	6.88	14.32	25.04	38.3	50.55	102.6	125	
42	Mo	7.099	16.15	27.16	46.4	61.2	68	126.8	153
43	Te	7.28	15.26	29.54					
44	Ru	7.37	16.76.	28.47					
45	Rh	7.46	18.08	31.06					
46	Pd	8.34	19.43	32.93					
47	Ag	7.576	21.49	34.83					
48	Cd	8.993	16.908	37.48					
49	In	5.786	18.869	28.03	54				
50	Sn	7.344	14.632	30.502	40.734	72.28	108		
51	Sb	8.641	16.53	25.3	44.2	56			
52	Te	9.009	18.6	27.96	37.41	58.75	70.7	137	
53	I	10.451	19.131	33					
54	Xe	12.130	21.21	32.1					
55	Cs	3.894	25.1						
56	Ba	5.212	10.004						
57	La	5.577	11.06	19.175					
58	Ce	5.47	10.85	20.20	36.72	57.45			
59	Pr	5.42	10.55	21.62	38.95				
60	Nd	5.49	10.72						
61	Pm	5.55	10.90						
62	Sm	5.63	11.07						
63	Eu	5.67	11.25						
64	Gd	6.14	12.1						
65	Tb	5.85	11.52						

Z	Element	I	II	III	IV	V	VI
66	Dy	5.93	11.67				
67	Ho	6.02	11.80				
68	Er	6.10	11.93				
69	Tm	6.18	12.05	23.71			
70	Yb	6.254	12.17	25.2			
71	Lu	5.426	13.9				
72	Hf	7.0	14.9	23.3	33.3		
73	Ta	7.89					
74	W	7.98					
75	Re	7.88					
76	Os	8.7					
77	Ir	9.1					
78	Pt	9.0	18.563				
79	Au	9.225	20.5				
80	Hg	10.437	18.756	34.2			
81	Tl	6.108	20.428	29.83			
82	Pb	7.416	15.032	31.937	42.32	68.8	88.3
83	Bi	7.289	16.69	25.56	45.3	56.0	
84	Po	8.48					
85	At						
86	Rn	10.748					
87	Fr						
88	Ra	5.279	10.147				
89	Ac	6.9	12.1				
90	Th	11.5	20.0	28.8			
91	Pa						
92	U						
93	Np						
94	Pu	5.8					
95	Am	6.0					

† C. E. Moore, "Ionization Potentials and Ionization Limits Derived from the Analyses of Optical Spectra," NSRDS-NBS 34, National Bureau of Standards, Washington, D.C., 1970.
‡ Italicized entries are estimates.

complete d shells. Even less pronounced, but distinct, maxima are observed for elements with half-filled valence p shells. Thus both nitrogen and phosphorus have first ionization potentials greater than those of the adjacent elements in the periodic table. The latter maxima occur because the next element in the periodic table has an np^4 valence-shell configuration, in which two electrons are paired in the same atomic orbital, with consequent loss of exchange energy and a relatively great increase in electronic repulsive energy.

The stability of completely filled shells of p, d, and f electrons is very important in controlling the chemistry of most of the elements in the periodic table. As a general rule, atoms tend to acquire filled electronic configurations of these types. For example, the $2s^2 2p^6$ configuration is achieved in C^{4-}, N^{3-}, O^{2-}, F^-, Ne, Na^+, Mg^{2+}, and Al^{3+}. The $3s^2 3p^6 3d^{10}$ configuration is achieved in Cu^+, Zn^{2+}, and Ga^{3+}. The $5s^2 5p^6 5d^{10} 6s^2$ configuration is achieved in Tl^+, Pb^{2+}, and Bi^{3+}. The $4s^2 4p^6 4d^{10} 4f^{14} 5s^2 5p^6$ configuration (with a just-completed $4f$ shell) is achieved in Yb^{2+}, Lu^{3+}, Hf^{4+}, and Ta^{5+}. A lesser tendency for atoms to achieve *half*-filled shells is shown by Eu^{2+}, Gd^{3+}, and Tb^{4+}, all of which have the $4s^2 4p^6 4d^{10} 4f^7 5s^2 5p^6$ configuration (with a half-filled $4f$ shell).

It should be noted that the ionization potential of a negative species such as F^- is seldom referred to as such, but rather as the "electron affinity" of the corresponding neutral species F. Thus, the electron affinity of the fluorine atom is equivalent to the "zeroth" ionization potential of fluorine. The periodic-table trends of electron affinities are similar to those of ionization potentials.

ATOMIC-SIZE TRENDS

Another atomic property which varies systematically throughout the periodic table is atomic size. Unfortunately, atomic size, unlike ionization potential, is an ill-defined quantity. We shall later discuss several methods for defining and measuring effective radii for atoms and ions, but at this point it will suffice to discuss qualitative trends in atomic size. When the principal quantum number of the valence electrons is increased while the valence configuration is kept constant, atomic size increases. Thus the bromine atom is bigger than the chlorine atom. This effect is due to the fact that an increase in principal quantum number corresponds to a decrease in electron binding energy and an increased average distance from the nucleus. Thus atomic size increases as one descends any column of the periodic table. On the other hand, when atomic number is increased by increasing the number of valence electrons in a given valence shell, atomic size decreases. This effect is a consequence of the increase in the effective nuclear charge and the contraction of the electron cloud. Thus atomic size decreases from left to right in any row of the periodic table, as long as the n or l values of the electrons added are not changed.

The contraction in size which occurs in the first series of transition metals (scandium through zinc) has interesting consequences in the succeeding elements, gallium, germanium, etc. The contraction causes gallium and germanium to be almost the same size as aluminum and silicon, respectively. Therefore, because chemical properties are highly dependent on atomic size, the chemical properties of aluminum and gallium are very similar, as are those of silicon and germanium. Similarly, the so-called lanthanide contraction which occurs from La to Lu, corresponding to the filling of the $4f$ shell, has important consequences in the series from Hf to Pt. The elements of the latter series are remarkably similar to the corresponding members of the second transition series, from Zr to Pd. Thus zirconium and hafnium are so similar that it is very difficult to separate these elements from each other. The block of six elements Ru, Rh, Pd, Os, Ir and Pt have many chemical similarities and are called the "platinum metals." Beyond platinum, however, the effect of the lanthanide contraction markedly peters out; thus cadmium is more like zinc than like mercury, and tin is more like germanium than like lead.

PROBLEMS

(Problems 1.4 to 1.15 should be answered without help of any sort.)

1.1 The probability of finding the electron of a hydrogen atom at a distance r from the proton is at a maximum for $r = 0.529$ Å. Does this statement contradict the fact that the electron density is greatest at the proton? Explain.

1.2 Describe the nodal surfaces of a $3p$ orbital.

1.3 Calculate the ninth ionization potential of fluorine.

1.4 How many unpaired electrons do each of the following atoms or ions have in their ground states: Al, S, Sc^{3+}, Cr^{3+}, Ir^{3+}, Dy^{3+}?

1.5 What $+2$ ion has just six $4d$ electrons in its ground state?

1.6 Write out the electronic configuration of arsenic.

1.7 On the basis of analogy with the lanthanides, which two actinides would you expect to show a stable $+2$ oxidation state?

1.8 List, in order of atomic numbers, the symbols of the elements between calcium and gallium.

1.9 Which element has the following electronic configuration: $1s^2 2s^2 2p^6 3s^2 3p^6 3d^{10} 4s^2 4p^6 4d^4 5s^1$?

1.10 Name the two elements whose chemical properties are very similar to those of chromium.

1.11 Which element has atomic number 25?

1.12   Which of the following gas-phase reactions can proceed spontaneously?

$$Kr + He^+ \rightarrow Kr^+ + He$$
$$Si + Cl^+ \rightarrow Si^+ + Cl$$
$$Cl^- + I \rightarrow I^- + Cl.$$

1.13   Why does potassium have a lower first ionization potential than lithium?

1.14   Why does copper have an outer electronic configuration $3d^{10}4s^1$ instead of $3d^9 4s^2$?

1.15   Arrange the ions of each group in order of increasing size:

a Y^{3+}, Ba^{2+}, Al^{3+}, Co^{3+}, Cs^+, La^{3+}, Ir^{3+}, Fe^{3+}

b Cl^-, I^-, Te^{2-}, Ar^+

1.16   Argue for the inclusion of B and Al in group IIIA, that is, in the same family with Sc, Y, La, and Ac.

COVALENT BONDING

THE LEWIS OCTET THEORY

The Lewis octet theory[1] is extremely useful for describing the bonding in practically all compounds of the nontransition elements. The theory is based on the stability of valence shells containing eight electrons in the s and p orbitals (s^2p^6). According to the theory, each atom in a compound achieves an octet of valence electrons by sharing electrons with the other atoms to which it is bonded. Of course, hydrogen atoms are excepted; their complete valence shells contain only two electrons.

Lewis octet structures for molecules can be represented in several ways. For example, the structure of ammonia can be written in the following ways:

$$
\begin{array}{ccc}
\text{H} & \text{H} & \text{H} \\
\text{:N:H} & \text{:N—H} & \text{N—H} \\
\text{H} & \text{H} & \text{H} \\
\text{I} & \text{II} & \text{III}
\end{array}
$$

---

[1] The modern concept of the Lewis octet theory has undergone a significant evolution from that originally outlined by G. N. Lewis in "Valence and the Structure of Atoms and Molecules," Chemical Catalog, New York, 1923.

The first way, in which all the valence electrons are indicated by dots, is somewhat cumbersome and is usually abandoned in the freshman chemistry course. The second and third ways are commonly used by practicing chemists; the choice between these methods is determined by whether or not one wishes to emphasize the presence of nonbonding electron pairs.

In writing octet structures, it is common practice to indicate the charges which atoms would have if each pair of bonding electrons were equally divided between the bonded atoms. These charges are called formal charges and should not be confused with the actual charges on the atoms. In NH_3, none of the atoms has a formal charge. However, many other molecules, and of course all ions, have formally charged atoms. The structures of a few of these are shown below.

$$
\begin{array}{cccc}
\overset{\displaystyle H}{\underset{\displaystyle H}{H-\overset{+}{N}}} - \overset{\displaystyle H}{\underset{\displaystyle H}{\overset{-}{B}-H}}
&
\overset{\displaystyle H}{:\overset{+}{O}-H}
&
\overset{\displaystyle O^-}{\underset{\displaystyle O^-}{}}
\end{array}
$$

$$ H-\overset{\overset{\textstyle H}{|}}{\underset{\underset{\textstyle H}{|}}{\overset{+}{N}}} - \overset{\overset{\textstyle H}{|}}{\underset{\underset{\textstyle H}{|}}{\overset{-}{B}}} -H \qquad :\overset{\overset{\textstyle H}{|}}{\underset{\underset{\textstyle H}{|}}{\overset{+}{O}}}-H \qquad ^-O-\overset{\overset{\textstyle O^-}{|}}{\underset{\underset{\textstyle O^-}{|}}{\overset{3+}{Cl}}}-O^- $$

All chemists should be proficient in the assignment of formal charges to the atoms of a Lewis octet structure. One simple method for making this assignment is based on the "total bond order" of an atom, i.e., the sum of the bond orders of the bonds to an atom. The total bond order has an easily remembered characteristic value for a given atom with zero formal charge. Thus it is 1 for a halogen atom, 2 for an atom of the oxygen family, 3 for an atom of the nitrogen family, etc. When the total bond order is 1 greater than the characteristic value, the formal charge is $+1$; when it is 2 greater, the formal charge is $+2$, etc. Similarly, when the total bond order is 1 less than the characteristic value, the formal charge is -1, etc. These rules[1] are summarized in Table 2.1.

Proficiency in the assignment of the formal charges can be very helpful in writing Lewis structures. For example, if the structure of nitrosyl chloride were mistakenly written

$$
\begin{array}{c}
N \\
\diagup \quad \diagdown \\
O \qquad Cl
\end{array}
$$

the structure would immediately be recognized as wrong, because the oxygen and nitrogen atoms would each have -1 formal charges, corresponding to a -2

---

[1] There is some difficulty in extending the rules to elements to the left of group IV. Thus it is impossible for boron to have both a complete octet and a zero formal charge.

ion. By introducing the N=O double bond,

a satisfactory structure for the neutral molecule is achieved.

Sometimes more than one structure can be written for a species with multiple bonds. In such a case the multiple bonding is distributed throughout the molecule, a situation often referred to as resonance[1] or delocalized π bonding. (A detailed discussion of π bonding is presented in Chap. 3.) Thus for ozone we may write either

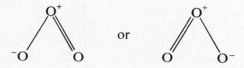

Neither structure is a satisfactory representation of the bonding because the two terminal oxygen atoms in ozone are known to be equivalent. The actual bonding is midway between these extreme "resonance structures." We may represent the actual molecule, which is a "resonance hybrid," by the structure

[1] L. Pauling, "The Nature of the Chemical Bond," 3d ed., Cornell University Press, Ithaca. N.Y., 1960.

Table 2.1 FORMAL CHARGES ON ATOMS WITH COMPLETE OCTETS

	Periodic table group					
Total bond order	III	IV	V	VI	VII	VIII
0	$\cdots$	$\cdots$	-3	-2	-1	0
1	$\cdots$	-3	-2	-1	0	$+1$
2	$\cdots$	-2	-1	0	$+1$	$+2$
3	-2	-1	0	$+1$	$+2$	$+3$
4	-1	0	$+1$	$+2$	$+3$	$+4$

Here each bond has a bond order of $1\frac{1}{2}$, and each terminal oxygen atom has a formal charge of $-\frac{1}{2}$, corresponding to the average values of these quantities in the two resonance structures. Similarly, we may represent the bonding in the nitrate ion by the hybrid structure

where each bond has a bond order of $1\frac{1}{3}$.

The two examples of resonance which we have just discussed involve resonance structures which are equivalent, i.e., structures which are indistinguishable upon the execution of a symmetry operation such as rotation about an axis of symmetry or reflection through a mirror plane. However, sometimes nonequivalent resonance structures can be written, as in the case of the cyanate ion:

$$^{-}O\!-\!C\!\equiv\!N \qquad O\!=\!C\!=\!N^{-} \qquad {}^{+}O\!\equiv\!C\!-\!N^{2-}$$

In such cases the best representation of the bonding is usually not just a simple average of the individual resonance structures but rather some weighted average of these structures. The third cyanate structure, with a negative formal charge on the nitrogen atom and a positive formal charge on the oxygen atom, has a charge distribution which contradicts the fact that oxygen is more electronegative (electron-attracting) than nitrogen. Therefore it is reasonable to suppose that the actual bonding is a resonance hybrid mainly of the first two structures, with negligible contribution from the third structure.

Let us consider a more complicated example, the $S_3N_2Cl^{+}$ ion, which is found in the salt $(S_3N_2Cl)Cl$.[1]

[1] A Zalkin, T. E. Hopkins, and D. H. Templeton, *Inorg. Chem.*, **5**, 1767 (1966).

Of the six different resonance structures, structures I to III are very unimportant because they involve adjacent atoms with formal charges of the same sign — an electrostatically unfavorable situation. Structure IV is the most favorable in terms of the electrostatic interaction of the formal charges. Indeed, the measured bond distances, given in the following diagram, are in accord with the main contribution from the structure IV. The shortest S—N bond corresponds to the

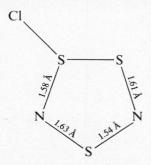

double bond of resonance structure IV.

A special kind of resonance, called "no-bond" resonance or "hyperconjugation," is occasionally employed to rationalize extraordinary bonding features. This special resonance involves an increase in the order of one bond at the expense of the order of another bond:

$$X-Y-Z \leftrightarrow {}^{+}X{=}Y \quad Z^{-}$$

For example, in ONF_3, the N—F bonds appear to be abnormally long, as determined by vibrational spectroscopy.[1] Resonance of the following type can explain this "abnormality."

$$
{}^{-}O{-}\overset{+}{N}{-}F \leftrightarrow O{=}\overset{+}{N} \quad F^{-}
$$

Similarly, the unexpectedly strong acid-base interaction between BH_3 and CO has been explained in terms of hyperconjugation of the following type:[2]

$$
H{-}\overset{-}{B}{-}C{\equiv}O^{+} \leftrightarrow {}^{+}H \quad B{=}C{=}O
$$

Although many structural rationalizations based on hyperconjugation are somewhat ad hoc, they are nevertheless reasonable insofar as the hyperconjugation

[1] C. Curtis, D. Pilipovich, and W. H. Moberly, *J. Chem. Phys.*, **46**, 2904 (1967).
[2] B. E. Douglas and D. H. McDaniel, "Concepts and Models of Inorganic Chemistry," pp. 213–214, Blaisdell, New York, 1965.

causes a change in formal charge distribution consistent with electronegativities. Thus in ONF_3 hyperconjugation causes a shift in negative formal charge from oxygen to fluorine, and in H_3BCO it causes a shift in positive formal charge from oxygen to hydrogen.

Some species have more valence electrons than can be accommodated by a Lewis octet structure involving only single bonds. For example, if we write single-bonded structures for I_3^- and SF_6, we find that the central atoms are required to have 10 and 12 valence electrons, respectively:

$$\left[\ddot{\ddot{I}} - \ddot{\ddot{I}} - \ddot{\ddot{I}} \colon \right]^- \qquad \underset{F \quad F \quad F}{\overset{F \quad F \quad F}{\diagdown \mid \diagup \atop S}}$$

Such species are "electron-rich," and the central atoms are referred to as "hypervalent." There are two common ways of handling this problem. One way is to invoke "no-bond" resonance structures or, what is equivalent, fractional bonds. Thus we may write the following Lewis structures for I_3^-:

$$I-I \quad I^- \qquad I^- \quad I-I$$

These correspond to a symmetric resonance hybrid which can be represented as follows:

$$^{-\frac{1}{2}}I \cdots I \cdots I^{-\frac{1}{2}}$$

In the latter structure each bond has a bond order of $\frac{1}{2}$. The molecule SF_6 may be similarly treated; the structure involves six $\frac{2}{3}$ bonds:

$$\begin{array}{ccc} & F^{-\frac{1}{3}} & \\ ^{-\frac{1}{3}}F & \mid & F^{-\frac{1}{3}} \\ & S^{2+} & \\ ^{-\frac{1}{3}}F & \mid & F^{-\frac{1}{3}} \\ & F^{-\frac{1}{3}} & \end{array}$$

Another way of handling hypervalent atoms is to invoke the use of valence-shell d orbitals. Then, instead of restricting atoms like iodine and sulfur to an s^2p^6 valence shell, we can permit electrons to be accepted by the empty d orbitals of the same principal quantum number. In principle, we can thereby accommodate hypervalency for any atom heavier than neon. Such d-orbital participation has been postulated even in species for which satisfactory Lewis octet structures can be written. Thus it has sometimes been assumed that the central atoms in $SO_4{}^{2-}$ and SiF_4 use $3d$ orbitals, and resonance structures of the following type have been written:

$$
\begin{array}{cc}
\text{O}^- & \text{F} \\
| & | \\
\text{O}=\text{S}=\text{O} & \text{F}-\text{Si}^-=\text{F}^+ \\
| & | \\
\text{O}^- & \text{F}
\end{array}
$$

However, the evidence for such valence-shell d-orbital participation in chemical bonding is relatively weak,[1] and in recent years the tendency has been to describe chemical bonding in terms of octet structures whenever possible.

Straightforward octet structures cannot be written for molecules with an odd number of electrons, such as NO. However, Linnett[2] devised a procedure in which the valence electrons of a molecule are split into two parts (corresponding to electrons of opposite spin) so that each atom achieves a *quartet* of each type of electron. By this "double quartet" theory, one indicates the separate quartets for NO as follows:

$$
{}^{\times}_{\times}\text{N}^{\times}_{\times}\text{O}^{\times}_{\times} \qquad {}_{\circ}\text{N}^{\circ}_{\circ}\text{O}_{\circ}
$$

The molecule can then be represented as follows:

$$
{}^{\circ\times}_{\circ\times}\text{N}^{\circ\times}_{\circ\times}\text{O}^{\times}_{\times}{}^{\circ} \qquad \text{or} \qquad {}^{\times}\!-\!\text{N}\!\equiv\!\text{O}\!-\!{}^{\times}
$$

The N—O bond order is 2.5, and, inasmuch as the molecule has one more electron of one spin than of the other, the molecule has a net spin of $\frac{1}{2}$. By a similar treatment of O_2, it is possible to rationalize the fact that this molecule has two unpaired electrons and a bond order of 2. For further applications of this method, the book by Linnett should be consulted.[2]

THE ISOELECTRONIC PRINCIPLE

Molecules that have the same number of electrons and the same number of heavy atoms (i.e., atoms heavier than hydrogen) usually have similar electronic structures, similar heavy-atom geometries, and similar chemical properties. This statement is the "isoelectronic principle," first discussed by Langmuir.[3] As

---

[1] T. B. Brill, *J. Chem. Educ.*, **50**, 392 (1973).
[2] J. W. Linnett, "The Electronic Structure of Molecules," Wiley, New York, 1964.
[3] I. Langmuir, *J. Am. Chem. Soc.*, **41**, 868, 1543 (1919).

an example of a set of isoelectronic species, the molecules and ions consisting of 3 heavy atoms and 22 electrons may be considered:

$H_2C=C=CH_2$
Allene

$H_2C=C=O$
Ketene

$HN=C=O$
Isocyanic acid

$O=C=O$
Carbon dioxide

$^-N=\overset{+}{N}=O$
Nitrous oxide

$^-N=\overset{+}{N}=N^-$
Azide ion

$H_2C=\overset{+}{N}=N^-$
Diazomethane

$O=\overset{+}{N}=O$
Nitryl ion

$F-C\equiv N$
Cyanogen fluoride

$H_3\bar{B}-C\equiv N$
Cyanoborohydride ion

$H_3C-C\equiv N$
Acetonitrile

$^-N=C=N^-$
Cyanamide ion

$H_3\bar{B}-C\equiv O^+$
Borane carbonyl

$O=\bar{B}=O$
Metaborate ion

The important common feature of these species is not that they have the same total number of electrons, but rather that they have the same number of *valence* electrons (in this case, 16). Indeed, the word isoelectronic is usually loosely interpreted as meaning *having the same number of valence electrons*. Thus we may include the following species in the preceding group of 3–heavy-atom, 16–valence-electron species:

$O=C=S$
Carbonyl sulfide

$S=C=S$
Carbon disulfide

$^-N=C=S$
Thiocyanate ion

$Br-C\equiv N$
Cyanogen bromide

etc.

One feature which all these species have in common is a linear heavy-atom skeleton. They also have similar values for their corresponding vibrational

Table 2.2 VIBRATIONAL FREQUENCIES FOR LINEAR TRI-ATOMIC SPECIES[†]

Molecule	Bond bending frequency, cm^{-1}	Symmetric stretching frequency, cm^{-1}	Asymmetric stretching frequency, cm^{-1}
NO_2^+	538	1400	2375
H_2CN_2	564	1170	2102
HNCO	572	1327	2274
H_2CCO	588	1120	2152
N_2O	589	1285	2223
BO_2^-	610	1070	1970
NCO^-	629	1205	2170
N_3^-	630	1348	2080
CO_2	667	1388	2349

[†] H. A. Bent, *J. Chem. Educ.*, **43**, 170 (1966).

frequencies, as shown by the data in Table 2.2. Many of the species undergo analogous reactions. Thus Lewis bases such as OH^-, H_2O, and NH_2^- react by attacking the middle atom:

$$H_2C=C=O \xrightarrow{\ OH^-\ } H_3C-C\overset{\displaystyle O^-}{\underset{\displaystyle O}{<}}$$

$$O=C=O \xrightarrow{\ OH^-\ } {}^-O-C\overset{\displaystyle OH}{\underset{\displaystyle O}{<}}$$

$$H_3\bar{B}-C\equiv O^+ \xrightarrow{\ OH^-\ } H_3\bar{B}-C\overset{\displaystyle OH}{\underset{\displaystyle O}{<}}$$

$$O=\overset{+}{N}=O \xrightarrow{\ OH^-\ } {}^-O-\overset{+}{N}\overset{\displaystyle OH}{\underset{\displaystyle O}{<}}$$

$$H_2N-C\equiv N \xrightarrow{\ H_2O\ } H_2N-\overset{\displaystyle O}{\overset{\|}{C}}-NH_2$$

$${}^-N=\overset{+}{N}=O \xrightarrow{\ NH_2^-\ } \left[{}^-O-\overset{+}{N}\overset{\displaystyle NH_2}{\underset{\displaystyle N^-}{<}} \right] \rightarrow \left[\bar{N}=\overset{+}{N}=\bar{N} \right]^- + H_2O$$

$$Br-C\equiv N \xrightarrow{\ 2OH^-\ } O=C=N^- + Br^- + H_2O$$

Many of the compounds, when irradiated with ultraviolet light, break into two fragments:

$$H_2CCO \xrightarrow{\ h\nu\ } CH_2 + CO$$
$$H_2CNN \xrightarrow{\ h\nu\ } CH_2 + N_2$$
$$HNNN \xrightarrow{\ h\nu\ } NH + N_2$$
$$OCO \xrightarrow{\ h\nu\ } O + CO$$
$$OCS \xrightarrow{\ h\nu\ } S + CO$$
etc.

The existence of isoelectronic analogs of an unknown compound has often served as the impetus for its first synthesis. For example, in 1971, the following isoelectronic compounds were known: $Ni(CO)_4$, $Co(CO)_3NO$, $Fe(CO)_2(NO)_2$, and $Mn(NO)_3CO$. The last member of this series, $Cr(NO)_4$, was unknown.

However, in 1972, several chemists[1] had sufficient faith in the isoelectronic principle to photolyze a solution of $Cr(CO)_6$ in the presence of NO, and thus they prepared the elusive compound.

For many years chemists unsuccessfully tried to prepare the perbromate ion BrO_4^-, which is analogous to the well-known perchlorate and periodate ions. The first successful synthesis of perbromate involved an isoelectronic species as the starting material. The synthesis involved the β decay of radioactive ^{83}Se incorporated in a selenate[2]:

$$^{83}SeO_4{}^{2-} \rightarrow {}^{83}BrO_4{}^- + \beta^-$$

The ^{83}Br is itself radioactive, and its coprecipitation with $RbClO_4$ was taken as evidence that it was in the form of BrO_4^-. Perbromate has since been prepared by more conventional methods.[3]

One must use reasonable caution in applying the isoelectronic principle. For example, because of the large difference in the electronegativities of fluorine and nitrogen, one should not expect FCN to be a close analog of CO_2. Indeed, the trimerization reaction of FCN has no known parallel with CO_2.

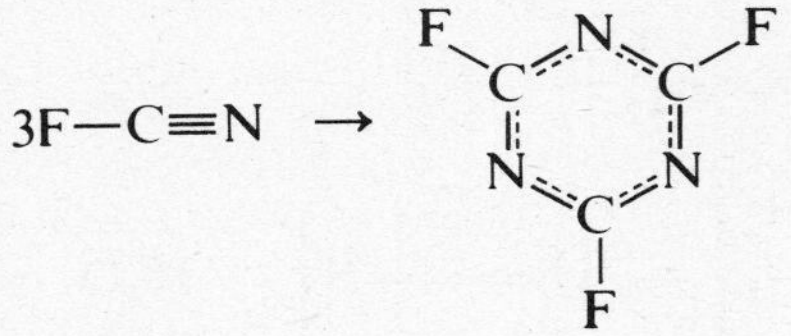

Similarly, borazine, $B_3N_3H_6$, is an imperfect analog of benzene, C_6H_6:

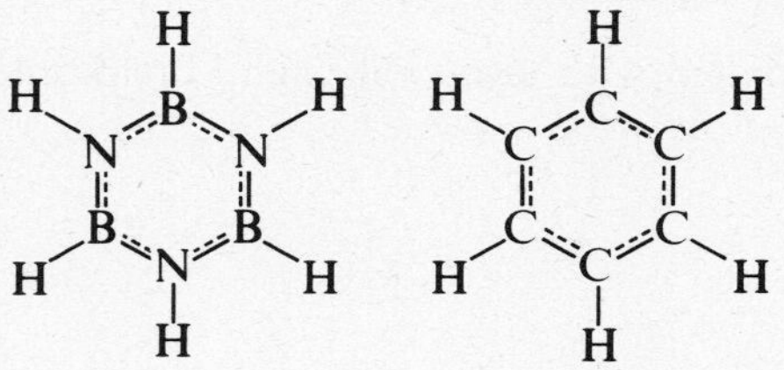

Borazine forms an adduct with 3 mol of HCl, suggesting protonation of the ring nitrogen atoms[4], whereas benzene is inert toward HCl under ordinary conditions. Although hexamethylborazine forms transition-metal complexes such as $B_3N_3(CH_3)_6Cr(CO)_3$ (analogous to complexes formed by the corresponding

[1] B. I. Swanson and S. K. Satija, *J. Chem. Soc. Chem. Commun.*, p. 40, 1973; M. Herberhold and A. Razavi, *Angew. Chem. Int. Ed.*, **11**, 1092 (1972).
[2] E. H. Appelman, *J. Am. Chem. Soc.*, **90**, 1900 (1968).
[3] E. H. Appelman, *Inorg. Syn.*, **13**, 1 (1972).
[4] E. Wiberg and A. Bolz, *Chem. Ber.*, **73**, 209 (1940).

benzene derivative), the B_3N_3 ring in these complexes is believed to be puckered, in contrast to the planar C_6 rings of the benzene derivative complexes.[1]

BOND DISTANCES

Interatomic distance are now known for many molecules with probable errors of ±0.03 Å or less.[2] One finds that the distance between two bonded atoms (say Si and Cl) is very nearly the same in different molecules, as long as the general nature of the bonding is the same in the various molecules. Some idea of the constancy of covalent bond distances can be obtained from Table 2.3, where the experimental Si—Cl bond distances for 10 different compounds are listed. The average value, 2.02 Å, is taken as the characteristic Si—Cl bond distance and may be used as an estimate of the Si—Cl bond distance in any other molecule containing such a bond. A brief listing of other characteristic covalent single-bond distances is given in Table 2.4.

The Si–Cl bond distances listed in Table 2.3 have practically the same value because the bonds not only have the same order but they also involve silicon atoms which have similar bonding geometries (tetrahedral) and similar hybrid

[1] J. L. Adcock and J. J. Lagowski, *Inorg. Chem.*, **12**, 2533 (1973).
[2] L. E. Sutton (ed.), "Tables of Interatomic Distances and Configuration in Molecules and Ions," Special Publ. 11, The Chemical Society, London, 1958; Supplement, Special Publ. 18, 1965.

Table 2.3 SILICON-CHLORINE BOND DISTANCES IN VARIOUS MOLECULES†

Compound	r_{Si-Cl}, Å
$SiClF_3$	2.00
$SiClH_3$	2.05
$SiCl_2H_2$	2.02
$SiCl_3H$	2.02
$SiCl_3SH$	2.02
$SiCl_4$	2.01
Si_2Cl_6	2.01
Si_2Cl_6O	2.02
Cl_3CSiCl_3	2.01
$C_6H_5SiCl_3$	2.00

† L. E. Sutton (ed.), "Tables of Interatomic Distances and Configuration in Molecules and Ions," Special Publ. 11, The Chemical Society, London, 1958; Supplement, Special Publ. 18, 1965.

Table 2.4 COVALENT SINGLE-BOND DISTANCES FOR THE NONMETALS, IN ANGSTROMS

Bracketed values are probably high because of lone pair repulsions or hyperconjugation, and parenthesized values are probably low because of multiple bonding.

	As	B	Br	C	Cl	F	Ge	H	I	N	O	P	S	Sb	Se	Si	Sn	Te
As	2.44		2.33	1.96	2.16	1.71		1.52	2.54		1.78		2.24			2.16	2.46	2.51
B		1.72	1.93	1.56	(1.74)	1.42		1.12		(1.42)	(1.36)							
Br			2.28	1.94	2.14	1.76	2.30	1.41	2.14	[2.14]			2.27	2.51				
C				1.54	1.77	1.35	1.94	1.09	2.32	1.47	1.43	1.84	1.82	2.20	1.98	1.87	2.15	
Cl					1.99	1.63	2.08	1.27		1.75	[1.70]	2.01	2.03	2.32		2.02	2.33	2.33
F						[1.42]	1.68	0.92		1.36	[1.42]	1.52	1.53			1.56		
Ge							2.44	1.53										
H								0.74	1.61	1.01	0.96	1.44	1.34	1.71	1.46	1.48	1.70	
I									2.67									
N										[1.45]	[1.41]		1.67					
O											[1.48]	1.66	1.5			1.63	2.05	
P												2.24	1.86		2.24			
S													2.05			2.15		
Sb														2.80				
Se															2.32			
Si																2.34		
Sn																	2.81	
Te																		2.86

bonding orbitals (probably very close to sp^3). (The concept of orbital hybridization will be discussed in Chap. 3.) A bond distance changes significantly upon changing the hybridization of one or both of the atomic orbitals involved in the bond.[1] This effect is quite obvious in the C—C bonds of hydrocarbons, as can be seen from the data in Table 2.5. As the fractional s character of the bonding orbitals increases from sp^3 to sp hybridization, the bond distance decreases. It can be shown that orbital overlap of carbon orbitals is at a maximum when they are sp-hybridized.[2] Indeed, it appears to be a good rule of thumb that bond distance decreases and bond strength increases as the bonding orbital hybridization changes from p to sp.

By use of the data in Table 2.4, it can be shown that, in most cases, the distance of a heteronuclear covalent bond, A—B, is approximately the arithmetic average of the distances of the corresponding homonuclear covalent bonds, A—A and B—B. Therefore it is possible to assign covalent radii to atoms and to use these to estimate unknown covalent bond distances. The covalent radii listed in Table 2.6 are, in most cases, simply one-half of the homonuclear single-bond distances. However, in some cases the values were chosen so as to give as good agreement as possible with the heteronuclear bond distances, especially with the bonds to carbon. Notice the correlation of covalent radius with position in the periodic table. The data bear out the trends discussed in Chap. 1.

[1] H. A. Bent, *Chem. Rev.*, **61**, 275 (1961).
[2] C. A. Coulson, "Valence," p. 199, Oxford University Press, New York, 1952.

Table 2.5 CARBON-CARBON SINGLE-BOND LENGTHS AND HYBRIDIZATIONS IN VARIOUS HYDROCARBONS

Bond type	Hybridization	r, Å
$\geq$C—C$\leq$	sp^3-sp^3	1.54
$\geq$C—C$\diagup$	sp^3-sp^2	1.51
$\geq$C—C≡	sp^3-sp	1.46
$\diagup$C—C$\diagup$	sp^2-sp^2	1.46
$\geq$C—C≡	sp^2-sp	1.44
≡C—C≡	sp-sp	1.37

As one might expect, bond distance decreases as bond order increases. This can be seen from the covalent radii for carbon, nitrogen, and oxygen in Table 2.7. This general correlation can be used to detect partial multiple-bond character, or delocalized π bonding, in molecules. For example, in SO_2 the S—O bond order is 1.5, and the observed bond distance, 1.43 Å, is 0.25 Å shorter than the sum of the single-bond covalent radii $(1.02 + 0.66 = 1.68$ Å$)$. In BF_3, the B—F bond order is 1.33, and the observed bond distance, 1.30 Å, is 0.14 Å shorter than the sum of the single-bond covalent radii $(0.86 + 0.58 = 1.44$ Å$)$. The correlation of N—O bond distance with bond order can be seen from the data in Table 2.8.

In SiF_4, the Si—F bond distance, 1.56 Å, is 0.19 Å shorter than the sum of the single-bond radii $(1.17 + 0.58 = 1.75$ Å$)$. The cause of this apparent bond shortening is a subject of some dispute. Schomaker and Stevenson[1] proposed that the shortening is a result of the relatively high ionic character of the bond. Others have proposed that the shortening of such bonds is a result of $p\pi \rightarrow d\pi$ bonding,[2] which is not feasible in the case of bonds involving only first-row atoms,

[1] V. Schomaker and D. P. Stevenson, *J. Am. Chem. Soc.*, **63**, 37 (1941).
[2] T. B. Brill, *J. Chem. Educ.*, **50**, 392 (1973).

Table 2.6 SINGLE-BOND COVALENT RADII, IN ANGSTROMS

				H 0.30
B 0.86	C 0.77	N 0.70	O 0.66	F 0.58
	Si 1.17	P 1.12	S 1.02	Cl 1.00
	Ge 1.22	As 1.22	Se 1.16	Br 1.14
	Sn 1.40	Sb 1.40	Te 1.43	I 1.34

Table 2.7 COVALENT RADII FOR CARBON, NITROGEN, AND OXYGEN

Atom	Single bond	Double bond	Triple bond
C	0.77	0.67	0.60
N	0.70	0.60	0.55
O	0.66	0.55	0.53

such as the C—F bond. Another possibility is that the Si—F bond distance is normal but that the C—F bond distance is extraordinarily long because of repulsion between the lone-pair electrons of the fluorine and the electrons of the other three bonds to the carbon atom. (We shall later have more to say about the repulsive interactions of nonbonding and bonding electrons.) Perhaps the apparent shortening of the Si—F bond is due to a combination of the effects that we have discussed. In any event, the phenomenon serves as an indication of the errors which one can make by incautious use of covalent atomic radii.

The covalent radius for hydrogen, which is consistent with the other covalent radii and the various X—H bond distances, is much less than one-half the H—H bond distance. Probably the H—H bond is abnormally long (and weak) because the 1s orbitals of the atoms cannot overlap adequately without undue proton-proton repulsion. On the other hand, when a hydrogen atom forms a bond to an atom which offers a highly directional bonding orbital (e.g., an orbital with appreciable p character), the hydrogen atom can *immerse itself* in the other atom's bonding orbital, and thus very strong orbital overlap can be achieved.

The covalent single-bond radii for fluorine, oxygen, and nitrogen in Table 2.6, which were calculated by subtracting the covalent radius of carbon from the C—F, C—O, and C—N bond distances, respectively, are considerably less than one-half the F—F, O—O, and N—N single-bond distances, respectively. It is

Table 2.8   NITROGEN-OXYGEN BOND DISTANCES

Compound	Formal bond order	Bond distance, Å
OṄ—OH, H₂Ṅ—OH	1	1.46
O₂N—OH	1	1.41
(CH₃)₃N⁺—O⁻	1	∼1.4
NO₃⁻ {N₂O₅	1.33	1.24
{NaNO₃	1.33	1.21
NO₂⁻	1.5	1.24
ClNO₂	1.5	1.24
CH₃NO₂	1.5	1.22
HONO₂ (nitro group)	1.5	1.22
N≡N⁺—O⁻ ↔ N⁻=N⁺=O	∼1.5	1.19
NO₂	1.75	1.19
ClN=O	2	1.14
NO₂⁺	2	1.15
NO	2.5	1.15
NO⁺	3	1.06

generally supposed that the latter distances are abnormally large because of strong repulsive interaction between the lone-pair electrons on adjacent atoms. Because the fluorine, oxygen, and nitrogen atoms are relatively small, lone pairs on adjacent atoms are close together and repel each other more strongly than they do in bonds such as Cl—Cl and S—S. The N—N bond distance in hydrazine $(H_2\ddot{N}$—$\ddot{N}H_2)$ is reduced from 1.47 to 1.40 Å by protonation of the two lone pairs and formation of the hydrazinium ion, $H_3\overset{+}{N}$—$\overset{+}{N}H_3$. The same sort of effect is observed on going from hydroxylamine, $H_2\ddot{N}$—$\ddot{O}H$, $(r_{N-O} = 1.46$ Å$)$ to trimethylamine oxide, $(CH_3)_3\overset{+}{N}$—$O^- (r_{N-O} = 1.36$ Å$)$.

It is rather interesting that the O—O bond distance in O_2F_2 is only 1.22 Å, almost as short as that in O_2 (1.21 Å).[1] This short bond has been explained in terms of hyperconjugation, which shifts nonbonding electrons from the oxygen atoms to the fluorine atoms.

$$F-O-O-F \quad \leftrightarrow \quad F^- \quad O=O^+-F \quad \leftrightarrow \quad F-O^+=O \quad F^-$$

As would be expected on the basis of such resonance, the O—F bond length is exceptionally long, 1.58 Å.

It should come as no surprise that bonds of order less than 1 are longer than single bonds. Thus in I_3^-, which has two bonds of order 0.5, the I—I bond distance is 2.91 Å, or 0.24 Å longer than that in I_2. The molecule PF_5 has a trigonal bipyramidal structure, with two axial fluorines, each 1.58 Å from the phosphorus, and three equatorial fluorines, each 1.53 Å from the phosphorus. These bond distances are slightly longer than the P—F bond distance in PF_3, 1.52 Å, and reflect the fact that the average bond order in PF_5 is 0.8, if we ignore the participation of phosphorus $3d$ orbitals in the bonding.

BOND STRENGTH

The most straightforward measure of the strength of the bond in a diatomic molecule is the dissociation energy, i.e., the energy required to break the molecule into its constituent atoms. A listing of dissociation energies for some important diatomic molecules is given in Table 2.9.[2] From these data it can be seen that, for isoelectronic molecules, the heavier analogs generally have lower dissociation

[1] R. H. Jackson, *J. Chem. Soc.*, p. 4285, 1962; J. K. Burdett, D. J. Gardiner, J. J. Turner, R. D. Spratley, and P. Tchir, *J. Chem. Soc. Dalton*, p. 1928, 1973.

[2] The dissociation energies, bond energies, and other thermodynamic data given in this chapter are taken from S. W. Benson, *J. Chem. Educ.*, **42**, 502 (1965); D. A. Johnson, "Some Thermodynamic Aspects of Inorganic Chemistry," Cambridge University Press, London, 1968; S. R. Gunn, *Inorg. Chem.*, **11**, 796 (1972); L. Pauling, "The Nature of the Chemical Bond," 3d ed., Cornell University Press, Ithaca, N. Y., 1960; L. Brewer, *Science*, **161**, 115 (1968).

energies. By a comparison of data for molecules such as N_2, O_2, and F_2, it is clear that the dissociation energy increases with increasing bond order. Thus the data as a whole indicate, as one might expect, that there is an inverse relationship between dissociation energy and bond length.

In the case of a polyatomic molecule, the energy required to break just one bond, with formation of two molecular fragments, is called a bond dissociation energy. A brief selection of bond dissociation energies is given in Table 2.10.[1] Bond dissociation energies are valuable data for thermochemical calculations, but their interpretation in terms of bond strengths is somewhat hazardous because of the structural and electronic rearrangements which occur in the molecular fragments that form when a bond in a molecule is broken. Unfortunately, the sum of the separate dissociation energies of the bonds of a molecule is not equal to the energy of dissociating the molecule completely into atoms. For example, consider the molecule OCS:

$$OCS(g) \rightarrow CO(g) + S(g) \qquad D(OC{=}S) = 74 \text{ kcal mol}^{-1}$$
$$OCS(g) \rightarrow O(g) + CS(g) \qquad D(O{=}CS) = 31 \text{ kcal mol}^{-1}$$
$$OCS(g) \rightarrow O(g) + C(g) + S(g) \qquad \Delta H^{\circ}_{\text{atom}}(OCS) = 331 \text{ kcal mol}^{-1}$$

[1] See footnote 2 on page 38.

Table 2.9 DISSOCIATION ENERGIES OF SOME DIATOMIC MOLECULES

Molecule	D, kcal mol^{-1}	Molecule	D, kcal mol^{-1}
H_2	104.2	O_2	119.2
D_2	106.0	SO	125
HF	135.8	S_2	101
HCl	103.0	NO	151.0
HBr	87.5	CO	256.9
HI	71.3	SiO	190
LiH	58	BeO	107
NaH	48	MgO	91
Li_2	26	N_2	226
Na_2	18	P_2	117
K_2	13	Sb_2	70
F_2	36.7	C_2	143
Cl_2	57.1	Si_2	81
Br_2	45.5	Pb_2	13
I_2	35.5	NaF	115
ClF	60.3	$NaCi$	98
BrF	59.4	$AgCl$	75
IF	66.2	$CuCl$	78
$BrCl$	51.5		
ICl	49.6		
IBr	41.9		

The sum of $D(OC{=}S)$ and $D(O{=}CS)$ is much less than the atomization energy of OCS. However, each of the sums $D(OC{=}S) + D(C{\equiv}O)$ and $D(O{=}CS) + D(C{\equiv}S)$ must equal the atomization energy of OCS.

Because of the nonadditivity of bond dissociation energies of molecules, chemists commonly measure bond strengths in terms of *bond energies* which, for polyatomic molecules, are hypothetical because they are calculated on the *assumption* that the sum of the bond energies of a molecule equals the atomization energy of the molecule. The usefulness of this concept lies in the fact that the calculated bond energy for a particular type of bond, say an S—H bond, is found to be approximately constant in different molecules containing that type of bond. We may demonstrate the approximate validity of the method by a simple calculation. From the atomization energy of cyclooctasulfur,

$$S_8(g) \rightarrow 8S(g) \qquad \Delta H^\circ_{atom}(S_8) = 508.4 \text{ kcal mol}^{-1}$$

we calculate the S—S bond energy, $E(S{-}S) = 508.4/8 = 63.6$ kcal mol^{-1}. From the atomization energy of hydrogen sulfide,

$$H_2S(g) \rightarrow 2H(g) + S(g) \qquad \Delta H^\circ_{atom}(H_2S) = 175.7 \text{ kcal mol}^{-1}$$

Table 2.10 SOME BOND DISSOCIATION ENERGIES OF MOLECULES AND RADICALS

Molecule	D, kcal mol^{-1}	Molecule	D, kcal mol^{-1}
HO—H	119	$H_2N{-}NH_2$	58
$CH_3COO{-}H$	112	HO—OH	51
$H_3CO{-}H$	102	$CH_3{-}Cl$	84
HOO—H	90	$NH_2{-}Cl$	60
$C_6H_5O{-}H$	85	HO—Cl	60
ClO—H	78	$CH_3{-}NH_2$	79
$H_3C{-}H$	104	$CH_3{-}OH$	91
$CH_3CH_2{-}H$	98	$CH_3{-}F$	108
$(CH_3)_3C{-}H$	91	$CH_3{-}I$	56
NC—H	130	HO—I	56
$Cl_3C{-}H$	96	O—H	102
$C_6H_5{-}H$	74	OO—H	47
$HC{\equiv}CH$	230	$CH_2O{-}H$	31
$HC{\equiv}N$	224	COO—H	12
$CH_2{=}CH_2$	163	$CH_2{-}H$	106
$CH_2{=}O$	175	CH—H	106
HN=O	115	C—H	81
$CH_2{=}NH$	154	$CH_2CH_2{-}H$	39
$CH_3{-}CH_3$	88		

we calculate $E(S-H) = 175.7/2 = 87.8\,\text{kcal mol}^{-1}$. Using these data, we calculate $63.6 + 2(87.8) = 239.3\,\text{kcal mol}^{-1}$ for the atomization of disulfane,

$$H_2S_2(g) \rightarrow 2H(g) + 2S(g)$$

The actual atomization energy for this molecule is 235 kcal mol^{-1}.

A wide variety of bond energies are listed in Table 2.11.[1] These approximate bond energies, in combination with the relatively accurately known heats of formation of the gaseous atoms in Table 2.12,[1] may be used to estimate the heat of formation of any nonresonant molecule containing these bonds.

A brief examination of the data in Table 2.11 will show that the bond energies for heteronuclear bonds are considerably greater than the average of those for the corresponding homonuclear bonds. Pauling[2] ascribed this enhancement of a heteronuclear bond energy to the ionic character of the bond and

[1] See footnote 2 on page 38.
[2] Pauling, op. cit.

Table 2.11 THERMOCHEMICAL (AVERAGE) BOND ENERGIES
(Values in kilocalories per mole)

Bond	kcal	Bond	kcal	Bond	kcal
As—As	40	Ge—Ge	38	S—S	63
As—F	116	Ge—F	113	S—Cl	65
As—Cl	74	Ge—Cl	81	S—Br	51
As—Br	61	Ge—Br	67	S—H	88
As—H	71	Ge—I	51	Sb—Sb	34
As—O	79	Ge—H	69	Sb—Cl	75
B—B	79	Ge—O	86	Sb—Br	63
B—F	154	Ge—N	61	Sb—H	61
B—Cl	106	N—N	38	Se—Se	38
B—Br	88	N=N	100	Se—Cl	58
B—I	65	N≡N	226	Se—H	73
B—H	91	N—F	67	Si—Si	46
B—O	125	N—Cl	45	Si—Ge	42
C—C	83	N—H	93	Si—F	143
C=C	147	N—O	39	Si—Cl	96
C≡C	194	N=O	142	Si—Br	79
C—F	117	O—O	34	Si—H	76
C—Cl	78	O=O	119	Si—O	111
C—Br	65	O—F	51	Si—N	80
C—I	57	O—Cl	49	Si—S	54
C—H	99	O—H	111	Si—C	73
C—O	86	P—P	47	Sn—Sn	36
C=O	174	P—F	119	Sn—Cl	75
C—N	73	P—Cl	79	Sn—Br	64
C=N	147	P—Br	64	Sn—H	60
C≡N	213	P—H	77	Sn—C	50
C=S	114	P—O	88		

showed that the bond energy in kilocalories per mole can be represented by the equation

$$E(A\!-\!B) = \tfrac{1}{2}[E(A\!-\!A) + E(B\!-\!B)] + 23(x_A - x_B)^2 \qquad (2.1)$$

where x_A and x_B refer to the empirically evaluated electronegativity values listed in Table 2.13. It is obvious from Eq. 2.1 that the bond energy for a bond between two atoms of equal electronegativity would be expected to be the arithmetic average of the homonuclear bond energies. It should be emphasized that Eq. 2.1 is a very approximate relation and should be used only when a very crude estimate will suffice. Nevertheless, it serves as the basis for a very useful generalization: The most stable arrangement of covalent bonds connecting a group of atoms is that arrangement in which the atom with the highest

Table 2.12   HEATS OF FORMATION OF THE GASEOUS ATOMS AT 25°C IN KILOCALORIES PER MOLE

Aluminum	78.7	Hydrogen	52.1	Radon	0
Americium	66	Indium	58	Rhenium	187
Antimony	63	Iodine	25.5	Rhodium	133
Argon	0	Iridium	160	Rubidium	19.5
Arsenic	72 ± 3	Iron	99.3	Ruthenium	155.5
Barium	42.5	Krypton	0	Samarium	49.4
Beryllium	77.5	Lanthanum	103	Scandium	90
Bismuth	50.1	Lead	46.6	Selenium	54
Boron	135	Lithium	38.6	Silicon	109
Bromine	26.7	Lutecium	102	Silver	68
Cadmium	26.7	Magnesium	35	Sodium	25.9
Calcium	42.5	Manganese	68	Strontium	39
Carbon	170.9	Mercury	14.5	Sulfur	66.6
Cerium	101 ± 3	Molybdenum	157	Tantalum	187
Cesium	18.7	Neodymium	77	Technetium	158
Chlorine	29.0	Neon	0	Tellurium	48
Chromium	95	Neptunium	105	Terbium	93
Cobalt	102.4	Nickel	102.8	Thallium	44
Copper	80.7	Niobium	172	Thorium	137.5
Dysprosium	71	Nitrogen	113.0	Thulium	55.5
Erbium	75.8 ± 1	Osmium	188	Tin	72
Europium	42.4	Oxygen	59.2	Titanium	112
Fluorine	18.9	Palladium	91	Tungsten	203
Gadolinium	95.0	Phosphorus	80 ± 10	Uranium	125
Gallium	66	Platinum	135	Vanadium	123
Germanium	89.5	Plutonium	84	Xenon	0
Gold	88	Polonium	34.5	Ytterbium	36.4
Hafnium	148	Potassium	21.5	Yttrium	101.5
Helium	0	Praeseodymium	85.0	Zinc	31.2
Holmium	71.9	Protactinium	126	Zirconium	145.5

electronegativity is bonded to the atom with the lowest electronegativity. There are many examples of spontaneous reactions, such as the following, which bear out this rule.

$$BBr_3 + PCl_3 \rightarrow BCl_3 + PBr_3$$
$$2HI + Cl_2 \rightarrow 2HCl + I_2$$
$$TiCl_4 + 4EtOH \rightarrow Ti(OEt)_4 + 4HCl$$

Another useful measure of the strength of a bond is the bond-stretching force constant. This is the constant of proportionality between the force tending to restore the equilibrium bond distance and an infinitesimal displacement from that distance. For small displacements, most bonds act as harmonic oscillators, and we can write the following relationship between the vibrational frequency v and the force constant k of a bond A—B:

$$v = \frac{1}{2\pi c}\sqrt{\frac{k}{\mu}}$$

Here μ is the reduced mass of A and B, $\mu = m_A m_B/(m_A + m_B)$. If we express v in wave numbers (reciprocal centimeters), k in millidynes per angstrom, and μ in the usual atomic-weight units, we obtain the relation $v^2 = (1.70 \times 10^6)k/\mu$. A diatomic molecule has only one vibrational frequency, which obviously corresponds to the bond stretching. However, polyatomic molecules have a number of vibrational frequencies, and these are not necessarily directly related to the

Table 2.13 THE PAULING SCALE OF ELECTRONEGATIVITIES

H 2.1																
Li 1.0	Be 1.5	B 2.0											C 2.5	N 3.0	O 3.5	F 4.0
Na 0.9	Mg 1.2	Al 1.5											Si 1.8	P 2.1	S 2.5	Cl 3.0
K 0.8	Ca 1.0	Sc 1.3	Ti 1.5	V 1.6	Cr 1.6	Mn 1.5	Fe 1.8	Co 1.8	Ni 1.8	Cu 1.9	Zn 1.6	Ga 1.6	Ge 1.8	As 2.0	Se 2.4	Br 2.8
Rb 0.8	Sr 1.0	Y 1.2	Zr 1.4	Nb 1.6	Mo 1.8	Tc 1.9	Ru 2.2	Rh 2.2	Pd 2.2	Ag 1.9	Cd 1.7	In 1.7	Sn 1.8	Sb 1.9	Te 2.1	I 2.5
Cs 0.7	Ba 0.9	La–Lu 1.1–1.2	Hf 1.3	Ta 1.5	W 1.7	Re 1.9	Os 2.2	Ir 2.2	Pt 2.2	Au 2.4	Hg 1.9	Tl 1.8	Pb 1.8	Bi 1.9	Po 2.0	At 2.2
Fr 0.7	Ra 0.9	Ac 1.1	Th 1.3	Pa 1.5	U 1.7	Np–No 1.3										

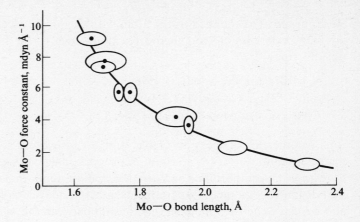

FIGURE 2.1
Plot of Mo—O force constants versus the corresponding bond lengths. [*Reproduced with permission from F. A. Cotton and R. M. Wing, Inorg. Chem.,* **4,** 867 (1965). *Copyright © by the American Chemical Society.*]

Table 2.14   SOME   BOND-STRETCHING FORCE CONSTANTS†

Bond	Force constant, mdyn $Å^{-1}$
N≡N	22.6
C≡O	18.7
HC≡CH	15.7
P≡P	5.5
As≡As	4.0
O=O	11.4
$H_2C=CH_2$	9.6
OC=S	7.5
H—H	5.1
H_3C—H	5.0
H_2N—H	6.4
H_2P—H	3.2
K—H	0.53
Li—I	0.77
Cl_2P—Cl	2.1
Br_2P—Br	1.6
Na—Na	1.7
H—F	8.8
H—Cl	4.8
H—Br	3.8
H—I	2.9

† T. L. Cottrell, "The Strengths of Chemical Bonds," 2d ed., Butterworth, London, 1958.

stretching of individual bonds. The calculation of bond force constants from the vibrational frequencies of a polyatomic molecule is a relatively complicated procedure, involving assumptions regarding the dependence of the energy of the molecule on the atomic coordinates. Such calculations have been made for a wide variety of polyatomic molecules, and some of the resulting bond force constants are included with similar data for diatomic molecules in Table 2.14.

For a series of related bonds, the force constant increases with increasing bond energy, with increasing bond order, and with decreasing bond length. The relationship between force constant and bond length for Mo—O bonds is shown in Fig. 2.1.

BOND POLARITY

The electrons of a heteronuclear bond are generally not shared equally by the two bonded atoms. The bonding electron density is almost always greater near one atom (the more electronegative atom) than the other. This polarization is depicted graphically for the gaseous LiH molecule in Fig. 2.2. This figure is a plot of the electron density of LiH minus the electron density of the hypothetical pair of neutral, noninteracting atoms, for a plane passing through the two nuclei.[1] The plot shows that bond formation causes electron density to leave the outer regions of the lithium atom, including a region in the direction of the hydrogen atom, and to enter the region around the hydrogen atom (especially in the bonding region) and the region near the lithium nucleus. This polarization is qualitatively what one would expect for a bond between the electropositive lithium atom and the relatively electronegative hydrogen atom. The electron flow is not complete, in the sense that the compound does not correspond to Li^+ and H^-, but the covalent bond between the atoms does have a lot of ionic character. The effective nuclear charge of the lithium atom is increased because the reduced electron density gives poorer shielding; hence some electron density crowds around the nucleus, and the atom is effectively smaller. Conversely, the hydrogen atom has greater electron density, and is effectively larger, than a neutral H atom.

One way of quantitatively expressing the ionic character of a bond is by assigning fractional charges to the atoms. The dipole moment μ of LiH is 5.88×10^{-18} esu cm = 5.88 Debye units; if we assume that this moment is due to two point charges, $+q$ and $-q$, separated a distance r equal to the Li—H

---

[1] A. Streitwieser, Jr., and P. H. Owens, "Orbital and Electron Density Diagrams," Macmillan, New York, 1973.

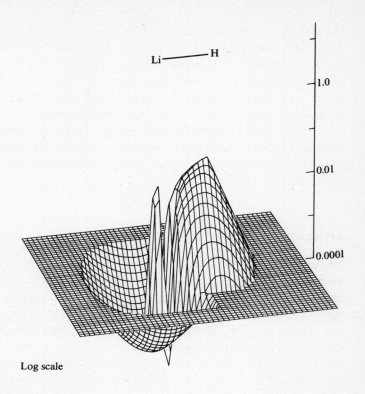

Li———H

1.0

0.01

0.0001

Log scale

FIGURE 2.2
Plot of electron density in LiH less electron density of neutral Li + H. *(Reproduced with permission from A. Streitwieser, Jr., and P. H. Owens, "Orbital and Electron Density Diagrams," Macmillan, New York, 1973.)*

bond distance (1.595 Å), it is an easy matter to calculate q from the relation $\mu = qer$.

$$q = \frac{\mu}{er} = \frac{5.88 \times 10^{-18}}{(4.8 \times 10^{-10}) \times (1.595 \times 10^{-8})} = 0.768$$

Thus the dipole moment may be accounted for by assigning a charge of $+0.768$ to the lithium atom and a charge of -0.768 to the hydrogen atom.

The treatment of LiH as a pair of point charges, or even as a pair of charged, spherically symmetric atoms, is an approximation. From Fig. 2.2 it can be seen that the charge distribution of the molecule would be better represented by a pair of charged atoms, each with a dipole moment. The atomic moments would be relatively small, however, and the representation of the overall dipole

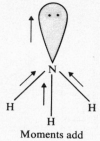

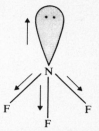

FIGURE 2.3
Explanation of the low dipole moment of
NF_3. Arrows indicate the direction from
+ to −.

Moments add

Moments almost cancel

moment with simple atomic charges is fairly reasonable in the case of LiH. However, in the case of a molecule with nonbonding valence electrons, such as HCl or NH_3, such a representation is extremely misleading because of the relatively large moments due to the nonbonding electrons. It is instructive to compare the dipole moments of the pyramidal molecules NF_3 ($\mu = 0.23$ Debye unit) and NH_3 ($\mu = 1.47$ Debye units). Although the bond angle of NF_3 is more acute than that of NH_3 and although the N—F bond would be expected to be at least as polar as the N—H bond (but with opposite sign), the dipole moment of NF_3 is much smaller than that of NH_3. The apparent anomaly is caused by the large moment of the lone pair of electrons on the nitrogen atom. In NH_3, the lone-pair–nitrogen moment adds to the resultant of the N—H moments. Thus NH_3 has a large dipole moment. However, in NF_3, the lone-pair–nitrogen moment partly cancels the resultant of the N—F moments. Thus NF_3 has a small dipole moment. Reference to Fig. 2.3 will make this clear.

Other physical properties of molecules are related to molecular charge distribution. Chemical shifts and spin-spin coupling constants from nuclear magnetic resonance (nmr) spectra[1] are sensitive to electron density at the nucleus. Thus in Fig. 2.4 is shown a correlation of proton chemical shift in methyl compounds with the electronegativity of the atom bonded to the methyl group. One can rationalize the data by arguing that the more electronegative atoms pull electrons away from the protons, thus deshielding them and allowing the protons to resonate at a lower applied magnetic field. Unfortunately, however, nmr parameters are affected by other features besides electron density, and these effects are not understood well enough to permit the use of nmr parameters as reliable measures of atomic charge.

---

[1] Nuclear magnetic resonance spectroscopy, nuclear quadrupole resonance spectroscopy, and Mössbauer spectroscopy are discussed in W. L. Jolly, "The Synthesis and Characterization of Inorganic Compounds," Prentice-Hall, Englewood Cliffs, N.J., 1970; R. S. Drago, "Physical Methods in Inorganic Chemistry," Reinhold, New York, 1965; and H. A. O. Hill and P. Day (eds.), "Physical Methods in Advanced Inorganic Chemistry," Wiley-Interscience, New York, 1968.

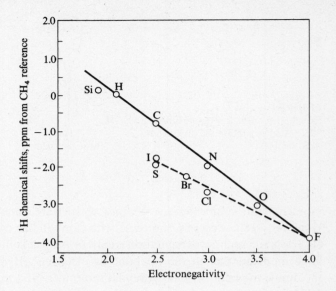

FIGURE 2.4
Plot of proton magnetic resonance chemical shifts for methane derivatives versus the Pauling electronegativity of the substituent atoms. [*Reproduced with permission from H. Spiesecke and W. G. Schneider, J. Chem. Phys.*, **35**, 722 (1961).]

Table 2.15   CHLORINE-35 NUCLEAR QUADRUPOLE COUPLING CONSTANTS†

Compound	Coupling constant, MHz
Free Cl^-	0.0
LiCl	−6.1
$(CH_3)_3CCl$	−62.1
$(CH_3)_2CHCl$	−64.1
C_2H_5Cl	−65.8
CH_3Cl	−68.1
CH_2Cl_2	−72.0
$CHCl_3$	−76.7
CCl_4	−81.2
ICl	−82.5
BrCl	−103.6
Cl_2	−109.0
ClF	−146.0

† J. E. Huheey, "Inorganic Chemistry," p. 171, Harper & Row, New York, 1972.

The nuclear quadrupole (nqr) coupling constant[1] of an atom is a measure of the asymmetry of the electric field at the nucleus. A free chloride ion, for example, should exhibit no quadrupole coupling. The greater the electronegativity of the atom bonded to a chlorine atom, the more asymmetric is the electric field at the chlorine nucleus, and the greater is the absolute magnitude of the nqr coupling constant. The trend of the chlorine-35 coupling constant with the electronegativity of the substituent is shown by the data in Table 2.15. Unfortunately the interpretation of these data in terms of chlorine's atomic charge is complicated by the fact that the electric field gradient is a function of the hybridization of the bonding orbital as well as the atomic charge. Because of the difficulty of estimating the s character of a chlorine bonding orbital, data such as those in Table 2.15 are restricted to qualitative interpretation.

In Mössbauer spectroscopy[1] one can measure shifts in the energy of a nuclear transition caused by changes in the electronic environment of the nucleus. These *isomer shifts* are caused by changes in s electron density at the nucleus and are essentially independent of the population and shape of other atomic orbitals. Thus isomer shifts can be meaningfully interpreted in terms of changes in atomic charge only for compounds having similar structures and orbital

[1] See footnote on page 47.

Table 2.16 MÖSSBAUER ISOMER SHIFTS FOR SOME HEXAHALOSTANNATES†

Compound	Isomer shift, mm s^{-1}‡
K_2SnF_6	-0.36
$(Et_4N)_2SnCl_4F_2$	$+0.29$
$(Et_4N)_2SnCl_6$	$+0.52$
$(Et_4N)_2SnBr_4F_2$	$+0.53$
$(Et_4N)_2SnCl_4Br_2$	$+0.67$
$(Et_4N)_2SnBr_4Cl_2$	$+0.77$
$(Et_4N)_2SnCl_4I_2$	$+0.78$
$(Et_4N)_2SnBr_6$	$+0.84$
$(Et_4N)_2SnBr_4I_2$	$+0.96$
$(Et_4N)_2SnI_4Br_2$	$+1.09$
$(Et_4N)_2SnI_6$	$+1.23$

† C. A. Clasen and M. L. Good, *Inorg. Chem.*, **9**, 817 (1970).
‡ Shifts are customarily expressed as the experimentally measured relative velocities of the sample and a gamma-ray source required to give resonance absorption.

hybridizations. For example, shifts for the tin nucleus in a series of hexahalo-stannate salts, listed in Table 2.16, can readily be correlated with the electron-withdrawing abilities of the coordinated halide ions.

Probably the most useful technique for the estimation of atomic charges is X-ray photoelectron spectroscopy (often referred to as ESCA).[1] We have already alluded to this technique in Chap. 1 in connection with the determination of the ionization potentials (binding energies) of the various kinds of electrons in argon. With respect to the estimation of atomic charges, *core*-electron binding energies are of principal interest. It has been shown that, for a given core level of an element in a set of compounds, the binding energies E_B may be fairly well represented by the following equation:

$$E_B = k_i Q_i + \sum_{j \neq i} Q_j e^2 / r_j + l_i \qquad (2.2)$$

In this expression, Q_i is the charge of the atom i for which the core binding energy is being calculated, Q_j is the charge on another atom, j, of the compound, r_j is the distance between atoms j and i, and k_i and l_i are empirical constants characteristic of the element. Inasmuch as the constant k_i is always positive, the first term in this expression shows that the core binding energy increases with increasing atomic charge. This relation is qualitatively reasonable; one would expect the difficulty of removing an electron from an atom to increase with the positive charge of the atom.

The xenon $3d_{5/2}$, xenon $3d_{3/2}$, and fluorine $1s$ spectra for elementary xenon, molecular fluorine, and a series of xenon fluorides are shown in Fig. 2.5. It can be seen that, as expected, the xenon core binding energies increase as the oxidation state of the xenon increases and that the fluorine $1s$ binding energy decreases on going from the free element to the fluorides. By assuming that there is an equation analogous to Eq. 2.2 for every atom in the set of xenon compounds and by using the fact that the sum of the atomic charges is zero for each compound, the binding-energy data can be used to estimate atomic charges for the atoms in the xenon compounds. These estimated charges are given in Table 2.17.

The electronegativities listed in Table 2.13, which were evaluated from thermodynamic data, were interpreted by Pauling as the tendencies of atoms in molecules to attract electrons. Ever since this original definition of electronegativity, many other definitions and methods of evaluation have been

[1] K. Siegbahn et. al., "ESCA; Atomic, Molecular and Solid State Structure by Means of Electron Spectroscopy," Almqvist and Wiksells, Uppsala, Sweden, 1967; K. Siegbahn et al., "ESCA Applied to Free Molecules," North-Holland, Amsterdam, 1969; J. M. Hollander and W. L. Jolly, *Acc. Chem. Res.*, **3**, 193 (1970).

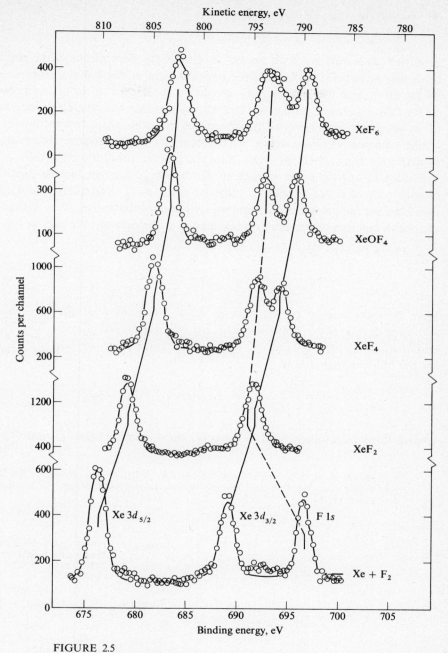

FIGURE 2.5
X-ray photoelectron spectra of Xe, F_2, XeF_2, XeF_4, $XeOF_4$, and XeF_6.
[*Reproduced with permission from T. X. Carroll et al., J. Am. Chem. Soc.,* **96**, 1989
(1974). *Copyright © by the American Chemical Society.*]

proposed. It is now generally agreed, in view of the dependence of the electron-attracting power of an atom on the hybridization of the bonding orbital used by the atom, that electronegativity should be considered as an *orbital* property rather than an *atomic* property.[1] It is also recognized that the electronegativity of an orbital is a function of the charge of the orbital (i.e., a function of the orbital occupancy). If we define electronegativity as the tendency of an atomic orbital to attract electrons from another atomic orbital with which it has combined to form a bond, it is clear that *in a molecule* the electronegativities of each pair of bonding orbitals must be equal as a consequence of electron flow between the orbitals. As a fair approximation, the energy of an orbital is a quadratic function of the electron occupancy of the orbital. We may write

$$E - E_0 = aq + \tfrac{1}{2}bq^2$$

$$\frac{dE}{dq} = a + bq$$

Here E is the energy of the system, E_0 is the energy of the zero-charged reference state, q is the positive charge on the orbital, and a and b are positive empirical constants. The energy of a pair of combining orbitals is minimized for the following condition:

$$\frac{d(E_1 + E_2)}{dq_1} = 0 = \frac{dE_1}{dq_1} - \frac{dE_2}{dq_2} = (a_1 + b_1 q_1) - (a_2 + b_2 q_2)$$

Charge flows from one orbital to the other until the quantities $\dfrac{dE_1}{dq_1}$ and $\dfrac{dE_2}{dq_2}$ are

[1] J. Hinze and H. H. Jaffé, *J. Am. Chem. Soc.*, **84**, 540 (1962); *J. Phys. Chem.*, **67**, 1501 (1963); J. E. Huheey, "Inorganic Chemistry, Principles of Structure and Reactivity," pp. 158–168. Harper & Row, New York, 1972.

Table 2.17   ATOMIC CHARGES DERIVED FROM CORE BINDING ENERGIES OF XENON FLUORIDES†

Compound	Q_{Xe}	Q_F	Q_O
XeF_2	0.48	−0.24	
XeF_4	0.97	−0.24	
XeF_6	1.43	−0.24	
$XeOF_4$	1.37	−0.24	−0.43

† T. X. Carroll, R. W. Shaw, T. D. Thomas, C. Kindle, and N. Bartlett, *J. Am. Chem. Soc.*, **96**, 1989 (1974).

equal. Thus it is logical to identify electronegativity with $\dfrac{dE}{dq}$ or with a quantity linearly related to $\dfrac{dE}{dq}$. Values of the parameters a and b for some orbitals of first-row atoms are listed in Table 2.18. The charge flow in a bond can be calculated as follows:

$$a_1 + b_1 q_1 = a_2 + b_2 q_2 = a_2 - b_2 q_1$$

$$q_1 = \frac{a_2 - a_1}{b_1 + b_2}$$

Thus for the molecule HF, if we assume sp^3 hybridization for the fluorine, we calculate $q_H = (16.97 - 7.17)/(12.85 + 16.48) = +0.33$. Somewhat more complicated calculations are required if one wishes to treat π-bonded systems and to account for the effects of formal charge and induction.[1]

[1] W. L. Jolly and W. B. Perry, *J. Am. Chem. Soc.*, **95**, 5442 (1973); *Inorg. Chem.*, **13**, 2686 (1974).

Table 2.18   SOME ORBITAL ELECTRO-
NEGATIVITY PARAMETERS†

Element	Orbital	a, eV/q	b, eV/q^2
H	s	7.17	12.85
He	s	9.7	29.8
Li	s	3.10	4.57
Be	sp	4.78	7.59
B	sp^3	5.99	8.90
	sp^2	6.33	9.91
C	sp^3	7.98	13.27
	sp^2	8.79	13.67
	sp	10.39	14.08
N	sp^3	11.54	14.78
	sp^2	12.87	15.46
O	p	9.65	15.27
	sp^3	15.25	18.28
F	p	12.18	17.36
	sp^3	16.97	16.48

† J. Hinze and H. H. Jaffé, *J. Am. Chem. Soc.*, **84**, 540 (1962); *J. Phys. Chem.*, **67**, 1501 (1963); J. E. Huheey, "Inorganic Chemistry, Principles of Structure and Reactivity," pp. 158–168, Harper & Row, New York, 1972.

PREDICTION OF MOLECULAR TOPOLOGY

The empirical formula of a polyatomic molecule tells nothing about the structure of the molecule. For example, the formula S_2O could correspond to a three-membered ring,

or a chain with either an oxygen atom or a sulfur atom in the middle,

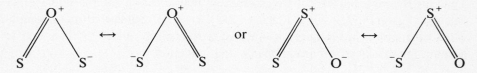

The chain structures could be either linear or bent. From this and other examples we see that a description of molecular structure consists of two stages. The first stage is simply a statement of which atoms are bonded to each other, i.e., a topological description of the bonding. The second stage is the specification of bond angles and bond distances. In this section we shall discuss methods for predicting bonding topologies; methods for predicting structural parameters will be discussed in the next section.

The following procedure may be used to predict the most stable of several possible bonding topologies for a molecule or ion. First write valence structures for the various possible topologies. Then discard the less stable structures according to the following rules, in the order given, until only one structure remains.

1 Discard structures which do not satisfy the Lewis octet rule. Thus for CH_3N discard the structure in which a central carbon atom is attached to each of the other four atoms in favor of the $HN{=}CH_2$ structure.

2 Discard structures with three-membered rings. Thus the ring structure for S_2O is discarded in favor of the chain structures.

3 Discard structures that require adjacent atoms to have formal charges of the same sign. For example, for S_2N_2 discard the $^-N{=}\overset{+}{S}{-}\overset{+}{S}{=}N^-$ structure in favor of the following ring structure:

$$
\begin{array}{ccc}
{}^-N & - & S^+ \\
| & & \| \\
S & - & N
\end{array}
$$

4 Discard structures in which nearby atoms have unlike formal charges contradicting electronegativities, or in which an atom has an unlikely formal charge (according to electronegativities). Thus for N_2O discard the $^-N\!\!=\!\!\overset{2+}{O}\!\!=\!\!N^-$ structure in favor of the $N\!\!\equiv\!\!N^+\!\!-\!\!O^-$ structure.

5 Discard structures for which relatively few stable resonance forms can be written. For example, we predict nitramide, $H_2N\!\!-\!\!NO_2$ (for which two equivalent stable resonance structures can be written), to be stable with respect to hyponitrous acid, $HON\!\!=\!\!NOH$ (for which only one stable structure can be written). This example illustrates the utility of the procedure for estimating the relative stabilities of structural isomers.

6 The most stable structure of those remaining is generally that for which the sum of the electronegativity differences of adjacent atoms is the greatest. Discard all the other structures. Thus the structure $H\!\!-\!\!O\!\!-\!\!Cl$ is predicted to be more stable than the structure $H\!\!-\!\!\overset{+}{Cl}\!\!-\!\!O^-$.

VALENCE-SHELL ELECTRON REPULSION[1]

The spatial arrangement of the bonds to an atom is strongly correlated with a quantity which we shall call the "total coordination number" of the atom. We define the total coordination number as the sum of the number of ligands and valence lone pairs. Thus the oxygen atom of H_2O has a total coordination number of 4 (two ligands and two lone pairs), and the boron atom of BF_3 has a total coordination number of 3 (three ligands and no lone pairs). The ligands and lone pairs of an atom are generally located at or near the vertices of a regular polyhedron which has as many vertices as the total coordination number of the atom. Such an arrangement is energetically favorable because it keeps all the valence-shell electron pairs as far apart as possible. When more than one arrangement of ligands and lone pairs is possible at the vertices of a polyhedron, the observed stable geometry is generally consistent with the following empirical rules: (1) Electron repulsions decrease in the order lone-pair–lone-pair repulsion > lone-pair–bonding-pair repulsion > bonding-pair–bonding-pair repulsion, and (2) repulsions between electrons at vertices which subtend an angle at the central atom greater than about $115°$ can be neglected. A summary of the stereochemistries of a wide variety of molecules is presented in Table 2.19. We shall discuss some specific examples in the following paragraphs.

[1] This valuable concept was introduced by R. J. Gillespie and R. S. Nyholm [*Q. Rev.*, **11**, 339 (1957)] and has been developed and popularized by R. J. Gillespie [*J. Chem. Educ.*, **47**, 18 (1970)]. The discussion of the concept in this chapter includes certain ideas which are quite different from those expressed by Gillespie.

Total Coordination Number 2

When only two ligands are coordinated to a central atom, with no nonbonding electrons on the central atom, the repulsions between the two ligands and between the bonding electrons are minimized when the ligands form a 180° bond angle with the central atom. Therefore a linear structure is generally observed in such cases. It is convenient to think of an s orbital and a p orbital in terms of their linear combinations, i.e., as two hybrid orbitals[1] directed 180° from each other. If we wish the hybrid orbitals to be equivalent, they are formed as the sum and difference of the atomic orbitals, as shown schematically in Fig. 2.6. In a linear molecule such as $BeCl_2$, $HgCl_2$, or CO_2, we may think of the central atom as using such hybrid sp orbitals in the formation of the σ bonds between the central atom and the ligands.

When the central atom is large and the ligand atoms are small, as in BaF_2, the molecule is V-shaped. This result may be rationalized by considering the molecule as a combination of a polarizable central cation and two anions. The two anions are not bound at an angle of 180°, because then the induced dipole moments in the central cation would be opposed and there would be no extra bonding energy due to polarization. However, if the angle between the anions is less than 180°, the central cation is polarized and the

---

[1] Hybridization is discussed in Chap. 3.

Table 2.19 PREDICTIONS OF THE VALENCE-SHELL ELECTRON REPULSION THEORY OF DIRECTED VALENCY

Total coordination no.	Arrangement of lone pairs and ligands	No. of ligands	No. of lone pairs	Shape of molecule	Examples
2	Linear	2	0	Linear	$BeCl_2$, $HgCl_2$, ZnI_2, CO_2
3	Equilateral	3	0	Planar triangular	BCl_3, NO_3^-
	Triangular	2	1	V-shaped	O_3, NO_2^-, $SnCl_2$
4	Tetrahedral	4	0	Tetrahedral	CH_4, Al_2Cl_6, ClO_4^-
		3	1	Pyramidal	NF_3, H_3O^+, $(TlOR)_4$, ClO_3^-
		2	2	V-shaped	H_2O, SCl_2, ClO_2^-
5	Trigonal bipyramidal	5	0	Trigonal bipyramidal	PCl_5, $PF_3(CH_3)_2$
		4	1	Irregular tetrahedral	SF_4, R_2TeCl_2
		3	2	T-shaped	ClF_3, $C_6H_5ICl_2$
		2	3	Linear	ICl_2^-, XeF_2
6	Octahedral	6	0	Octahedral	SF_6, PCl_6^-, S_2F_{10}
		5	1	Square pyramidal	BrF_5, $XeOF_4$
		4	2	Square planar	ICl_4^-, XeF_4

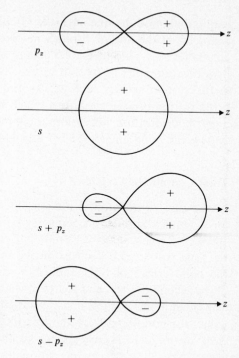

FIGURE 2.6
Formation of *sp* hybrid orbitals by combination of an *s* and a *p* orbital. Notice that, for simplicity, a 1*s* orbital is combined with a 2*p* orbital.

system is stabilized by the interaction between the resultant induced dipole and the anions. As the bond angle decreases, the polarization stabilization increases and the anion-anion electrostatic repulsion increases. The observed bond angle corresponds to a balance of these effects.

Total Coordination Number 3

When three ligands are coordinated to a central atom, with no nonbonding electrons on the central atom, repulsions are minimized when the ligands are disposed at the corners of an equilateral triangle, with 120° bond angles. In such cases it is often convenient to consider the σ bonds as being formed by the overlap of sp^2 hybrid orbitals on the central atom with appropriate orbitals of the ligands.

If one of the ligands is replaced by a lone pair, the lone pair occupies the site of the replaced ligand, and the resulting molecule is V-shaped. Because of the relative stability of *s* orbitals compared with *p* orbitals, a lone pair is more stable the greater the *s* character of its orbital. This effect is somewhat counterbalanced by the fact that bond strength increases with increasing s character of the atomic orbitals involved (at least up to a hybridization of sp). Nevertheless, replacement of a ligand by a lone pair almost always causes re-hybridization such as to shift more s character to the lone-pair orbital. Thus, in the case of species such as O_3 and $NO_2{}^-$, the σ bonds are formed using orbitals from the central atoms which have more p character than corresponds to sp^2 hybridization. Therefore the bond angles are less than 120°. For O_3 and $NO_2{}^-$, the bond angles are 117 and 115°, respectively.

Total Coordination Number 4

In the case of an atom with four identical ligands, repulsions are minimized for a regular tetrahedral geometry with sp^3 hybridization and bond angles of 109.47°. Replacement of ligands by lone pairs causes a shift of s character to the lone pairs and a decrease in the bond angles. It is interesting to compare the bond angles of NH_3 and H_2O with those for the corresponding hydrides of the second-, third-, and fourth-row elements; the data are given in Table 2.20. The bond angles for the first-row hydrides are slightly less than the tetrahedral value; the bond angles for the hydrides of the heavier elements are remarkably close to 90°, which corresponds to the use of pure p orbitals by the central atom. These results can be rationalized as follows: In the heavier hydrides, the central atoms are so big that there is very little repulsion between valence electrons. Thus bonding is accomplished using the p orbitals of the central atoms, without the need of the promotional energy required by hybridization. In the first-row hydrides, 90° bond angles would cause excessive crowding of the bonding electron pairs. Hybridization, with the introduction of s character into the bonding orbitals almost to the extent of sp^3 hybridization, permits the lone pairs and bonding pairs to separate, with reduction of the repulsion.

The bond angle of OF_2 is 102°, significantly less than that of H_2O. Presumably the bond angle is reduced because the highly electronegative fluorine atoms withdraw much more of the bonding electron density than the hydrogen

Table 2.20   BOND ANGLES OF SOME HYDRIDES

NH_3	106.8°	H_2O	104.5°
PH_3	93.3°	H_2S	92.2°
AsH_3	91.5°	H_2Se	91.0°
SbH_3	91.3°	H_2Te	89.5°

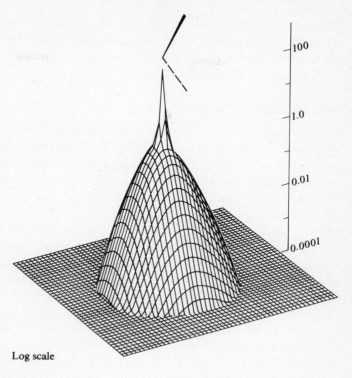

Log scale

FIGURE 2.7
Electron density plot for H_2O in the plane of the lone pairs. (*Reproduced with permission from A. Streitwieser, Jr., and P. H. Owens, "Orbital and Electron Density Diagrams," Macmillan, New York, 1973.*)

atoms do. Such electron withdrawal is energetically favored by increasing the p character of the oxygen orbitals used in the bonding. On the other hand, the bond angle of OCl_2 is $\sim 111°$, appreciably greater than that of H_2O. Apparently the chlorine atoms are so big that repulsion between nonbonding electrons on separate chlorine atoms overwhelms the electronegativity effect.

The lone-pair electrons of a molecule such as water are often illustrated as separate lobes or "rabbit ears," as shown in the following structure:

Although such structures are helpful for illustrating the stereochemical influence of lone pairs, they are very misleading representations of valence-electron density. The electron density plot for H_2O, given in Fig. 2.7, shows that the lone pairs overlap and that the separate lobes are fictitious.[1]

Total Coordination Number 5

When an atom with no lone pairs is bonded to five ligands, the ligands are almost always located at the vertices of a trigonal bipyramid. Such a structure is obtained in the case of PF_5, which is illustrated in Fig. 2.8. There are two stereochemically distinct types of ligands in a trigonal bipyramidal MX_5 molecule; two axial ligands and three equatorial ligands. For convenience, let us define the threefold axis of the molecule as the z axis of the central atom. Then the central atom's p_z orbital lies along the threefold axis and the p_x and p_y orbitals lie in the equatorial plane. Because the axial bonds must share the p_z orbital, each axial bond has available one-half of a p orbital, and because the equatorial bonds must share the p_x and p_y orbitals, each of these bonds has available two-thirds of a p orbital. The spherically symmetric s orbital could, in principle, be distributed in any proportion between the axial and equatorial bonding orbitals, although the two axial orbitals must have the same s character, and the three equatorial orbitals must have the same s character. At one extreme, there could be a full sp hybrid orbital for each axial bond and a $\frac{2}{3}$ p orbital for each equatorial bond. At the other extreme, there could be a $\frac{1}{2}$ p orbital for each axial bond and a full sp^2 hybrid orbital for each equatorial bond. We shall assume, as a reasonable working hypothesis, that the s orbital is equally distributed among all five bonds. Thus we obtain bond orders of 0.7 for the axial bonds and 0.867 for the equatorial bonds, and formal charges of -0.3 for the axial ligand atoms and -0.133 for the equatorial ligand atoms. Because the equatorial bond order is greater than the axial bond order, the equatorial bond lengths are shorter than the axial bond lengths. (For PF_5, $r_{ax} = 1.58$ Å and $r_{eq} = 1.53$ Å; for PCl_5, $r_{ax} = 2.19$ Å and $r_{eq} = 2.04$ Å.) Because, on the basis of our structural hypothesis, the formal charge on the axial ligand atoms is more negative than that on the equatorial ligand atoms, we would expect that, when more than one type of ligand is bonded to the same central atom, the more electronegative ligands would preferentially occupy the axial positions. This effect is clearly seen in the structures of PF_4CH_3 and $PF_3(CH_3)_2$, illustrated in Fig. 2.8. Because CH_3 groups are more electropositive than the F atoms, the CH_3 groups occupy equatorial positions.

[1] Streitwieser and Owens, op. cit.

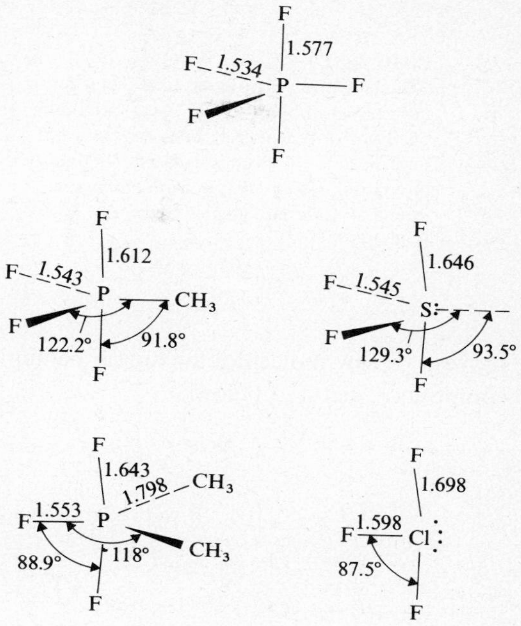

FIGURE 2.8
Structures of PF_5, PF_4CH_3, $PF_3(CH_3)_2$, SF_4, and ClF_3. [*Reproduced with permission from R. J. Gillespie, J. Chem. Educ.*, **47**, 18 (1970).]

The trigonal bipyramidal configuration for molecules with five ligands and no lone pairs on the central atom is not exceedingly stable with respect to other configurations. Distortions of the bond angles readily occur, and it is believed that many pentacoordinate molecules undergo rapid interconversion between a trigonal bipyramidal configuration and a square pyramidal configuration, even though the former configuration may be the stable, equilibrium configuration. For example, the ^{19}F nmr spectrum of PF_5 shows a signal apparently due to only one kind of fluorine atom, even though one would expect separate signals due to the axial and equatorial fluorine atoms. Apparently the axial and equatorial atoms are scrambled at a rate greater than the difference in their nmr frequencies, and therefore the nmr signal corresponds to a weighted average signal of the axial and equatorial atoms. The mechanism shown in Fig. 2.9 has been postulated to explain the scrambling.[1,2] It will be noted that the molecule passes through an intermediate square pyramidal configuration in the scrambling

[1] S. Berry, *J. Chem. Phys.*, **32**, 933 (1960).
[2] R. J. Gillespie, *Angew. Chem. Int. Ed.*, **6**, 819 (1967).

FIGURE 2.9
The mechanism postulated to account for
fluorine scrambling in PF_5. (*a*) Original
positions of F atoms, ∘; final positions of
F atoms, • (*b*) Pseudo-rotated trigonal
bipyramid. [*Reproduced with permission
from R. J. Gillespie, Angew. Chem. Int. Ed.,*
6, 819 (1967).]

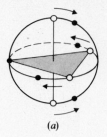

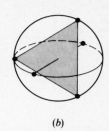

(*a*)   (*b*)

process. A few molecules, including pentaphenylantimony[1] and the phosphorus
compound[2] pictured below,

are so distorted in the solid state that their configurations are closer to square
pyramidal than to trigonal bipyramidal. These distortions are not understood.
They may be a consequence of solid-state crystal packing forces; perhaps the
free molecules would have trigonal bipyramidal configurations.

When one, two, or three of the ligands in a trigonal bipyramidal molecule
are replaced by lone pairs, the lone pairs always occupy equatorial positions,
presumably to minimize lone-pair–bonding-pair repulsions. Thus molecules of
this type always have two ligands in the axial positions. The structures of SF_4
and ClF_3 are shown in Fig. 2.8.

Total Coordination Number 6

In a 6-coordinate molecule such as SF_6 the ligand atoms are situated at the
vertices of a regular octahedron. If we arbitrarily place the ligand atoms on the
x, y, and z axes of the central atom, it is easy to see that each bonding orbital
of the central atom consists of a combination of $\frac{1}{2}$ p orbital and $\frac{1}{6}$ s orbital.
Neglecting any contributions of the valence-shell d orbitals, each bond has a

[1] A. L. Beauchamp, M. J. Bennett, and F. A. Cotton, *J. Am. Chem. Soc.*, **90,** 6675 (1968).
[2] J. A. Howard, D. R. Russell, and S. Trippett, *J. Chem. Soc. Chem. Commun.*, p. 856, 1973.

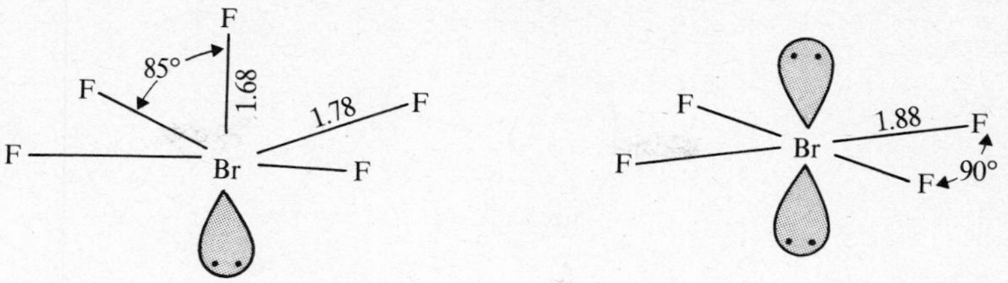

FIGURE 2.10
Structures of BrF_5 and BrF_4^-.

bond order of $\frac{2}{3}$. If we replace one of the ligands with a lone-pair, the resulting molecule has a square pyramidal configuration, with the axial ligand trans to the lone pair. As an illustration of this type of bonding, the structure of BrF_5 is shown in Fig. 2.10. Because of the relatively high lone-pair–bonding-pair repulsions, the bromine atom lies below the basal plane of the four equivalent fluorine atoms, and the bond angles are less than 90°. In order to replace a ligand of a 6-coordinate molecule by a lone pair, it is necessary to extract electron density from the other bonds. (The average bond order drops from $\frac{2}{3}$ to $\frac{3}{5}$.) Because of the repulsion of the lone pair by the bonding electrons of the four basal bonds, it is to be expected that more electron density will be extracted from these basal bonds than from the axial bond. Hence the axial bond order is greater than the basal bond order, in qualitative agreement with the corresponding bond distances (1.68 and 1.79 Å, respectively, in BrF_5). If we replace two of the ligands in a 6-coordinate molecule with lone pairs, it is reasonable to place these in trans positions to minimize the strong lone-pair–lone-pair repulsion. Thus we expect the ligands to occupy a square planar configuration. Indeed such a configuration is found for all tetracoordinate species with two lone pairs on the central atom (for example, BrF_4^-, shown in Fig. 2.10, and XeF_4). In such species the bond order is $\frac{1}{2}$, with one-half of a p orbital contributing to each bonding orbital of the central atom.

Total Coordination Number 7

When the total coordination number exceeds 6, there are usually various symmetric, polyhedral configurations which have very similar energies, and one often finds examples of each of these configurations in actual compounds. It does not seem to be possible to predict structures with certainty in such cases, particularly if the ligands are not equivalent or if they are chelated. For

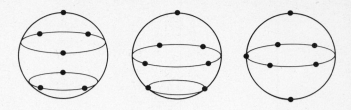

FIGURE 2.11
Three ways of achieving sevenfold coordination. [*Reproduced with permission from R. J. Gillespie, Angew. Chem. Int. Ed.,* **6**, 819 (1967).]

example, let us consider total coordination number 7. Three plausible configurations are shown in Fig. 2.11. The first structure is a monocapped octahedron, the second is a monocapped trigonal prism, and the third is a pentagonal bipyramid. The first structure has been observed in the case of $NbOF_6^{3-}$ and ZrF_7^{3-}, the second structure in the case of NbF_7^{2-} and TaF_7^{2-}, and the third structure in the case of $UO_2F_5^{3-}$ and IF_7. If one of the ligands is replaced with a lone pair, one might expect to obtain the structure in which the lone pair has the minimum number of nearest neighbors, i.e., the first structure, with the lone pair in the unique axial position. This configuration has been postulated[1,2] for XeF_6, but the exact structure of this molecule is not yet known. On the other hand, the structure of the trisoxalatoantimonate(III) ion, $Sb(C_2O_4)_3^{3-}$, has been shown to be based on a pentagonal bipyramid with the lone pair presumably occupying an axial position.[3] The structure of this ion is shown in Fig. 2.12. Two oxalate ions lie in the equatorial plane, and one bridges an equatorial site and an axial site.

When the ligands are large, repulsions between ligands can be of primary importance in determining stereochemistry. For example, $TeCl_6^{2-}$, $TeBr_6^{2-}$, and $SbBr_6^{3-}$ have regular octahedral structures even though the central atoms have nonbonding electrons. Apparently the lone pair is forced inside the shell of ligands into a spherical s orbital.

Higher coordination numbers are known, but for none of these is the geometry known to be affected by the presence of lone pairs on the central atom. Therefore we shall defer consideration of such structures until we discuss transition-metal complexes.

Theoretical chemists have been slow to appreciate the significance of the valence-shell electron-repulsion theory, even though it fairly reliably predicts

[1] R. J. Gillespie, *J. Chem. Educ.,* **47**, 18 (1970).
[2] R. J. Gillespie, *Angew. Chem. Int. Ed.,* **6**, 819 (1967).
[3] M. C. Poore and D. R. Russell, *Chem. Commun.,* p. 18, 1971.

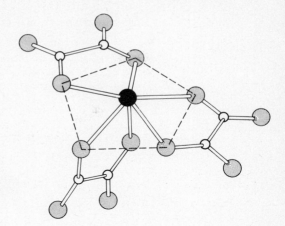

FIGURE 2.12
Perspective view of the $Sb(C_2O_4)_3^{3-}$ ion.
(*Reproduced with permission from M. C. Poore and D. R. Russell, Chem. Commun., p. 18, 1971.*)

structure for a wide variety of molecules. However, in recent years theoretical studies have shown that the theory is in good agreement with the results of *ab initio* quantum-mechanical calculations for simple molecules.[1]

BOND CONFORMATIONS

Many molecules have groups of atoms joined by a bond such that rotation of the groups with respect to one another around the bond axis causes a continuous change in the energy of the molecule. Rotation about a double bond causes the molecule to pass through conformations of such high energy that rotation is strongly resisted. However, in the case of a single bond, the rotational barriers are low (usually less than 4 kcal mol^{-1}), and rotation is relatively unrestricted at ordinary temperatures. Nevertheless the stable conformation for a single bond is of considerable interest both because of its relation to the dipole moment of the molecule and because of the theoretical problems posed. A conformation is generally described in terms of the dihedral angle between two atoms or lone pairs on the adjacent atoms, i.e., the angle projected by the atoms or lone pairs on a plane perpendicular to the bond axis. Conformational data for several molecules are given in Table 2.21. Such data are difficult to rationalize. Probably one must consider nucleus-electron attractions, nucleus-nucleus repulsions, and electron-electron repulsions; and no simple theory seems to be adequate. For a discussion of the theoretical aspects of bond conformations, the reader is referred to a paper by Wolfe.[2]

[1] G. W. Schnuelle and R. G. Parr, *J. Am. Chem. Soc.*, **94**, 8974 (1972); C. A. Naleway and M. E. Schwartz, *J. Am. Chem. Soc.*, **95**, 8235 (1973).
[2] S. Wolfe, *Acc. Chem. Res.*, **5**, 102 (1972).

Table 2.21 STABLE CONFORMATIONS OF SEVERAL MOLE-
CULES†

Compound	Projected conformation	Dihedral angle (ϕ), deg
N_2H_4 P_2H_4		90–95 90–100
H_2O_2		111
H_2S_2 O_2F_2 S_2F_2		91 88 88
FCH_2OH		60

† S. Wolfe, *Acc. Chem. Res.*, **5**, 102 (1972).

PROBLEMS

2.1   Write Lewis octet structures for the following species. (Show all nonbonding valence electrons, and indicate formal charges.) (a) Al_2Cl_6, (b) $SnCl_3^-$, (c) XeF_2, (d) BrF_4^-, (e) NS^+, (f) SO_3F^-, (g) HOClO, (h) $S_4N_3^+$ (S—S—N—S—N—S—N).

2.2   The S—O bond distance in SO_2F_2 is 1.37 Å, whereas that in SO_2 is 1.43 Å. Explain in terms of octet structures.

2.3   How can you account for CH_5^+ in terms of an octet structure?

2.4   For each of the following, name three well-known isoelectronic species containing the same number of atoms: (a) O_3, (b) OH^-, (c) BH_4^-, (d) NO_3^-.

2.5   How might you estimate or predict (a) the ultraviolet spectrum of CO_2^-, (b) the bond distance in BF, (c) the dissociation energy of F_2, (d) the symmetric stretching frequency of BF_2^+, (e) the bond angles in NH_3^+, (f) the structure of solid $AlPO_4$, and (g) the infrared spectrum of ClO_2^+?

2.6 What are the values of n and m in the anionic species $Fe(CO)_n{}^{-m}$? Hint: Consider $Ni(CO)_4$.

2.7 The Mössbauer tin isomer shifts, relative to $IS = 0.00$ for grey tin metal, are positive for tin(II) salts such as $SnCl_2$, and negative for tin(IV) compounds such as $SnCl_4$. Explain.

2.8 The Mn $2p_{3/2}$ binding energy of MnF_2 is greater than that of MnO_2, even though the manganese oxidation states in these compounds are $+2$ and $+4$, respectively. Explain.

2.9 Using the electronegativity parameters of Table 2.18, calculate atomic charges for NH_3 and CF_4. (Assume sp^3 hybridization for N, C, and F.)

2.10 Which of the following structures for $N_2F_3{}^+$ is more stable? (The lines do not necessarily indicate *single* bonds; they merely indicate bonds.)

$$\left[\begin{array}{c} F \\ | \\ N-N-F \\ | \\ F \end{array}\right]^+ \qquad \left[\begin{array}{c} F \\ N-N \\ F \qquad F \end{array}\right]^+$$

2.11 Draw valence-bond octet-satisfying structures for the following species, showing all nonbonding electron pairs and formal charges. Indicate when other resonance structures are important. Indicate the shapes of the molecules.

$$HNO_3 \quad NOF \quad NSCl \quad NOF_3 \quad NSF_3 \quad ClO_3{}^- \quad N_3{}^- \quad PH_2{}^- \quad SbCl_5{}^{2-} \quad IO_2F_2{}^-$$

2.12 Which of the following molecules would you expect to have permanent dipole moments?

$$SO_2 \quad CO_2 \quad BF_3 \quad NF_3 \quad SiF_4 \quad XeF_4 \quad SF_4 \quad SF_2 \quad C_2H_2$$
$$N_2F_4 \quad \textit{trans-}N_2F_2 \quad SiH_3Cl \quad O_3 \quad PF_5 \quad BrF_5 \quad S_2Cl_2$$

2.13 Predict structures, including estimated bond distances and bond angles, for the molecules ClO_2F, N_2O_5, and N_4O.

2.14 Estimate the heats of formation of the gaseous molecules N_3H_5 and $(GeH_3)_2Se$.

2.15 Rationalize the following bond-angle data for alkaline-earth dihalide *molecules*.

MX_2	$\angle$ XMX, degrees
MgF_2	158°
CaF_2	140°
SrF_2	108°
BaF_2	$\sim 100°$
$CaCl_2$	$\sim 180°$
$SrCl_2$	120°
$BaCl_2$	$\sim 100°$

3

MOLECULAR ORBITAL THEORY

SIMPLE LCAO THEORY

In Chap. 1, we discussed the electronic orbitals associated with an atomic nucleus, that is, atomic orbitals. In this chapter we shall discuss the orbitals associated with a set of two or more atomic nuclei, that is, molecular orbitals. We shall make the reasonable approximation that a molecular orbital (MO) is a linear combination of atomic orbitals (LCAO). The approximation is based on the assumption that, as a valence electron circulates among the nuclei, it will at any one time be much closer to one nucleus than any other and the MO wave function will be fairly well represented by a valence atomic orbital for that nucleus. To illustrate this idea we shall develop the MOs and their energies for the case of a single valence electron associated with two atoms. The results are applicable to molecule ions such as H_2^+ (in which two $1s$ atomic orbitals are involved in the bonding), Li_2^+ (two $2s$ orbitals), and LiH^+ (a $1s$ orbital and a $2s$ orbital). (In the case of Li_2^+ and LiH^+ we make the further very good approximation that the $1s^2$ closed shells are not involved in bonding.)

The Schrödinger equation which was introduced in Chap. 1 may be rearranged as follows:

$$\left[V - \frac{h^2}{8\pi^2 m} \left(\frac{\partial^2 \psi}{\partial x^2} + \frac{\partial^2 \psi}{\partial y^2} + \frac{\partial^2 \psi}{\partial z^2} \right) \right] \psi = E\psi$$

The left side of the equation can be considered the action of an operator (called the hamiltonian operator) on ψ, where ψ is now a molecular orbital. Therefore we can abbreviate the expression as

$$H\psi = E\psi \tag{3.1}$$

where H is the hamiltonian operator. Just as in the case of atomic wave functions, the MO wave function ψ can be either positive or negative, and ψ^2 is a quantity proportional to electron density. If we multiply both sides of Eq. 3.1 by ψ and integrate over all space, we obtain

$$\int \psi H \psi \, d\tau = E \int \psi^2 \, d\tau \tag{3.2}$$

In Eq. 3.2 we have allowed a single integral sign to stand for a triple integral sign and have made the substitution $d\tau = dx \, dy \, dz$. By simple rearrangement we obtain the following expression for the MO energy:

$$E = \frac{\int \psi H \psi \, d\tau}{\int \psi^2 \, d\tau} \tag{3.3}$$

The wave function ψ is represented by the following linear function,

$$\psi = c_1 \phi_1 + c_2 \phi_2 \tag{3.4}$$

where ϕ_1 and ϕ_2 are the atomic orbital wave functions of atoms 1 and 2, and c_1 and c_2 are coefficients to be determined. Combination of Eqs. 3.3 and 3.4 yields

$$E = \frac{\int (c_1 \phi_1 + c_2 \phi_2) H (c_1 \phi_1 + c_2 \phi_2) \, d\tau}{\int (c_1 \phi_1 + c_2 \phi_2)^2 \, d\tau}$$

$$= \frac{\int (c_1 \phi_1 H c_1 \phi_1 + c_1 \phi_1 H c_2 \phi_2 + c_2 \phi_2 H c_1 \phi_1 + c_2 \phi_2 H c_2 \phi_2) \, d\tau}{\int (c_1{}^2 \phi_1{}^2 + 2 c_1 c_2 \phi_1 \phi_2 + c_2{}^2 \phi_2{}^2) \, d\tau}$$

Now we shall accept, without proof,[1] the facts that

$$H c_1 \phi_1 = c_1 H \phi_1$$

and

$$\int \phi_1 H \phi_2 \, d\tau = \int \phi_2 H \phi_1 \, d\tau$$

[1] M. C. Day, Jr., and J. Selbin, "Theoretical Inorganic Chemistry," 2d ed., pp. 165–166, 571–574, Reinhold, New York, 1969.

Therefore we may write

$$E = \frac{c_1{}^2 \int \phi_1 H\phi_1 \, d\tau + 2c_1 c_2 \int \phi_1 H\phi_2 \, d\tau + c_2{}^2 \int \phi_2 H\phi_2 \, d\tau}{c_1{}^2 \int \phi_1{}^2 \, d\tau + 2c_1 c_2 \int \phi_1 \phi_2 \, d\tau + c_2{}^2 \int \phi_2{}^2 \, d\tau}$$

For simplification, we make the following substitutions:

$$H_{11} = \int \phi_1 H\phi_1 \, d\tau$$

$$H_{22} = \int \phi_2 H\phi_2 \, d\tau$$

$$H_{12} = \int \phi_1 H\phi_2 \, d\tau$$

$$S_{11} = \int \phi_1{}^2 \, d\tau$$

$$S_{22} = \int \phi_2{}^2 \, d\tau$$

$$S_{12} = \int \phi_1 \phi_2 \, d\tau$$

Hence

$$E = \frac{c_1{}^2 H_{11} + 2c_1 c_2 H_{12} + c_2{}^2 H_{22}}{c_1{}^2 S_{11} + 2c_1 c_2 S_{12} + c_2{}^2 S_{22}} \tag{3.5}$$

We are interested in determining the minimum value of E, corresponding to the equations

$$\left(\frac{\partial E}{\partial c_1}\right)_{c_2} = 0 \quad \text{and} \quad \left(\frac{\partial E}{\partial c_2}\right)_{c_1} = 0$$

Differentiation yields the equations

$$c_1(H_{11} - ES_{11}) + c_2(H_{12} - ES_{12}) = 0$$
$$c_1(H_{12} - ES_{12}) + c_2(H_{22} - ES_{22}) = 0$$

These are called the "secular equations." A nontrivial solution to these equations can be expressed in terms of the "secular determinant":

$$\begin{vmatrix} H_{11} - ES_{11} & H_{12} - ES_{12} \\ H_{12} - ES_{12} & H_{22} - ES_{22} \end{vmatrix} = 0 \tag{3.6}$$

The terms H_{11} and H_{22} are called "coulomb integrals." From our previous definition and Eq. 3.3, we see that a coulomb integral is approximately the energy of an electron in the valence atomic orbital, α. At least this approximation is reasonable for a neutral molecule, in which electron-electron and nucleus-nucleus repulsions somewhat compensate. Hence we may write $H_{11} = \alpha_1$ and $H_{22} = \alpha_2$. The term H_{12} is called the "resonance integral" and is essentially the interaction energy of the two atomic orbitals, β. Both α and β have negative values.

If we assume that the atomic orbital wave functions ϕ_1 and ϕ_2 of Eq. 3.4 are "normalized," then

$$S_{11} = \int \phi_1{}^2 \, d\tau = S_{22} = \int \phi_2{}^2 \, d\tau = 1 \tag{3.7}$$

Equation 3.7 simply states that the probability of finding an electron in the orbital is exactly unity. The term S_{12} is called the "overlap integral" because it is a measure of the extent to which orbitals 1 and 2 overlap. For simplification we shall omit the subscripts and write S for the overlap integral. The secular determinant reduces to

$$\begin{vmatrix} \alpha_1 - E & \beta - ES \\ \beta - ES & \alpha_2 - E \end{vmatrix} = 0 \tag{3.8}$$

For a homonuclear species such as $H_2{}^+$, we may substitute $\alpha_1 = \alpha_2 = \alpha$. The determinantal equation then corresponds to

$$(\alpha - E)^2 = (\beta - ES)^2$$

The two solutions of this equation are

$$\alpha - E = -(\beta - ES) \quad \text{or} \quad E = \frac{\alpha + \beta}{1 + S} \tag{3.9}$$

and

$$\alpha - E = (\beta - ES) \quad \text{or} \quad E = \frac{\alpha - \beta}{1 - S} \tag{3.10}$$

By appropriate substitution in the first secular equation, we obtain

$$c_1(\alpha - E) + c_2(\beta - ES) = 0$$

or

$$c_1 = -\frac{\beta - ES}{\alpha - E} c_2$$

From this relation it can be seen that, when $E = (\alpha + \beta)/(1 + S)$, then $c_1 = c_2$, and when $E = (\alpha - \beta)/(1 - S)$, then $c_1 = -c_2$. Thus the molecular orbital wave function can be written as follows:

$$\psi = c_1\phi_1 \pm c_1\phi_2$$

To evaluate c_1, we must normalize the wave function:

$$\int \psi^2 \, d\tau = c_1^2 \int \phi_1^2 \, d\tau \pm 2c_1^2 \int \phi_1\phi_2 \, d\tau + c_1^2 \int \phi_2^2 \, d\tau = 1$$

$$= c_1^2 S_{11} \pm 2c_1^2 S + c_1^2 S_{22} = 1$$

Hence

$$c_1^2(2 \pm 2S) = 1$$

and

$$c_1 = \pm \frac{1}{\sqrt{2 \pm 2S}}$$

The $+$ sign under the radical sign corresponds to $c_1 = c_2$, and the $-$ sign under the radical sign corresponds to $c_1 = -c_2$. Obviously the following wave functions are normalized.

$$\psi_B = \frac{1}{\sqrt{2 + 2S}} (\phi_1 + \phi_2) \tag{3.11}$$

$$\psi_A = \frac{1}{\sqrt{2 - 2S}} (\phi_1 - \phi_2) \tag{3.12}$$

The valence-electron density is obtained by squaring these functions:

$$\psi_B^2 = \frac{1}{2 + 2S} (\phi_1^2 + \phi_2^2 + 2\phi_1\phi_2)$$

$$\psi_A^2 = \frac{1}{2 - 2S} (\phi_1^2 + \phi_2^2 - 2\phi_1\phi_2)$$

ψ_B^2 shows an increase in electron density in the region of overlap between the atoms over that of the individual atoms. Such an electron distribution stabilizes the system, and we refer to ψ_B as the "bonding" MO. The energy level of ψ_B is given by $E = (\alpha + \beta)/(1 + S)$. ψ_A^2 shows a decrease in electron density in the overlap region, and the system is unstable relative to the separate atoms. We refer to ψ_A as the "antibonding" MO, for which $E = (\alpha - \beta)/(1 - S)$.

In Fig. 3.1 is shown a plot of ϕ_1^2, ϕ_2^2, ψ_B^2, and ψ_A^2 along the internuclear line. The dashed lines indicate ϕ_1^2 and ϕ_2^2, that is, the electron density of the individual atomic orbitals. The lower solid line indicates ψ_A^2, the electron density of the antibonding MO, and the upper solid line indicates ψ_B^2, the electron density of the bonding MO. Figure 3.2 is an energy level diagram which graphically indicates the energies of the two MOs which arise from the interaction of two atomic orbitals. Overlap integrals are generally fairly small (often in the range

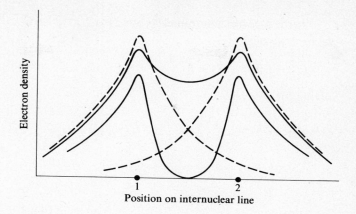

FIGURE 3.1
Plot of electron densities for the orbitals ϕ_1 and ϕ_2 (dashed lines), ψ_B (upper solid line), and ψ_A (lower solid line) along the internuclear axis of H_2^+.

0.2 to 0.3); hence the antibonding MO is destabilized approximately the same amount that the bonding MO is stabilized. As a matter of fact, in simple LCAO theory, it is often assumed that $S = 0$. This assumption simplifies the calculations and is actually not as drastic an approximation as it might appear.[1] With this approximation, the energy levels of ψ_B and ψ_A are $\alpha + \beta$ and $\alpha - \beta$, respectively.

[1] A. Streitwieser, Jr., "Molecular Orbital Theory for Organic Chemists," pp. 101–103, Wiley, New York, 1961.

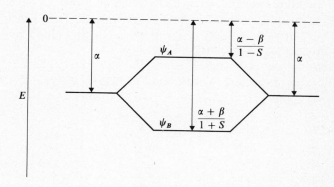

FIGURE 3.2
Energy level diagram for the molecular orbitals formed from similar atomic orbitals in a homonuclear molecule.

In the case of a heteronuclear bond (such as in LiH^+), the secular determinant yields, if we neglect S,

$$(\alpha_1 - E)(\alpha_2 - E) = \beta^2$$

Solving for E gives

$$E = \frac{\alpha_1 + \alpha_2}{2} \pm \frac{1}{2}\sqrt{\alpha_1{}^2 + \alpha_2{}^2 + 2\alpha_1\alpha_2 - 4\alpha_1\alpha_2 + 4\beta^2}$$

$$= \frac{\alpha_1 + \alpha_2}{2} \pm \frac{1}{2}\sqrt{(\Delta\alpha)^2 + 4\beta^2} \tag{3.13}$$

The corresponding energy level diagram is given in Fig. 3.3. Notice that, according to the approximate LCAO method that we are now employing, the energy of the bonding MO is depressed from that of the more stable atomic orbital by the same amount that the energy of the antibonding MO is raised from that of the less stable atomic orbital. If $|\beta|$ is very small, the energy spread between the bonding and antibonding levels is very little more than the separation between α_1 and α_2, and the MOs are essentially slightly perturbed atomic orbitals.

The energy level diagrams of Figs. 3.2 and 3.3 can be used to predict the electronic configurations of systems containing one to four valence electrons. The MOs are filled in the order of increasing energy. Table 3.1 gives the electron configurations for four species with $1s$ valence atomic orbitals to illustrate the filling of the ψ_B ($1s\sigma_B$) and ψ_A ($1s\sigma^*$) MOs. Asterisks are used to designate antibonding MOs, except in the case of ψ_A.[1] In MO theory, bond order is defined as one-half the number of bonding electrons minus one-half the number of antibonding electrons. Hence the species in Table 3.1 have the bond orders indicated in the third column. Note that the molecule He_2, which has never been observed, would have a bond order of zero, corresponding to no net bonding between the atoms.

---

[1] The symbol ψ^* is reserved for the complex conjugate of ψ.

Table 3.1

Species	Electron configuration	Bond order
$H_2{}^+$	$(1s\sigma_B)^1$	$\frac{1}{2}$
H_2	$(1s\sigma_B)^2$	1
HHe	$(1s\sigma_B)^2(1s\sigma^*)^1$	$\frac{1}{2}$
He_2	$(1s\sigma_B)^2(1s\sigma^*)^2$	0

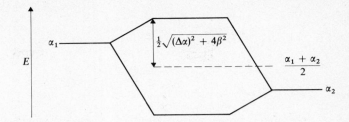

FIGURE 3.3
Energy level diagram for the molecular orbitals formed from dissimilar atomic orbitals in a heteronuclear molecule. The overlap integral has been neglected.

CRITERIA FOR STABLE MOLECULAR ORBITALS

The energy $-\beta$ may be taken as a measure of the covalent bond energy. It is commonly assumed in simple MO theory that $-\beta$ is proportional to the corresponding overlap integral S and to the average energy of the atomic orbitals. Thus stable covalent bonds can form whenever the overlap integral is large. The following three criteria must be met to ensure large values of S.

1 The energies of the atomic orbitals must be comparable. A small negative value for α corresponds to a large, diffuse atomic orbital, and a large negative value for α corresponds to a small, compact atomic orbital. Such orbitals overlap poorly, as indicated by the following diagram:

Poor overlap; low S Good overlap; high S

Examples of diatomic molecules which have low overlap integrals because of a mismatch of atomic orbital energies are KF and the hypothetical molecule NeS. In spite of a low β value, the KF molecule is stable because of the large amount of energy released on transferring a valence electron from potassium to the bonding MO, which has an energy slightly lower than that of the fluorine atomic orbital. The bonding in KF is principally ionic. The NeS molecule has very low stability because very little energy is released on transferring two electrons from the neon atomic orbital to the bonding MO.

2 The atomic orbitals must be positioned so that good overlap can occur. For example, consider internal hydrogen bonding in nitrophenol. Such bonding is very extensive in the ortho isomer but essentially nonexistent in the meta isomer. For another example, consider the C—C bonds in cyclopropane and cyclohexane. The C—C—C bond angles in cyclopropane are 60°, which is much lower than the tetrahedral angle or even the 90° angle corresponding to use of pure p orbitals. Obviously the C—C bonds in cyclopropane must be "bent"; i.e., the overlapping atomic orbitals meet at an angle less than 180°. Such overlap is poorer (that is, S is lower) than in the case of normal "straight" bonds such as probably exist in cyclohexane, in which the C—C—C bond angles are essentially the tetrahedral angle. Evidence for the strain in the cyclopropane configuration is found in the high exothermicity of the conversion of cyclopropane into cyclohexane:

$$2C_3H_6 \rightarrow C_6H_{12} \qquad \Delta H^\circ = -54.9 \text{ kcal mol}^{-1}$$

3 The atomic orbitals must have the same, or approximately the same, symmetry with respect to the bond axis. (Symmetry is briefly discussed in Appendix B.) For example, consider the overlap between two orbitals on atoms whose z axis lie on the bond axis. In the case of an s orbital and p_y orbital, the overlapping volume element on one side of the p_y orbital has the same magnitude but opposite sign as that on the other side of the p_y orbital. This fact is obvious from Fig. 3.4. Thus the integration over all space is exactly zero; that is, $S = 0$. The atomic orbitals are said to be "orthogonal." On the other hand, the overlap of an s orbital and a p_z orbital is finite, and $S \neq 0$. Such orbitals are "nonorthogonal." As another example, consider the overlap of a p_y orbital on one atom with a d orbital on another atom. In the case of d_{xz}, d_{xy}, $d_{x^2-y^2}$, and d_{z^2} orbitals, the overlap integral is exactly zero. However, in the case of the d_{yz} orbital, S is finite. The reader is invited to ponder the orthogonalities of all the other possible combinations of s, p, and d orbitals.

Molecular orbitals which are symmetric with respect to rotation around the internuclear axis (as, for example, those formed from $s + s$, $s + p_z$, $p_z + p_z$, $p_z + d_{z^2}$, etc.) are called sigma (σ) MOs. Those which have a nodal plane containing the internuclear axis (for example, $p_x + p_x$, $p_y + d_{yz}$, etc.) are called pi (π) MOs. The relatively rare MOs with two nodal planes along the internuclear axis (for example, $d_{xy} + d_{xy}$) are called delta (δ) MOs.

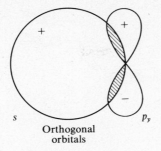 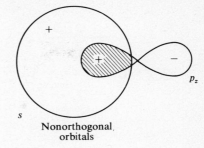

Orthogonal orbitals

Nonorthogonal orbitals

FIGURE 3.4

Overlap between s and p orbitals. In the case of the p_y orbital, $S = 0$; in the case of the p_z orbital, $S > 0$.

DIATOMIC MOLECULES

Let us now consider all the MOs which form as a consequence of the overlap of the valence-shell s and p orbitals in a homonuclear diatomic molecule. The two s orbitals yield an $s\sigma_B$ and an $s\sigma^*$ MO, and the six p orbitals yield a $p\sigma_B$ and a $p\sigma^*$ MO, and two equivalent $p\pi_B$ and $p\pi^*$ MOs. Because the $p\pi$ overlap is poorer than the $p\sigma$ overlap, $|\beta_{p\sigma}| > |\beta_{p\pi}|$. If there were no interaction between the s and p_z orbitals, the energy level diagram would look like that shown in Fig. 3.5a. However, in the molecules of the first-row elements, a significant amount of $2s$-$2p_z$ overlap occurs, causing the energies of the $2s\sigma$ orbitals to be depressed and the energies of the $2p\sigma$ orbitals to be raised. Consequently the $2p\pi_B$ levels lie below the $2p\sigma_B$ level, as shown in Fig. 3.5b.

Just as the electronic configurations of the elements can be deduced by appropriate filling of the atomic orbitals, the electronic configurations of the first-row diatomic molecules can be deduced by filling the molecular orbitals

Table 3.2 DIATOMIC SPECIES OF FIRST-ROW ELEMENTS

Species	Valence electron configuration	Unpaired electrons	Bond order	D, kcal mol^{-1}	r, Å
Li_2	$(s\sigma_B)^2$	0	1	26	2.67
Be_2	$(s\sigma_B)^2(s\sigma^*)^2$	0	0		
B_2	$(s\sigma_B)^2(s\sigma^*)^2(p\pi_B)^2$	2	1	69	1.59
C_2	$(s\sigma_B)^2(s\sigma^*)^2(p\pi_B)^4$	0	2	143	1.31
N_2	$(s\sigma_B)^2(s\sigma^*)^2(p\pi_B)^4(p\sigma_B)^2$	0	3	226	1.10
NO	$(s\sigma_B)^2(s\sigma^*)^2(p\pi_B)^4(p\sigma_B)^2(p\pi^*)^1$	1	2.5	151	1.15
O_2	$(s\sigma_B)^2(s\sigma^*)^2(p\pi_B)^4(p\sigma_B)^2(p\pi^*)^2$	2	2	119	1.21
O_2^-	$(s\sigma_B)^2(s\sigma^*)^2(p\pi_B)^4(p\sigma_B)^2(p\pi^*)^3$	1	1.5	$\cdots$	1.28
F_2	$(s\sigma_B)^2(s\sigma^*)^2(p\pi_B)^4(p\sigma_B)^2(p\pi^*)^4$	0	1	37	1.42
Ne_2	$(s\sigma_B)^2(s\sigma^*)^2(p\pi_B)^4(p\sigma_B)^2(p\pi^*)^4(p\sigma^*)^2$	0	0		

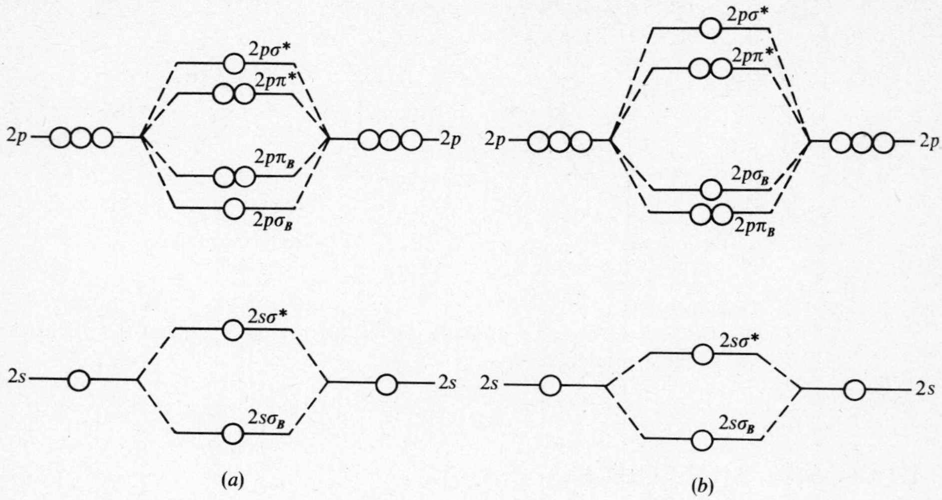

FIGURE 3.5
Energy level diagrams for a homonuclear diatomic molecule containing first-row atoms. (a) corresponds to no interaction between $2s$ and $2p$ levels; (b) to substantial $2s$-$2p$ interaction.

shown in Fig. 3.5b. The configurations for the homonuclear molecules from Li_2 to Ne_2, as well as those of NO and O_2^-, are given in Table 3.2. It should be noted that, as expected, dissociation energy increases and bond distance decreases with increasing bond order. The series O_2^+ ($r = 1.122$ Å), O_2 ($r = 1.21$ Å), O_2^- ($r = 1.28$ Å), and O_2^{2-} ($r = 1.49$ Å) also shows the expected inverse correlation of bond order and bond distance.

Probably the most significant feature of the simple MO treatment of the diatomic molecules is the ease with which the paramagnetism of B_2 and O_2 is accounted for. In each of these molecules, a pair of degenerate $p\pi$ MOs contain two electrons. According to Hund's rule, these electrons occupy separate orbitals and have parallel spins. Thus, although each of these molecules has an even number of electrons, each has a permanent magnetic moment and is paramagnetic. Another valuable achievement of the simple MO treatment is the prediction of zero bond order (that is, no net bonding) in Be_2. This result is not at all obvious from simple valence pictures.

In heteronuclear molecules such as CO, NO, CS, SiC, etc., the atomic orbital energies are not equal. Nevertheless, most data indicate that the order of MO levels in such molecules is very similar to that indicated in Fig. 3.5b, and the bonding can be discussed in terms of these MOs.

SYMMETRY AND POLYATOMIC MOLECULES

In symmetric molecules, there are sets of equivalent atoms and atomic orbitals. When considering the molecular orbitals of such molecules, it is helpful to form linear combinations of the equivalent atomic orbitals, called "group orbitals." These group orbitals can then be treated just like individual atomic orbitals; they can be combined with other atomic orbitals or with other group orbitals to form molecular orbitals. However, no matter what types of orbitals are combined to form MOs, they must have the same symmetry properties and they must have satisfactory overlap.[1]

To illustrate the use of group orbitals, we shall discuss the MOs of the triiodide ion, I_3^-. From valence-shell electron repulsion theory (Chap. 2) we know that this is a linear species. The middle atom can be considered as having three nonbonding electron pairs in sp^2 hybrid orbitals in a plane normal to the bonds. Each of the terminal atoms also has a set of three nonbonding electron pairs, but the orientation of these pairs is not certain; they probably occupy orbitals having somewhat more p character than sp^2. Thus we have three collinear atomic orbitals which can be involved in σ bonding: a p orbital on the middle atom, and hybrid orbitals on the terminal atoms. The two equivalent hybrid orbitals on the terminal atoms, ϕ_1 and ϕ_3, can be combined into two group orbitals:

$$\phi_G = \frac{1}{\sqrt{2}}(\phi_1 \pm \phi_3)$$

These combinations are pictured below:

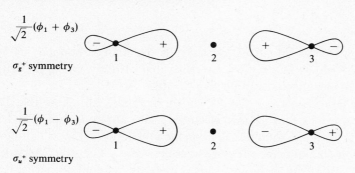

$\frac{1}{\sqrt{2}}(\phi_1 + \phi_3)$

σ_g^+ symmetry

$\frac{1}{\sqrt{2}}(\phi_1 - \phi_3)$

σ_u^+ symmetry

---

[1] Group theoretical labels such as σ_g^+, b_1, t_2, etc., are useful for designating the symmetry properties of orbitals and combinations of orbitals. A full understanding of their significance is not necessary to follow the discussions of this text. A brief introduction to the subject is given in Appendix B.

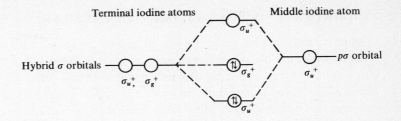

FIGURE 3.6
Energy level diagram for the sigma molecular orbitals of $I_3{}^-$ formed from the atomic orbitals lying on the bonding axis. Valence-shell d orbitals are ignored.

Note that the $\sigma_g{}^+$ group orbital[1] is symmetric with respect to inversion through the center of the $I_3{}^-$ ion and that the $\sigma_u{}^+$ group orbital is antisymmetric with respect to inversion. The p orbital on the middle atom is antisymmetric with respect to inversion; like the $\phi_1 - \phi_3$ group orbital, it has $\sigma_u{}^+$ symmetry.[1]

The overlap integral of the symmetric group orbital and the middle p orbital is exactly zero. Thus only the antisymmetric group orbital can interact with the middle p orbital to form a bonding and an antibonding MO. These MOs are represented as follows:

$$\psi = \frac{1}{\sqrt{2}}\,\phi_2 \pm \frac{1}{2}\,(\phi_1 - \phi_3)$$

The symmetric group orbital $(1/\sqrt{2})(\phi_1 + \phi_3)$, is essentially a nonbonding MO. The reader may wish, as an exercise, to prove that all three MOs, as written, are normalized (under the assumption that $S = 0$).

An energy level diagram for $I_3{}^-$ is shown in Figure 3.6. Note that these orbitals are occupied by four valence electrons. Two are nonbonding and two are bonding. Obviously we should look upon the I—I bonds in $I_3{}^-$ as half-bonds, in agreement with simple octet structure considerations. The other 18 valence electrons occupy the 9 nonbonding orbitals that we referred to earlier. The nonbonding orbital of Fig. 3.6 is shared by the terminal atoms; thus $3\frac{1}{2}$ lone pairs are located on each terminal atom.

---

[1] See footnote on page 79 .

The model of I_3^- which we have just described completely ignores valence-shell d-orbital participation. If we choose to have the z axis of the middle atom lie along the bonds, we see that the d_{z^2} orbital of the middle atom has the same symmetry as that of the symmetric group orbital $(1/\sqrt{2})(\phi_1 + \phi_3)$. Now it is likely that the overlap between this d_{z^2} orbital and the terminal-atom hybrid orbitals is very poor; the orbital energies are probably quite far apart. However, if we are willing to admit a weak interaction between the middle-atom d_{z^2} orbital and the symmetric group orbital, we may draw a modified energy level diagram as shown in Fig. 3.7. Most evidence indicates that d-p bonding of the type shown is very weak; for all practical purposes the σ_g^+ orbital is nonbonding.[1]

Let us now consider the π bonding in the ozone molecule. Assume that this V-shaped molecule lies in the yz plane; we are then concerned with the overlap of the p_x orbitals on each oxygen atom. The two terminal p_x orbitals are equivalent; their corresponding group orbitals are

$$\phi_G = \frac{1}{\sqrt{2}}(\phi_1 \pm \phi_3)$$

These may be shown as the following projections:

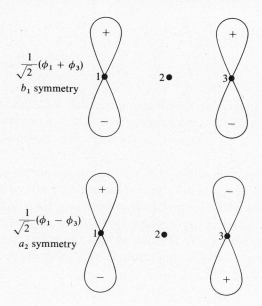

$\dfrac{1}{\sqrt{2}}(\phi_1 + \phi_3)$

b_1 symmetry

$\dfrac{1}{\sqrt{2}}(\phi_1 - \phi_3)$

a_2 symmetry

[1] T. B. Brill, *J. Chem. Educ.*, **50**, 392 (1973); C. D. Cornwell and R. S. Yamasaki, *J. Chem. Phys.*, **27**, 1060 (1957).

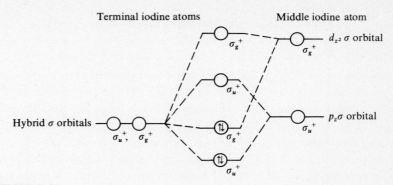

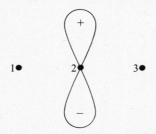

FIGURE 3.7
Energy level diagram for the sigma molecular orbitals of I_3^- formed from the atomic orbitals lying on the bonding axis. Valence-shell d_{z2} orbital is included.

The p_x orbital on the middle oxygen atom has the same symmetry, b_1, as the $(1/\sqrt{2})(\phi_1 + \phi_3)$ group orbital.

Therefore these orbitals combine to form antibonding and bonding MOs, and the group orbital of a_2 symmetry is essentially a nonbonding MO. The corresponding energy level diagram is shown in Fig. 3.8. Inasmuch as there are only two bonding π electrons, the π bond order in O_3 is $\frac{1}{2}$.

Finally we shall consider the σ bonding in a tetrahedral species such as BH_4^-, CH_4, NH_4^+, ClO_4^-, PCl_4^+, etc. The central atom has an s orbital and three p orbitals in its valence shell which are involved in bonding. Each ligand atom has one atomic orbital involved in the bonding (a $1s$ orbital in the case of a hydrogen atom, and a hybrid orbital in the case of an oxygen or chlorine atom). The group orbital wave function which can combine with the central-atom s orbital is

$$\phi_G = \frac{1}{\sqrt{4}} (\phi_1 + \phi_2 + \phi_3 + \phi_4)$$

where ϕ_1, ϕ_2, ϕ_3, and ϕ_4 are the individual ligand orbital wave functions.

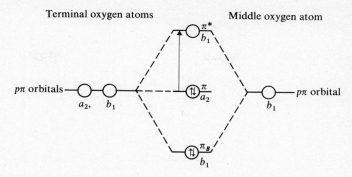

FIGURE 3.8
Energy level diagram for the pi molecular orbitals of ozone, O_3. The vertical arrow indicates $\pi \to \pi^*$ transition energy.

The corresponding MO wave functions are

$$\psi = \frac{1}{\sqrt{2}}\phi_s \pm \frac{1}{2\sqrt{2}}(\phi_1 + \phi_2 + \phi_3 + \phi_4)$$

In order to see which combinations of ligand orbitals can combine with the central-atom p orbitals, it is helpful to refer to Fig. 3.9. Figure 3.9a shows the bonding combination,

$$\psi_B(p_z) = \frac{1}{\sqrt{2}}\phi_{p_z} + \frac{1}{2\sqrt{2}}(\phi_1 + \phi_2 - \phi_3 - \phi_4)$$

and Fig. 3.9b shows the antibonding combination,

$$\psi_A(p_z) = \frac{1}{\sqrt{2}}\phi_{p_z} - \frac{1}{2\sqrt{2}}(\phi_1 + \phi_2 - \phi_3 - \phi_4)$$

Similarly, for the p_x and p_y orbitals, we write

$$\psi_B(p_x) = \frac{1}{\sqrt{2}}\phi_{p_x} + \frac{1}{2\sqrt{2}}(-\phi_1 + \phi_2 - \phi_3 + \phi_4)$$

$$\psi_A(p_x) = \frac{1}{\sqrt{2}}\phi_{p_x} - \frac{1}{2\sqrt{2}}(-\phi_1 + \phi_2 - \phi_3 + \phi_4)$$

$$\psi_B(p_y) = \frac{1}{\sqrt{2}}\phi_{p_y} + \frac{1}{2\sqrt{2}}(-\phi_1 + \phi_2 + \phi_3 - \phi_4)$$

$$\psi_A(p_y) = \frac{1}{\sqrt{2}}\phi_{p_y} - \frac{1}{2\sqrt{2}}(-\phi_1 + \phi_2 + \phi_3 - \phi_4)$$

Because of the symmetry of the molecule, the p_x, p_y, and p_z orbitals are equivalent. Therefore $\psi_B(p_x)$, $\psi_B(p_y)$, and $\psi_B(p_z)$ are degenerate; i.e., they have the same

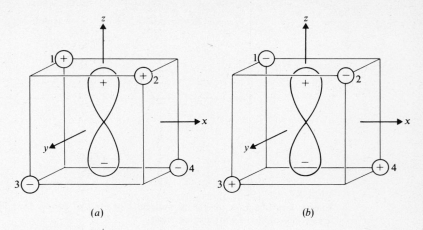

(a)   (b)

FIGURE 3.9
(a) Bonding and (b) antibonding combinations of the hydrogen s orbitals and a carbon $2p_z$ orbital in methane. Similar combinations can be drawn for the degenerate $2p_x$ and $2p_y$ orbitals.

energy levels. The same is true of $\psi_A(p_x)$, $\psi_A(p_y)$, and $\psi_A(p_z)$. Thus the energy level diagram for this system, in Fig. 3.10, shows that the eight bonding electrons occupy a single orbital (of a_1 symmetry) and three degenerate orbitals (of t_2 symmetry). Although the four σ bonds are equivalent, there are essentially two kinds of bonding electrons, with entirely different ionization potentials. The separate ionization potentials corresponding to the two MOs are obvious in the ultraviolet photoelectron spectrum of Fig. 3.11.

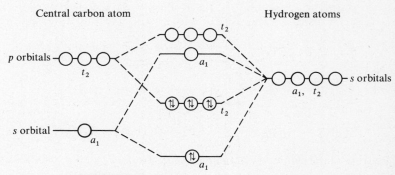

FIGURE 3.10
Energy level diagram for the valence molecular orbitals of methane.

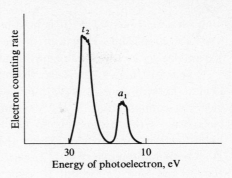

FIGURE 3.11
Ultraviolet photoelectron spectrum of methane. The photon energy was 41 eV; the abscissa is graduated in the kinetic energy of the ejected electron. [*Reproduced with permission from W. C. Price et al., in D. A. Shirley (ed.), "Electron Spectroscopy," pp. 187–198, (North-Holland, Amsterdam, 1972.*]

HYBRIDIZATION[1]

Although the concept of hybridization is strictly a part of valence bond theory, we have postponed discussion of this topic until now, when the reader is familiar with the concept of linear combinations of atomic orbitals and can directly compare the application of hybridization theory and MO theory to a particular molecule.

We know from experiment that the methane molecule is tetrahedral and that the four C—H bonds are equivalent. If we assume that the C—H bonds are single bonds, formed by combination of the hydrogen $1s$ orbitals with the carbon $2s$ and $2p$ orbitals, then each bond must involve, or "use," one-quarter of the carbon $2s$ orbital and one-quarter of the three carbon $2p$ orbitals. If we wish to describe the bonding of methane in terms of localized two-electron bonds we can assume that the one $2s$ and three $2p$ orbitals of carbon "mix" to form four equivalent hybrid orbitals directed toward the hydrogen atoms. Then each bond can be considered the result of the overlap of an sp^3 hybrid orbital of carbon and a $1s$ orbital of hydrogen. If we let the symbols s, p_x, p_y and p_z stand for the valence atomic orbitals of carbon, the four hybrid orbitals can be represented as the following linear combinations:

$$\phi_1 = \tfrac{1}{2}(s + p_x + p_y + p_z)$$
$$\phi_2 = \tfrac{1}{2}(s + p_x - p_y - p_z)$$
$$\phi_3 = \tfrac{1}{2}(s - p_x - p_y + p_z)$$
$$\phi_4 = \tfrac{1}{2}(s - p_x + p_y - p_z)$$

Similar sets of hybrid orbitals can be constructed from other combinations of atomic orbitals. Some of these types of hybridization are listed in Table 3.3.

---

[1] L. Pauling, "The Nature of the Chemical Bond," 3d ed., pp. 111–123, Cornell University Press, Ithaca, N.Y., 1960; J. E. Huheey, "Inorganic Chemistry," pp. 114–120, Harper & Row, New York, 1972; F. A. Cotton and G. Wilkinson, "Advanced Inorganic Chemistry," 3d ed., pp. 87–97, Wiley-Interscience, New York, 1972; F. A. Cotton, "Chemical Applications of Group Theory," 2d ed., chap. 8, Wiley-Interscience, New York, 1971.

In each case, except dsp^3, it is possible to construct the hybrid orbitals so that they are equivalent. However, in the case of trigonal bipyramidal and square pyramidal geometries, such equivalence is obviously impossible.

It is instructive to discuss the bonding in the water molecule in terms of orbital hybridization and valence-shell electron repulsion. The ground-state electron configuration of the oxygen atom is $2s^2 2p^4$, with two unpaired $2p$ electrons. If this atom were to react with two hydrogen atoms without any hybridization, one would expect the resulting water molecule to have an H—O—H bond angle of 90°. In such a molecule the lone pairs would occupy the $2s$ and one $2p$ orbital or, what is precisely equivalent, two sp hybrid orbitals directed at right angles to the plane of the molecule. Obviously in such a molecule there would be particularly strong repulsion between the lone-pair electrons and the bonding electrons. Now let us assume that we use an oxygen atom in which the two lone pairs occupy hybrid orbitals which have more than 50 percent p character and in which the unpaired electrons occupy orbitals which have less than 100 percent p character. Energy is required to promote a ground-state oxygen atom to this hypothetical hybridized state, because the promotion corresponds to the excitation of a fraction of an electron from a $2s$ orbital to a $2p$ orbital. (If a complete electron were excited, one would obtain an atom with the configuration $2s^1 2p^5$.) In the hybridized oxygen atom, the angle between the lone pairs would be less than 180°, and the promotion energy can be partly ascribed to the resultant extra electron-electron repulsion. The hybridized atom would react with two hydrogen atoms to form a molecule in which the H—O—H bond angle is greater than 90° and in which the angle between either lone pair and either O—H bond is greater than 90°. The promotion energy would be compensated by the reduction in the electron-electron repulsions and by the fact that an increase in the s character of an orbital causes an increase in the bond energy. Obviously in the actual water molecule these energy terms are exactly compensating. Such balancing of promotion energy, electron repulsion energy, and bond energy is a common feature of orbital hybridization schemes.

It should be emphasized that the concept of hybridization is essentially a

Table 3.3 SOME COMMON HYBRIDIZATIONS AND THEIR GEOMETRIES

sp	Linear
sp^2	Trigonal (120° bond angles)
sp^3, sd^3	Tetrahedral (109.5° bond angles)
d^2sp^3	Octahedral
dsp^2	Square planar
dsp^3	Trigonal bipyramidal or square pyramidal

mathematical operation which we use mainly for convenience in the description of bonding. One must be careful not to draw incorrect conclusions regarding the excited states of molecules for which the ground states can be described in terms of hybridized orbitals. For example, in the case of methane, if we consider only the four equivalent bonding orbitals corresponding to the use of hybrid sp^3 orbitals, it is not apparent that there should be two different ionization potentials for the bonding electrons. However, evidence for these two ionizations is seen in the photoelectron spectrum of Fig. 3.11, which shows two separate photolines.

CALCULATIONS

Mulliken et al.[1] calculated values of the overlap integral S for various pairs of atomic orbitals, such as $(1s, 1s)$, $(1s, 2p\sigma)$, $(2p\pi, 2p\pi)$, etc. For the atomic orbital wave functions with $n = 1$ to 3, they used approximate Slater[2] functions,

$$\phi = N_{nl} r^{n-1} e^{-\mu r/a_H} \tag{3.14}$$

where N_{nl} is a normalization constant, r is the electron-nucleus distance, a_H is the Bohr radius of the hydrogen atom (0.529 Å), and $\mu = Z^*/n$ (i.e., the effective nuclear charge calculated by the rules outlined in Chap. 1 divided by the principal quantum number). The values of μ for several atoms are given in Table 3.4. Values of S_{ij}, the overlap integral for orbitals i and j, are listed in Tables 3.5 to 3.12 as a function of two parameters, p and t.[1] These parameters are defined as follows,

$$p = 0.946(\mu_i + \mu_j)R \tag{3.15}$$

$$t = \frac{\mu_i - \mu_j}{\mu_i + \mu_j} \tag{3.16}$$

where R is the internuclear distance in angstroms. When the atomic orbitals have different n values, i corresponds to the orbital of smaller n; when the atomic orbitals have the same n value, i corresponds to the larger μ value. A more complete set of tables may be found in the original literature.[1]

[1] R. S. Mulliken, C. A. Rieke, D. Orloff, and H. Orloff, *J. Chem. Phys.*, **17**, 1248 (1949).
[2] J. C. Slater, *Phys. Rev.*, **36**, 57 (1930).

Table 3.4   SLATER μ VALUES FOR VALENCE-SHELL ns, np ATOMIC ORBITALS

H	1.00	C^+	1.80
Li	0.65	N	1.95
Be	0.975	N^+	2.125
B^-	1.125	O^-	2.10
B	1.30	O	2.275
C^-	1.45	O^+	2.45
C	1.625	F	2.60
		Na	0.733

Master Tables for Slater-AO Overlap Integrals[1]

Table 3.5

p	$t = 0.0$	$t = 0.1$	$t = 0.2$	$S(1s, 1s)$ $t = 0.3$	$t = 0.4$	$t = 0.5$	$t = 0.6$	$t = 0.7$	$t = 0.8$
0.0	1.000	0.985	0.941	0.868	0.770	0.650	0.512	0.364	0.216
0.5	0.960	0.946	0.905	0.837	0.744	0.630	0.499	0.357	0.213
1.0	0.858	0.847	0.812	0.756	0.678	0.580	0.465	0.337	0.205
1.2	0.807	0.797	0.766	0.715	0.644	0.554	0.447	0.327	0.200
1.3	0.780								
1.4	0.753	0.744	0.717	0.671	0.608	0.526	0.428	0.316	0.196
1.5	0.725								
1.6	0.697	0.689	0.666	0.626	0.570	0.498	0.409	0.305	0.191
1.7	0.669								
1.8	0.641	0.635	0.615	0.581	0.533	0.469	0.388	0.293	0.185
1.9	0.614								
2.0	0.586	0.581	0.565	0.536	0.495	0.439	0.368	0.281	0.180
2.1	0.560								
2.2	0.533	0.529	0.515	0.493	0.458	0.410	0.348	0.269	0.175
2.3	0.508								
2.4	0.483	0.479	0.469	0.451	0.423	0.382	0.328	0.256	0.169
2.5	0.458								
2.6	0.435	0.432	0.425	0.411	0.388	0.355	0.308	0.244	0.164
2.7	0.412								
2.8	0.390	0.388	0.383	0.373	0.356	0.329	0.289	0.233	0.158
2.9	0.369								
3.0	0.349	0.348	0.344	0.338	0.325	0.304	0.271	0.221	0.153
3.2	0.310	0.310	0.309	0.305	0.297	0.281	0.254	0.210	0.148
3.4	0.275	0.276	0.276	0.274	0.270	0.259	0.237	0.200	0.143
3.5	0.259								
3.6	0.244	0.244	0.246	0.247	0.245	0.238	0.221	0.189	0.138
3.8	0.215	0.216	0.218	0.221	0.222	0.219	0.206	0.180	0.133
4.0	0.189	0.190	0.194	0.198	0.201	0.201	0.192	0.170	0.128
4.2	0.166	0.167	0.171	0.176	0.182	0.184	0.179	0.161	0.123
4.4	0.146	0.147	0.151	0.157	0.164	0.168	0.166	0.152	0.119
4.5	0.136								
4.6	0.127	0.129	0.133	0.140	0.148	0.154	0.155	0.144	0.114
5.0	0.097	0.098	0.103	0.110	0.120	0.129	0.134	0.129	0.106
5.5	0.068	0.069	0.074	0.082	0.091	0.102	0.111	0.112	0.097
6.0	0.047	0.049	0.053	0.060	0.070	0.081	0.092	0.097	0.088
6.5	0.032	0.034	0.037	0.044	0.053	0.064	0.076	0.084	0.080
7.0	0.022	0.023	0.026	0.032	0.040	0.050	0.063	0.073	0.072
7.5	0.015	0.016	0.018	0.023	0.030	0.040	0.052	0.063	0.066
8.0	0.010	0.011	0.013	0.017	0.023	0.031	0.043	0.054	0.059
9.0	0.005	0.005	0.006	0.009	0.013	0.019	0.029	0.040	0.049
10.0	0.002	0.002	0.003	0.004	0.007	0.012	0.019	0.030	0.040

[1] Reproduced with permission from R. S. Mulliken et al., *J. Chem. Phys.* **17**, 1248 (1949).

Table 3.6

$S(1s, 2s)$

p	$t=-0.5$	$t=-0.4$	$t=-0.3$	$t=-0.2$	$t=-0.1$	$t=0.0$	$t=0.1$	$t=0.2$	$t=0.3$	$t=0.4$	$t=0.5$	$t=0.6$	$t=0.7$	$t=0.8$	$t=0.9$
0.0	0.844	0.933	0.977	0.978	0.938	0.866	0.768	0.652	0.526	0.400	0.281	0.177	0.095	0.037	0.007
0.5	0.829	0.916	0.959	0.960	0.923	0.854	0.760	0.647	0.525	0.401	0.284	0.180	0.097	0.039	0.007
1.0	0.787	0.866	0.906	0.907	0.875	0.814	0.730	0.628	0.516	0.400	0.288	0.186	0.102	0.042	0.008
1.5	0.722	0.790	0.823	0.825	0.799	0.749	0.679	0.593	0.496	0.393	0.290	0.192	0.109	0.046	0.009
2.0	0.644	0.697	0.722	0.723	0.702	0.664	0.610	0.542	0.463	0.376	0.285	0.195	0.114	0.050	0.011
2.2	0.611	0.657	0.679	0.679	0.661	0.627	0.579	0.519	0.447	0.367	0.282	0.196	0.116	0.051	0.011
2.4	0.577	0.617	0.635	0.635	0.619	0.589	0.547	0.494	0.430	0.357	0.277	0.195	0.117	0.053	0.012
2.6	0.543	0.578	0.592	0.591	0.576	0.550	0.514	0.468	0.411	0.345	0.272	0.194	0.119	0.054	0.012
2.8	0.510	0.539	0.550	0.547	0.534	0.512	0.481	0.441	0.392	0.333	0.265	0.192	0.119	0.055	0.012
3.0	0.478	0.501	0.508	0.505	0.493	0.474	0.448	0.414	0.372	0.320	0.258	0.190	0.120	0.057	0.013
3.2	0.446	0.464	0.468	0.464	0.453	0.437	0.416	0.388	0.352	0.306	0.251	0.187	0.120	0.058	0.013
3.4	0.416	0.428	0.430	0.425	0.415	0.402	0.384	0.361	0.331	0.292	0.243	0.184	0.120	0.058	0.014
3.6	0.386	0.395	0.394	0.388	0.379	0.368	0.354	0.336	0.311	0.278	0.234	0.180	0.119	0.059	0.014
3.8	....	....	....	....	....	0.335	0.325	0.311	0.291	0.263	0.225	0.176	0.119	0.060	0.014
4.0	0.332	0.333	0.327	0.320	0.312	0.305	0.297	0.287	0.272	0.249	0.216	0.172	0.118	0.060	0.015
4.2	....	....	....	....	....	0.276	0.271	0.264	0.253	0.235	0.207	0.167	0.116	0.061	0.015
4.4	....	....	....	....	....	0.250	0.246	0.242	0.235	0.221	0.198	0.162	0.115	0.061	0.016
4.5	0.272	0.265	0.256	0.247	0.241										
4.6	....	....	....	....	....	0.225	0.223	0.221	0.217	0.208	0.189	0.157	0.113	0.061	0.016
5.0	0.221	0.209	0.197	0.188	0.183	0.181	0.182	0.184	0.185	0.182	0.170	0.147	0.109	0.062	0.017
5.5	....	....	....	....	....	0.136	0.139	0.144	0.149	0.152	0.149	0.134	0.104	0.061	0.017
6.0	0.143	0.127	0.114	0.105	0.101	0.101	0.105	0.111	0.119	0.126	0.128	0.120	0.098	0.060	0.018
6.5	....	....	....	....	....	0.074	0.078	0.085	0.094	0.104	0.110	0.108	0.092	0.059	0.018
7.0	0.091	0.075	0.063	0.056	0.053	0.054	0.058	0.065	0.074	0.085	0.093	0.096	0.086	0.058	0.019
7.5	....	....	....	....	....	0.039	0.042	0.048	0.058	0.068	0.079	0.085	0.079	0.056	0.019
8.0	0.057	0.044	0.034	0.029	0.027	0.027	0.030	0.036	0.044	0.055	0.066	0.075	0.073	0.054	0.020
9.0	....	....	....	....	....	0.014	0.016	0.020	0.026	0.035	0.046	0.057	0.061	0.050	0.020
10.0	....	....	....	....	....	0.007	0.008	0.010	0.015	0.022	0.032	0.043	0.051	0.046	0.020

Table 3.7

$S(1s, 2p\sigma)$

p	t = −0.5	t = −0.4	t = −0.3	t = −0.2	t = −0.1	t = 0.0	t = 0.1	t = 0.2	t = 0.3	t = 0.4	t = 0.5	t = 0.6	t = 0.7	t = 0.8
0.0	0.000	0.000	0.000	0.000	0.000	0.000	0.000	0.000	0.000	0.000	0.000	0.000	0.000	0.000
0.5	0.117	0.155	0.189	0.216	0.234	0.240	0.235	0.218	0.191	0.157	0.119	0.080	0.046	0.022
1.0	0.208	0.276	0.336	0.385	0.417	0.429	0.421	0.394	0.348	0.288	0.220	0.151	0.087	0.038
1.5	0.264	0.348	0.423	0.484	0.525	0.544	0.538	0.507	0.454	0.381	0.296	0.206	0.121	0.054
2.0	0.287	0.375	0.454	0.518	0.564	0.586	0.585	0.558	0.506	0.433	0.344	0.245	0.148	0.067
2.2	0.289	0.375	0.453	0.517	0.563	0.587	0.587	0.563	0.515	0.444	0.355	0.256	0.156	0.071
2.4	0.287	0.371	0.447	0.509	0.554	0.579	0.582	0.561	0.517	0.449	0.363	0.264	0.164	0.076
2.6	0.282	0.363	0.435	0.495	0.540	0.565	0.570	0.553	0.513	0.450	0.368	0.271	0.170	0.079
2.8	0.275	0.352	0.420	0.477	0.520	0.546	0.553	0.540	0.505	0.447	0.369	0.275	0.175	0.083
3.0	0.266	0.338	0.402	0.456	0.497	0.523	0.532	0.522	0.492	0.441	0.368	0.278	0.179	0.086
3.2	0.255	0.323	0.382	0.432	0.471	0.497	0.508	0.502	0.477	0.431	0.364	0.279	0.182	0.089
3.4	0.244	0.306	0.361	0.407	0.443	0.468	0.481	0.479	0.459	0.419	0.359	0.278	0.184	0.091
3.6	0.232	0.289	0.338	0.380	0.414	0.439	0.453	0.454	0.439	0.406	0.351	0.276	0.186	0.093
3.8	.....	.....	.....	.....	.....	0.409	0.424	0.428	0.418	0.390	0.342	0.273	0.186	0.095
4.0	0.207	0.253	0.293	0.327	0.356	0.379	0.394	0.401	0.396	0.374	0.332	0.269	0.186	0.097
4.2	.....	.....	.....	.....	.....	0.349	0.365	0.374	0.373	0.357	0.321	0.264	0.186	0.098
4.4	.....	.....	.....	.....	.....	0.320	0.337	0.348	0.350	0.339	0.310	0.258	0.185	0.099
4.5	0.176	0.210	0.239	0.264	0.286									
4.6	.....	.....	.....	.....	.....	0.293	0.309	0.322	0.328	0.321	0.298	0.252	0.183	0.100
5.0	0.147	0.171	0.190	0.208	0.225	0.241	0.258	0.273	0.284	0.286	0.272	0.238	0.179	0.101
5.5	.....	.....	.....	.....	.....	0.186	0.202	0.219	0.234	0.243	0.241	0.219	0.172	0.102
6.0	0.099	0.108	0.115	0.122	0.130	0.141	0.155	0.172	0.189	0.204	0.210	0.199	0.163	0.101
6.5	.....	.....	.....	.....	.....	0.105	0.118	0.133	0.151	0.169	0.181	0.179	0.154	0.100
7.0	0.064	0.066	0.066	0.068	0.071	0.078	0.088	0.102	0.120	0.139	0.155	0.160	0.144	0.098
7.5	.....	.....	.....	.....	.....	0.057	0.065	0.077	0.094	0.113	0.132	0.142	0.134	0.095
8.0	0.041	0.039	0.037	0.036	0.037	0.041	0.048	0.058	0.073	0.092	0.111	0.126	0.124	0.092
9.0	.....	.....	.....	.....	.....	0.021	0.025	0.032	0.043	0.059	0.078	0.097	0.104	0.086
10.0	.....	.....	.....	.....	.....	0.010	0.013	0.017	0.025	0.037	0.054	0.073	0.087	0.078

Table 3.8

$S(1s, 3s)$

p	$t = -0.4$	$t = -0.3$	$t = -0.2$	$t = -0.1$	$t = 0.0$	$t = 0.1$	$t = 0.2$	$t = 0.3$	$t = 0.4$
0.0	0.954	0.928	0.857	0.754	0.632	0.505	0.381	0.269	0.175
0.5	0.945	0.920	0.851	0.751	0.632	0.507	0.384	0.273	0.179
1.0	0.915	0.893	0.832	0.739	0.628	0.510	0.393	0.284	0.190
1.5	0.864	0.847	0.792	0.713	0.615	0.508	0.399	0.295	0.203
2.0	0.794	0.780	0.736	0.670	0.588	0.495	0.399	0.304	0.215
2.5	0.712	0.701	0.665	0.612	0.546	0.470	0.389	0.306	0.224
3.0	0.623	0.612	0.584	0.544	0.494	0.436	0.370	0.299	0.227
3.2	0.587	0.575	0.551	0.515	0.471	0.419	0.360	0.295	0.226
3.4	0.551	0.539	0.516	0.486	0.447	0.402	0.347	0.289	0.225
3.6	0.516	0.504	0.483	0.456	0.423	0.383	0.334	0.283	0.223
3.8	0.481	0.469	0.450	0.426	0.398	0.364	0.320	0.275	0.220
4.0	0.448	0.435	0.418	0.397	0.373	0.344	0.309	0.267	0.217
4.2	0.416	0.403	0.386	0.369	0.348	0.324	0.295	0.257	0.213
4.4	0.385	0.371	0.356	0.341	0.324	0.305	0.280	0.248	0.207
4.6	0.355	0.341	0.327	0.314	0.301	0.285	0.265	0.238	0.202
4.8	0.327	0.312	0.300	0.288	0.279	0.266	0.249	0.227	0.195
5.0	0.301	0.287	0.274	0.265	0.256	0.248	0.235	0.217	0.190
5.5	0.242	0.227	0.216	0.210	0.206	0.204	0.200	0.190	0.173
6.0	0.193	0.178	0.168	0.164	0.163	0.165	0.166	0.164	0.155
6.5	0.152	0.137	0.129	0.126	0.127	0.132	0.137	0.140	0.137
7.0	0.119	0.105	0.098	0.096	0.098	0.104	0.111	0.117	0.120
7.5	0.092	0.080	0.073	0.072	0.074	0.080	0.089	0.098	0.104
8.0	0.071	0.060	0.054	0.053	0.056	0.062	0.070	0.080	0.089
9.0	0.042	0.033	0.029	0.028	0.030	0.035	0.043	0.053	0.064
10.0	0.024	0.018	0.015	0.015	0.016	0.020	0.025	0.034	0.044

Table 3.9

p	$S(2s, 2s)$						
	$t = 0.0$	$t = 0.1$	$t = 0.2$	$t = 0.3$	$t = 0.4$	$t = 0.5$	$t = 0.6$
0.0	1.000	0.975	0.903	0.790	0.647	0.487	0.328
0.5	0.986	0.962	0.892	0.782	0.642	0.486	0.328
1.0	0.948	0.926	0.862	0.760	0.629	0.481	0.329
1.5	0.890	0.871	0.815	0.725	0.608	0.472	0.329
2.0	0.815	0.799	0.753	0.678	0.578	0.458	0.327
2.5	0.729	0.717	0.679	0.622	0.540	0.438	0.322
3.0	0.637	0.629	0.603	0.559	0.496	0.413	0.312
3.2	0.600	0.593	0.571	0.533	0.477	0.402	0.308
3.4	0.563	0.557	0.539	0.507	0.457	0.389	0.302
3.6	0.527	0.522	0.507	0.480	0.438	0.377	0.296
3.8	0.491	0.487	0.476	0.454	0.417	0.364	0.290
4.0	0.456	0.454	0.445	0.427	0.397	0.350	0.283
4.2	0.423	0.421	0.415	0.402	0.377	0.336	0.276
4.4	0.390	0.389	0.386	0.376	0.357	0.323	0.268
4.6	0.360	0.359	0.358	0.352	0.337	0.309	0.260
4.8	0.330	0.331	0.331	0.328	0.318	0.295	0.252

p	$S(2s, 2s)$						
	$t = 0.0$	$t = 0.1$	$t = 0.2$	$t = 0.3$	$t = 0.4$	$t = 0.5$	$t = 0.6$
5.0	0.302	0.307	0.305	0.305	0.299	0.281	0.244
5.5	0.240	0.242	0.247	0.252	0.255	0.248	0.223
6.0	0.188	0.190	0.197	0.206	0.215	0.216	0.203
6.5	0.145	0.148	0.155	0.167	0.179	0.187	0.182
7.0	0.111	0.113	0.121	0.133	0.148	0.160	0.163
7.5	0.083	0.086	0.093	0.106	0.121	0.136	0.145
8.0	0.062	0.064	0.071	0.083	0.099	0.116	0.128
8.5	0.046	0.048	0.054	0.065	0.080	0.097	0.112
9.0	0.034	0.035	0.041	0.050	0.064	0.082	0.098
9.5	0.024	0.026	0.030	0.039	0.051	0.068	0.086
10.0	0.018	0.019	0.023	0.030	0.041	0.057	0.075
10.5	0.013	0.014	0.017	0.023	0.032	0.047	0.065
11.0	0.009	0.010	0.012	0.017	0.026	0.039	0.056
11.5	0.006	0.007	0.009	0.013	0.020	0.032	0.048
12.0	0.004	0.005	0.007	0.010	0.016	0.026	0.041

Table 3.10

$S(2s, 3s)$

p	$t=-0.6$	$t=-0.5$	$t=-0.4$	$t=-0.3$	$t=-0.2$	$t=-0.1$	$t=0.0$	$t=0.1$	$t=0.2$	$t=0.3$	$t=0.4$	$t=0.5$	$t=0.6$
0.0	0.479	0.667	0.826	0.937	0.989	0.979	0.913	0.801	0.659	0.505	0.354	0.222	0.120
0.5	0.477	0.663	0.820	0.929	0.979	0.970	0.905	0.796	0.657	0.505	0.356	0.225	0.122
1.0	0.473	0.651	0.801	0.904	0.952	0.943	0.884	0.782	0.650	0.505	0.361	0.231	0.127
1.5	0.466	0.634	0.771	0.865	0.909	0.902	0.849	0.758	0.638	0.504	0.367	0.240	0.136
2.0	0.457	0.611	0.733	0.815	0.853	0.847	0.803	0.725	0.620	0.499	0.372	0.250	0.146
2.5	0.445	0.582	0.688	0.756	0.789	0.785	0.747	0.682	0.594	0.488	0.373	0.258	0.156
3.0	0.430	0.549	0.637	0.692	0.716	0.712	0.684	0.632	0.560	0.472	0.371	0.265	0.165
3.5	0.411	0.512	0.582	0.623	0.639	0.636	0.614	0.575	0.519	0.448	0.362	0.268	0.173
4.0	0.390	0.472	0.524	0.552	0.562	0.557	0.541	0.514	0.473	0.418	0.349	0.267	0.178
4.5	0.365	0.429	0.466	0.483	0.485	0.481	0.469	0.451	0.423	0.384	0.329	0.261	0.181
5.0	0.339	0.387	0.408	0.415	0.413	0.407	0.395	0.389	0.372	0.347	0.307	0.252	0.182
5.5	0.312	0.344	0.354	0.352	0.346	0.339	0.334	0.329	0.322	0.308	0.282	0.239	0.180
6.0	0.285	0.304	0.303	0.295	0.285	0.278	0.275	0.275	0.275	0.270	0.255	0.225	0.175
6.5	0.259	0.266	0.258	0.245	0.233	0.226	0.225	0.227	0.232	0.234	0.229	0.209	0.170
7.0	0.232	0.230	0.216	0.200	0.187	0.180	0.179	0.184	0.191	0.199	0.201	0.191	0.162
7.5	0.207	0.198	0.180	0.162	0.149	0.142	0.142	0.147	0.157	0.168	0.176	0.173	0.153
8.0	0.184	0.169	0.148	0.130	0.117	0.111	0.111	0.117	0.127	0.140	0.152	0.156	0.144
9.0	0.143	0.121	0.099	0.082	0.071	0.065	0.066	0.071	0.081	0.095	0.111	0.123	0.124
10.0	0.110	0.085	0.065	0.050	0.041	0.037	0.037	0.042	0.050	0.062	0.079	0.095	0.104

Table 3.11

p	S(3s, 3s) t = 0.0	p	S(3s, 3s) t = 0.0
0.0	1.000	6.4	0.328
0.5	0.992	6.6	0.306
1.0	0.968	6.8	0.285
1.5	0.932	7.0	0.264
2.0	0.885	7.2	0.245
2.5	0.832	7.4	0.227
3.0	0.772	7.6	0.209
3.5	0.708	7.8	0.193
4.0	0.641	8.0	0.177
4.5	0.572	8.5	0.141
5.0	0.504	9.0	0.114
5.2	0.477	9.5	0.089
5.4	0.451	10.0	0.070
5.6	0.425	10.5	0.054
5.8	0.400	11.0	0.041
6.0	0.375	11.5	0.031
6.2	0.351	12.0	0.024

Table 3.12

p	S(2pπ, 2pπ)						
	t = 0.0	t = 0.1	t = 0.2	t = 0.3	t = 0.4	t = 0.5	t = 0.6
0.0	1.000	0.975	0.903	0.790	0.647	0.487	0.328
0.5	0.976	0.951	0.882	0.772	0.633	0.478	0.323
1.0	0.907	0.887	0.823	0.723	0.596	0.453	0.308
1.5	0.809	0.790	0.737	0.652	0.542	0.416	0.287
2.0	0.695	0.680	0.638	0.568	0.477	0.372	0.261
2.5	0.578	0.567	0.535	0.481	0.410	0.325	0.233
3.0	0.468	0.460	0.437	0.398	0.345	0.279	0.205
3.2	0.427	0.420	0.401	0.367	0.320	0.262	0.194
3.4	0.389	0.383	0.366	0.337	0.297	0.245	0.183
3.6	0.352	0.348	0.334	0.309	0.274	0.228	0.173
3.8	0.318	0.315	0.303	0.283	0.253	0.213	0.163
4.0	0.287	0.284	0.275	2.258	0.232	0.198	0.153
4.2	0.258	0.255	0.248	0.234	0.213	0.183	0.144
4.4	0.231	0.229	0.224	0.213	0.195	0.170	0.135

p	S(2pπ, 2pπ)						
	t = 0.0	t = 0.1	t = 0.2	t = 0.3	t = 0.4	t = 0.5	t = 0.6
4.6	0.207	0.205	2.201	0.193	0.179	0.157	0.127
4.8	0.184	0.183	0.181	0.174	0.163	0.145	0.119
5.0	0.164	0.163	0.162	0.157	0.149	0.134	0.111
5.2	0.146						
5.4	0.129	0.122	0.122	0.121	0.117	0.109	0.094
5.5	0.121						
5.6	0.114						
5.8	0.101	0.089	0.091	0.092	0.092	0.089	0.079
6.0	0.089						
6.2	0.078						
6.4	0.069	0.065	0.067	0.069	0.072	0.071	0.066
6.5	0.064	0.047	0.049	0.052	0.055	0.057	0.055
7.0	0.046	0.033	0.036	0.039	0.043	0.046	0.046
7.5	0.033						

Bond Dissociation Energies

Equation 3.9, which gives the energy of a bonding MO, may be rearranged as follows:

$$E = \frac{\alpha}{1 + S} + \frac{\beta}{1 + S}$$

For a two-electron bond,

$$E = \frac{2\alpha}{1 + S} + \frac{2\beta}{1 + S}$$

If we identify $2\alpha/(1 + S)$ with an energy of the separate atoms or radicals, then the dissociation energy can be equated to $-2\beta/(1 + S)$. Mulliken[1] proposed that $-\beta$ is approximately proportional to the product of S and the average of the ionization potentials of the atomic orbitals. He writes $-2\beta = AS(\alpha_1 + \alpha_2)/2$, where A is an empirically evaluated constant. Hence, we obtain the following expressions for the dissociation energy of a one-electron bond and an electron-pair bond, respectively:

$$D = \frac{-\frac{1}{2}AS(\alpha_1 + \alpha_2)}{2(1 + S)} \tag{3.17}$$

$$D = \frac{-AS(\alpha_1 + \alpha_2)}{2(1 + S)} \tag{3.18}$$

For s-s bonds, as in H_2^+ and LiH, a value of 0.7 has been suggested[1] for A. Values of S can be obtained from Tables 3.5 to 3.12, following the procedure described in the preceding paragraph. Values of α can be obtained from Table 3.13.[2]

[1] R. S. Mulliken, *J. Phys. Chem.*, **56**, 295 (1952).
[2] J. Hinze and H. Jaffé, *J. Am. Chem. Soc.*, **84**, 540 (1962).

Table 3.13 VALENCE STATE IONIZATION POTENTIALS†

Atom	Electron configuration	$-\alpha$, eV	Atom	Electron configuration	$-\alpha$, eV
H	$\underline{s}$	13.6	C	$\underline{s}ppp$	21.0
				$s\underline{p}pp$	11.3
Li	$\underline{s}$	5.4	N	$\underline{s}p^2pp$	26.9
Be	$\underline{s}p$	9.9		$sp^2\underline{p}p$	14.4
	$s\underline{p}$	6.0	O	$\underline{s}p^2p^2p$	36.1
B	$\underline{s}pp$	14.9		$sp^2p^2\underline{p}$	18.5
	$s\underline{p}p$	8.4	F	$\underline{s}p^2p^2p^2$	38.2
				$s^2p^2p^2\underline{p}$	20.9

† J. Hinze and H. Jaffé, *J. Am. Chem. Soc.*, **84**, 540 (1962).

Let us first apply this approximate method to the estimation of the dissociation energy of H_2. From Table 2.4 we obtain the bond distance, 0.74 Å. Hence $p = 0.946 \times 2.00 \times 0.74 = 1.40$ and $t = 0$. From Table 3.5, we obtain $S = 0.753$, and from Table 3.13, we obtain $\alpha_1 = \alpha_2 = -13.6$ eV. We then calculate

$$D(H_2) = 0.7 \times 0.753 \times \frac{13.6}{1.753} = 4.09 \text{ eV}$$

This may be compared with the experimental value 4.52 eV. By an analogous procedure we can calculate the dissociation energy of LiH. The bond distance is 1.595 Å. We calculate

$$p = 0.946 \times 1.65 \times 1.595 = 2.49$$

$$t = \frac{0.35}{1.65} = 0.212$$

Hence, from Table 3.6, $S = 0.475$.

$$D(\text{LiH}) = 0.7 \times 0.475 \times \frac{13.6 + 5.4}{2 \times 1.475} = 2.15 \text{ eV}$$

The experimental value is 2.52 eV.

When this method is applied to bonds involving other than s-s orbital overlap, different values of A must be used, corrections must be made for electron-electron repulsion, and hybridization promotion energy must be included. For details, the original literature should be consulted.[1]

Orbital Energies

When one wishes to obtain a rough value of β for the purpose of estimating orbital energies (or better, *differences* in orbital energies), the following relation may be used:[2]

$$\beta = 1.75 S_{1,2} \frac{\alpha_1 + \alpha_2}{2} \tag{3.19}$$

Note that the proportionality constant 1.75 differs considerably from that used for the estimation of dissociation energies. We may illustrate the use of Eq. 3.19 in the estimation of the $\pi \to \pi^*$ transition energy of ozone (see Fig. 3.8). It should be remembered that the middle oxygen atom of ozone has a formal charge of $+1$ (corresponding to $\mu = 2.45$) and that each terminal atom has a

[1] Mulliken, loc. cit.
[2] M. Wolfsberg and L. Helmholz, *J. Chem. Phys.*, **20**, 837 (1952).

formal charge of $-\frac{1}{2}$ (corresponding to $\mu = 2.18$). The O—O bond distance is 1.28 Å. Hence we calculate $p = 0.946 \times 4.63 \times 1.28 = 5.61$ and $t = 0.27/4.63 = 0.058$ and obtain (from Table 3.12) $S = 0.114$. Because of this low value of S, we shall be justified in the use of wave functions and orbital-energy expressions which assume $S = 0$. As we have already seen, the group orbital $(1/\sqrt{2})(\phi_1 + \phi_3)$ and the atomic orbital ϕ_2 combine to form the bonding and antibonding MOs. The resonance integral $H_{(1+3),\,2}$ is calculated as follows:

$$H_{(1+3),\,2} = \frac{1}{\sqrt{2}} \int (\phi_1 + \phi_3) H \phi_2 \, d\tau = \frac{1}{\sqrt{2}} (H_{12} + H_{23})$$

$$= \frac{1}{\sqrt{2}} (\beta + \beta) = \sqrt{2}\,\beta$$

By substitution in Eq. 3.6, we obtain

$$\begin{vmatrix} \alpha - E & \sqrt{2}\,\beta \\ \sqrt{2}\,\beta & \alpha - E \end{vmatrix} = 0$$

Hence $E = \alpha \pm \sqrt{2}\,\beta$, and we see that the $\pi \to \pi^*$ transition energy is $-\sqrt{2}\,\beta$, or

$$\Delta E_{\pi \to \pi^*} = 1.414 \times 1.75 \times 0.114 \times 18.5 = 5.2 \text{ eV}$$

Transitions of the $\pi \to \pi^*$ type are generally observed as very intense bands in ultraviolet or visible absorption spectra. Thus our calculation suggests that ozone should show a strong absorption band at a wavelength corresponding to

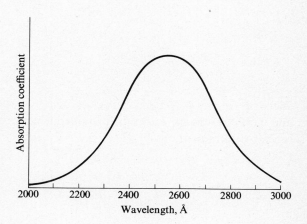

FIGURE 3.12

Absorption spectrum of ozone in the region 2000 to 3000 Å. [*Adapted with permission from E. C. Y. Inn and Y. Tanaka, J. Oct. Soc. Am.,* **43**, 870 (1953).]

5.2 eV, that is, 2400 Å. This prediction is borne out by the absorption spectrum shown in Fig. 3.12.[1]

PI BONDING IN PLANAR SYSTEMS

The P_4 molecule has a regular tetrahedral structure, with each atom bonded to three other atoms. The structure is remarkable in view of the strained P—P—P bond angles (60°) and the fact that simple valence bond structures suggest that a π-bonded square planar structure should be stable:

$$
\begin{array}{ccc}
\text{P=P} & & \text{P—P} \\
| \quad | & \leftrightarrow & \| \quad \| \\
\text{P=P} & & \text{P—P}
\end{array}
$$

Let us apply simple MO theory to the π bonding in this hypothetical square planar form of P_4. The secular determinant of Eq. 3.6 corresponds to a linear combination of two atomic orbitals, ϕ_1 and ϕ_2. In the general case, where $\psi = c_1\phi_1 + c_2\phi_2 + \cdots + c_n\phi_n$, the secular determinant is

$$
\begin{vmatrix}
H_{11} - ES_{11} & H_{21} - ES_{21} & \cdots & H_{1n} - ES_{1n} \\
H_{12} - ES_{12} & H_{22} - ES_{22} & \cdots & H_{2n} - ES_{2n} \\
H_{13} - ES_{13} & H_{23} - ES_{23} & \cdots & H_{3n} - ES_{3n} \\
\cdots\cdots\cdots\cdots\cdots\cdots\cdots\cdots\cdots\cdots\cdots\cdots\cdots \\
H_{1n} - ES_{1n} & H_{2n} - ES_{2n} & \cdots & H_{nn} - ES_{nn}
\end{vmatrix} = 0
$$

Consider the p orbitals which are perpendicular to the plane of the P_4 molecule:

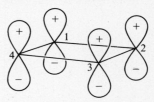

We write $\psi = c_1\phi_1 + c_2\phi_2 + c_3\phi_3 + c_4\phi_4$. If we ignore all overlap integrals and ignore resonance integrals for nonadjacent orbitals, we obtain the following secular determinant:

$$
\begin{vmatrix}
\alpha - E & \beta & 0 & \beta \\
\beta & \alpha - E & \beta & 0 \\
0 & \beta & \alpha - E & \beta \\
\beta & 0 & \beta & \alpha - E
\end{vmatrix} = 0
$$

[1] E. C. Y. Inn and Y. Tanaka, *J. Opt. Soc. Am.*, **43**, 870 (1953). For a discussion of the MOs, see A. D. Walsh, *J. Chem. Soc.*, p. 2266, (1953) or H. J. Maris, D. Larson, M. E. McCarville, and S. P. McGlynn, *Acc. Chem. Res.*, **3**, 368 (1970).

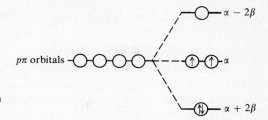

$p\pi$ orbitals

FIGURE 3.13
Energy level diagram for the π MOs in
the hypothetical planar P_4 molecule.

For simplification, we divide through by β and replace $(\alpha - E)/\beta$ by x.

$$\begin{vmatrix} x & 1 & 0 & 1 \\ 1 & x & 1 & 0 \\ 0 & 1 & x & 1 \\ 1 & 0 & 1 & x \end{vmatrix} = 0$$

This equation is equivalent to $x^2(x^2 - 4) = 0$, whose roots are -2, 0, 0, and $+2$. Hence there are two nonbonding MOs of energy α, a bonding MO of energy $\alpha + 2\beta$, and an antibonding MO of energy $\alpha - 2\beta$. These energy levels are shown in Fig. 3.13. Notice that there are only two bonding π electrons, corresponding to an average π bond order of only $\frac{1}{4}$. Thus simple MO theory predicts a total of five bonding electron pairs in planar P_4, whereas tetrahedral P_4 has six bonding electron pairs. On this basis one would predict the tetrahedral structure to be stable with respect to the planar structure, and the result may be considered a triumph of MO theory over valence bond theory.

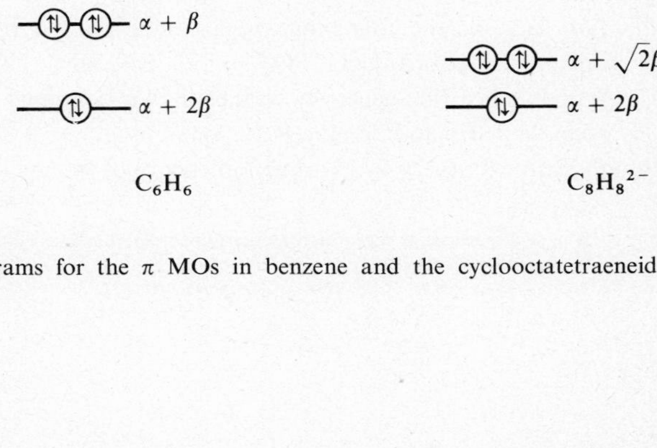

$\alpha - 2\beta$

$\alpha - \beta$

$\alpha + \beta$

$\alpha + 2\beta$

C_6H_6

$\alpha - 2\beta$

$\alpha - \sqrt{2}\beta$

α

$\alpha + \sqrt{2}\beta$

$\alpha + 2\beta$

$C_8H_8{}^{2-}$

FIGURE 3.14
Energy level diagrams for the π MOs in benzene and the cyclooctatetraeneide ion.

On the other hand, one might have expected that the strained bond angles in the tetrahedral structure would nullify the effect of the extra bond. Apparently the strain is not sufficiently pronounced.

The π bonding in cyclobutadiene, C_4H_4, is analogous to that in the hypothetical planar P_4. Cyclobutadiene has been observed only as an unstable transient species; however, it can be stabilized in the form of complexes with transition metals, as in $C_4H_4Fe(CO)_3$. (See Chap. 17.) The square planar forms of C_4H_4 and P_4 are examples of "antiaromatic" compounds.[1] These compounds do not satisfy the Hückel criterion for aromaticity, which is that *cyclic coplanar systems which contain $(4n + 2)$ π electrons are "aromatic" and possess relative electronic stability*. Note that species such as benzene (6 π electrons) and the cyclooctatetraeneide anion, $C_8H_8{}^{2-}$ (10 π electrons), do conform to the $4n + 2$ rule. MO energy level diagrams for these species are shown in Fig. 3.14.

PROBLEMS

3.1   What must be the heavy-atom geometry of $(SiH_3)_3N$ if there is to be significant π bonding using the d orbitals of the silicon atoms and a p orbital of the nitrogen atom?

3.2   Calculate by simple MO theory whether the linear $[H \cdots H \cdots H]^+$ or triangular

$$\left[\begin{array}{c} H \\ H \cdots \cdots H \end{array} \right]^+$$

form of $H_3{}^+$ is the more stable.

3.3   Calculate by simple MO theory the dissociation energies of Li_2, Na_2, and LiNa. The interatomic distances are 2.67, 3.08, and 2.86 Å, respectively.

3.4   Calculate by simple MO theory the $\pi \rightarrow \pi^*$ transition energy of $NO_2{}^-$. The N—O bond distance is 1.24 Å. Compare the calculated energy with the experimental value, 6.0 eV.

3.5   Construct a qualitative MO energy level diagram for the linear ion $HF_2{}^-$, involving only the hydrogen $1s$ orbital and the two fluorine hybrid orbitals that lie on the bonding axis.

3.6   For each of the following species, indicate the bond order and the number of unpaired electrons: NeO^+, $O_2{}^+$, CN^+, BN, SiF^+, NO^-, PCl, $I_2{}^+$, NeH^+.

3.7   Why is π bonding generally weaker than σ bonding?

3.8   Propose a structure for the $P_4{}^{2+}$ ion.

3.9   Explain why the $\pi \rightarrow \pi^*$ transition energy of $N_3{}^-$ is lower than that of CO_2.

---

[1] For a discussion of the antiaromaticity of cyclobutadiene and similar molecules, see R. Breslow, *Acc. Chem. Res.*, **6**, 393 (1973).

SOME REACTIONS OF MOLECULAR HYDROGEN

Hydrogen is by far the most abundant element in the universe, and there are probably more known compounds of hydrogen than of any other element. Therefore the reactions of hydrogen and its compounds are important, and we shall use them as examples in discussions of several types of gas-phase reaction mechanisms and the application of molecular orbital symmetry rules in kinetics.

REACTIONS WITH HALOGENS

Molecular hydrogen is capable of reacting with many elements and compounds, but at room temperature most of these reactions are extremely slow. This seeming inertness is undoubtedly related to the high dissociation energy of the molecule, 104.2 kcal mol^{-1}. Obviously, when an H_2 molecule reacts, the H—H bond must be broken. However, it is not necessary to supply the entire 104 kcal mol^{-1}

to the reacting molecules as activation energy in order to obtain reaction. Usually when an H_2 molecule reacts with another molecule at least one of the hydrogen atoms forms a bond to another atom at the same time that the $H-H$ bond is broken. Thus the bond-breaking contribution to the activation energy is partly compensated by the bond-forming contribution.

For example, consider the reaction of hydrogen with bromine vapor:

$$H_2 + Br_2 \rightarrow 2HBr$$

The reaction rate is negligible in the absence of light (which catalyzes the reaction) unless the temperature is raised above 200°C. Kinetic studies[1-3] of the reaction have shown that the rate is given by the following expression:

$$\frac{d[HBr]}{dt} = \frac{k[H_2][Br_2]^{1/2}}{1 + k'[HBr]/[Br_2]} \tag{4.1}$$

where k and k' are constants at a given temperature. This rate law is consistent with the following chain mechanism:

$$Br_2 \underset{2}{\overset{1}{\rightleftharpoons}} 2Br$$

$$Br + H_2 \underset{4}{\overset{3}{\rightleftharpoons}} HBr + H$$

$$H + Br_2 \overset{5}{\longrightarrow} HBr + Br$$

Reaction 1, in which radicals are produced, is the chain-initiating step, and reaction 2, in which radicals are consumed, is the chain-terminating step. Reactions 3, 4, and 5 are called chain-propagating steps; the number of radicals is unchanged by these reactions. Note that the sum of reactions 3 and 5 corresponds to the net reaction of H_2 and Br_2 to form two molecules of HBr. If a finite concentration of H and Br atoms is maintained, these steps can proceed as long as the supply of H_2 and Br_2 molecules lasts.

It is not difficult to show that the mechanism is consistent with the experimental rate law. For every molecule of HBr produced by reaction 5, another is produced by reaction 3. Therefore the overall rate of formation of HBr is equal to twice the rate of reaction 5:

$$\frac{d[HBr]}{dt} = 2k_5[H][Br_2] \tag{4.2}$$

[1] M. Bodenstein and S. C. Lind, *Z. Physik. Chem.*, **57,** 168 (1907).
[2] A. A. Frost and R. G. Pearson, "Kinetics and Mechanism. A Study of Homogeneous Chemical Reactions," 2d ed., pp. 236–241, Wiley, New York, 1961.
[3] S. W. Benson, "Foundations of Chemical Kinetics," McGraw-Hill, New York, 1960.

If we assume that a low, steady-state concentration of H atoms is achieved, then we may write

$$\frac{d[H]}{dt} = 0 = k_3[Br][H_2] - k_4[HBr][H] - k_5[H][Br_2]$$

Solving for [H], we obtain

$$[H] = \frac{k_3[Br][H_2]}{k_5[Br_2] + k_4[HBr]} \tag{4.3}$$

Substitution in Eq. 4.2 yields

$$\frac{d[HBr]}{dt} = \frac{2k_3 k_5[H_2][Br_2][Br]}{k_5[Br_2] + k_4[HBr]} \tag{4.4}$$

In the steady state, the rates of reactions 1 and 2 are equal; that is, the Br_2 molecules are in equilibrium with the Br atoms. We may write

$$\frac{[Br]^2}{[Br_2]} = \frac{k_1}{k_2}$$

and

$$[Br] = \left(\frac{k_1}{k_2}\right)^{1/2} [Br_2]^{1/2}$$

Substitution in Eq. 4.4, followed by rearrangement, gives

$$\frac{d[HBr]}{dt} = \frac{2k_3(k_1/k_2)^{1/2}[H_2][Br_2]^{1/2}}{1 + (k_4/k_5)[HBr]/[Br_2]}$$

This equation has the same form as Eq. 4.1; it can be seen that $k' = k_4/k_5$ and $k = 2k_3(k_1/k_2)^{1/2}$.

A rate constant may be expressed as $Ae^{-E_a/RT}$, and an equilibrium constant may be expressed as $e^{\Delta S^\circ/R}e^{-\Delta H^\circ/RT}$. Hence the last equation, in which k is represented as a function of the rate constant k_3 and the equilibrium constant k_1/k_2, may be rewritten as follows:

$$Ae^{-E_a/RT} = 2A_3 e^{-E_a(3)/RT}e^{\Delta S_1^\circ/2R}e^{-\Delta H_1^\circ/2RT}$$

where E_a is the activation energy of the overall reaction, $E_a(3)$ is the activation energy of reaction 3, and ΔH_1° is the heat of reaction 1 (that is, the dissociation energy of bromine). Hence

$$E_a = E_a(3) + \tfrac{1}{2}\Delta H_1^\circ \tag{4.5}$$

From studies of the $H_2 + Br_2$ reaction as a function of temperature it is known that k' is temperature-independent and that the overall activation energy E_a is

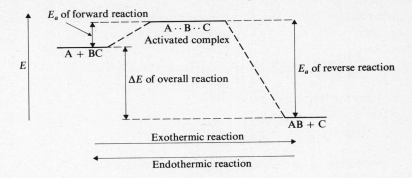

FIGURE 4.1
Energy level diagram for the reaction $A + BC \rightleftharpoons AB + C$, where A and C are atoms or radicals.

40.2 kcal mol^{-1}. From Table 2.9, we find that ΔH_1° is 45.5 kcal mol^{-1}. Thus we calculate

$$E_a(3) = 40.2 - 22.8 = 17.4 \text{ kcal mol}^{-1}$$

From data in Table 2.9 we calculate that the energy of reaction 3 is 16.7 kcal mol^{-1}, a value just slightly less than E_a. As a matter of fact, this result is expected. Most exothermic reactions involving radicals are very fast and have low activation energies—typically a few kilocalories per mole. A schematic plot of the energies of the species involved in such a reaction is shown in Fig. 4.1. From the plot we see that the activation energy of an *endothermic* reaction of this type is just slightly greater than the energy of the reaction.

Although reactions 1 to 5 satisfactorily account for the rate law of the reaction, it is of interest to consider why the following reactions are not involved in the reaction mechanism.

$$Br + HBr \xrightarrow{6} H + Br_2$$

$$HBr \underset{8}{\overset{7}{\rightleftharpoons}} H + Br$$

$$H_2 \underset{10}{\overset{9}{\rightleftharpoons}} 2H$$

From data in Table 2.9, we calculate that the activation energies of reactions 6, 7, and 9 are greater than 42.0, 87.5, and 104.2 kcal mol^{-1}, respectively. The activation energy of reaction 6 is much greater than that of any of the analogous chain-propagating steps, reaction 3, 4, or 5. Similarly, the activation energies of reactions 7 and 9 are much greater than that of the analogous chain-initiating step, reaction 1. Hence the rate constants of reactions 6, 7, and 9 are negligible compared with those of the analogous reactions in the actual mechanism. Rearrangement of Eq. 4.3 gives

$$\frac{[H]}{[Br]} = \frac{k_3[H_2]}{k_5[Br_2] + k_4[HBr]}$$

In the early stages of the reaction, when [HBr] is low, and when H_2 and Br_2 have comparable concentrations, $[H]/[Br] \approx k_3/k_5$. At a typical reaction temperature, say 300°C, $k_3/k_5 \approx e^{-17.4/RT} \approx 10^{-7}$. Therefore reaction of H and Br atoms (reaction 8) would be $\sim 10^{-7}$ as likely as recombination of two Br atoms (reaction 2), and recombination of two H atoms would be $\sim 10^{-14}$ as likely.

The chain reaction of hydrogen and chlorine, which proceeds by a mechanism analogous to reactions 1 to 5, is faster than that of hydrogen and bromine, whereas the chain reaction of hydrogen and iodine is slower. These qualitative results are readily predictable from the energetics of the steps in the general mechanism. If, as a rough approximation, we equate the energy of the $X + H_2 \rightarrow HX + H$ reaction to the activation energy of that reaction, then from Eq. 4.5 we obtain, for the *overall* activation energy,

$$E_a \approx D(H_2) - D(HX) + \tfrac{1}{2}D(X_2)$$

For chlorine, we calculate

$$E_a \approx 104.2 - 103.0 + 28.5 = 29.7 \text{ kcal mol}^{-1}$$

and, for iodine,

$$E_a \approx 104.2 - 71.3 + 17.8 = 50.7 \text{ kcal mol}^{-1}$$

Notice that the main factor in establishing the trend in the reactivity of the halogens with hydrogen is the trend in the H—X dissociation energy. In the case of chlorine, most of the energy required to break the H—H bond is compensated by the energy released upon formation of the H—Cl bond. In the case of iodine, a relatively small amount of energy is provided by the formation of the H—I bond.

The reaction of hydrogen with iodine by the chain mechanism is so slow that, under ordinary conditions, an entirely different mechanism predominates. The rate law has long been known to be

$$\frac{d[HI]}{dt} = k[H_2][I_2]$$

For many years this rate law was interpreted in terms of a simple bimolecular process with a four-center activated complex:

$$H_2 + I_2 \rightarrow 2HI$$

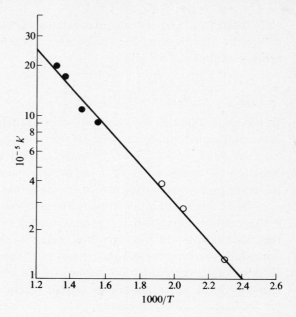

FIGURE 4.2

Plot of log $(10^{-5}\,k')$ versus $1000/T$ for the $H_2 + I_2$ reaction. Solid circles correspond to k/K values; open circles correspond to directly measured k' values. [*Reproduced with permission from J. H. Sullivan, J. Chem. Phys.,* **46,** 73 (1967).]

However, relatively recent experimental data show that the following, entirely different, mechanism is responsible for the reaction:[1]

$$I_2 \rightleftharpoons 2I \qquad K = [I]^2/[I_2]$$

$$2I + H_2 \xrightarrow{\;k'\;} 2HI$$

In the latter mechanism, I_2 molecules are in equilibrium with I atoms, and the latter species react with H_2 by a termolecular reaction (or perhaps two bimolecular reactions) to form the product. The corresponding rate law is indistinguishable from that for the simple bimolecular process:

$$\frac{d[HI]}{dt} = k'[H_2][I]^2 = k'K[H_2][I_2] = k[H_2][I_2]$$

The termolecular rate constant k' has been measured directly at low temperatures, using photochemically generated I atoms. The same rate constant can also be calculated from the relation $k' = k/K$, using values of k obtained from kinetic data on the $H_2 + I_2$ reaction at high temperatures and values of K obtained from

[1] J. H. Sullivan, *J. Chem. Phys.,* **46,** 73 (1967).

equilibrium thermodynamic data. Both types of k' values have been plotted in the log k' versus $1/T$ plot of Fig. 4.2. Both sets of points fall on the same straight line, indicating that both sets of data yield the same calculated rate constant for any given temperature. This result proves that the simple bimolecular process must be relatively unimportant; otherwise k' values calculated from kinetic data on the $H_2 + I_2$ reaction would be higher than k' values obtained from direct measurement of the $2I + H_2$ reaction rate.

ORBITAL SYMMETRY EFFECTS

In recent years chemists have come to realize the importance of molecular orbital symmetry in determining the rates of reactions.[1,2] We now know, for example, that the simple bimolecular $H_2 + I_2$ reaction mechanism involving a trapezoidal activated complex, $\begin{smallmatrix} H\text{---}H \\ I\text{----------}I \end{smallmatrix}$, is symmetry-forbidden, whereas a mechanism involving the preliminary dissociation of iodine is symmetry-allowed.

For a bimolecular reaction, the important orbitals are the highest occupied molecular orbital (HOMO) of one molecule and the lowest unoccupied molecular orbital (LUMO) of the other molecule.[1] In the course of the reaction, electrons flow from the HOMO to LUMO. This process will occur (i.e., the reaction will be allowed) if the following conditions are fulfilled. (1) As the reactants approach one another, the HOMO and LUMO must have a net positive overlap. (2) The energy of the LUMO must be lower than, or no more than about 6 eV higher than, the energy of the HOMO. (3) The HOMO must be either a bonding MO of a bond to be broken or an antibonding MO of a bond to be formed; and vice versa for the LUMO.

Let us apply these rules to the bimolecular $H_2 + I_2$ reaction. The HOMO of H_2 is the $s\sigma_B$ MO, and the LUMO of I_2 is the $p\sigma^*$ MO. These orbitals are illustrated below; the shading indicates electron occupancy. It is obvious that if the H_2 and I_2 approach each other broadside the net overlap of these orbitals

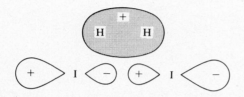

[1] R. G. Pearson, *Chem. Eng. News*, Sept. 28, 1970, pp. 66–72; *J. Am. Chem. Soc.*, **94**, 8287 (1972).
[2] R. G. Woodward and R. Hoffman, "The Conservation of Orbital Symmetry," Academic, New York, 1969.

is zero and the reaction is forbidden. An alternative approach is to consider the HOMO of I_2 (the $p\pi$* MO) and the LUMO of H_2 (the $s\sigma$* MO):

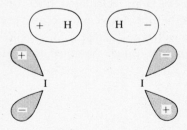

Although these orbitals have appropriate symmetries, i.e., the net overlap is positive, electron flow cannot occur for two reasons: The electron flow would correspond to the *strengthening* rather than the weakening of the I—I bond, and the electron flow from electronegative atoms to relatively electropositive atoms is energetically unfeasible. On the other hand, the reactions of an H_2 molecule with one or two iodine atoms, as in alternative mechanisms of the reaction, are symmetry- and energy-allowed processes:

For another example, let us consider the hydrogenation of ethylene to give ethane:

$$H_2 + C_2H_4 \rightarrow C_2H_6$$

If we assume a symmetric approach to H_2 to one side of the C_2H_4 plane, neither the flow of electrons from the HOMO of H_2 to the LUMO of C_2H_4 nor the flow of electrons from the HOMO[1] of C_2H_4 to the LUMO of H_2 is allowed by symmetry. In each case the net orbital overlap is zero.

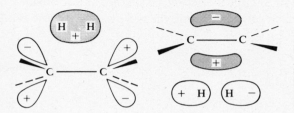

[1] The HOMO of C_2H_4 is a π MO, not a σ MO. The σ MOs associated with the C—H bonds have quite low energies and do not strongly interact with the σ MO of the C—C bond. Thus one does not observe the type of σ and π orbital ordering found in the first-row diatomic molecules (see Fig. 3.5b).

Thus the reaction is symmetry-forbidden. The nonreactivity of hydrogen with ethylene is in striking contrast to the reaction of hydrogen with B_2Cl_4, which proceeds readily even below room temperature.

$$3H_2 + 3B_2Cl_4 \rightarrow 4BCl_3 + B_2H_6$$

The first step in this complex reaction probably involves the broadside attack of an H_2 molecule on a B_2Cl_4 molecule to form intermediate $BHCl_2$ molecules:

$$H_2 + B_2Cl_4 \rightarrow 2BHCl_2$$

A series of exchange processes could then lead to the observed products:

$$2BHCl_2 \rightarrow BH_2Cl + BCl_3$$
$$BH_2Cl + BHCl_2 \rightarrow BH_3 + BCl_3$$
$$2BH_3 \rightarrow B_2H_6$$

It is easy to show that the first step of the proposed mechanism is both symmetry- and energy-allowed. Electrons can flow from the $s\sigma_B$ MO of H_2 to the empty $p\pi_B$ MO of B_2Cl_4:

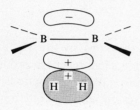

EXPLOSIONS[1]

The temperature of a reaction mixture is stable when the rate of loss of heat by conduction and convection equals the rate of heat production by the reaction. If the rate of heat loss cannot compensate for the rate of heat production, the reaction temperature and rate may increase rapidly, causing a "thermal explosion." Such an explosion can occur for highly exothermic reactions when the reactant concentrations are high. For example, an equimolar mixture of hydrogen and chlorine at a pressure of 50 torr in a glass vessel will explode at temperatures above 400°C. At lower temperatures the reaction rate is so slow that the heat of reaction can be carried off by conduction and convection. Explosion can occur at a lower temperature if the pressure of the

[1] Benson, op. cit.

reaction mixture is increased, and a higher temperature is required for explosion if the pressure is reduced.

It is interesting that the addition of only 0.5 torr of nitric oxide to 50 torr of $H_2 + Cl_2$ mixture lowers the critical explosion temperature from 400 to 270°C. This sensitization of the explosion is probably caused by an increase in the Cl atom concentration by the reaction

$$NO + Cl_2 \rightarrow NOCl + Cl$$

The latter reaction has an activation energy of only 22 kcal mol^{-1}, whereas the $Cl_2 \rightarrow 2Cl$ reaction has an activation energy of at least 57 kcal mol^{-1}.

The reaction of hydrogen with oxygen is the classic example of a "branching-chain" explosion:

$$2H_2 + O_2 \rightarrow 2H_2O$$

Like the $H_2 + Br_2$ reaction, this reaction proceeds by a chain mechanism. At temperatures above 400°C, there is good evidence that the chain propagation includes the following steps:

$$H + O_2 \xrightarrow{11} OH + O$$
$$O + H_2 \xrightarrow{12} OH + H$$

These reactions differ from the chain-propagating reactions in the $H_2 + Br_2$ reaction (and in most other chain reactions) in that each reaction produces more chain carriers (OH, O, or H) than it consumes. If these reactions can predominate, the overall reaction rate will increase essentially without bound; that is, an explosion will occur. Whether or not explosion takes place depends on whether or not the chain-terminating steps of the mechanism can compensate for these chain-branching steps. The chain-terminating steps are surface reactions of the following type:

$$H + \text{wall} \xrightarrow{13} \text{stable species}$$
$$OH + \text{wall} \xrightarrow{14} \text{stable species}$$

At low temperatures and pressures, these reactions prevent reactions 11 and 12 from taking place, and essentially no reaction can occur. However, because reactions 11 and 12 are of higher kinetic order than reactions 13 and 14, an increase in pressure causes reactions 11 and 12 to become faster relative to reactions 13 and 14. If the temperature of the reaction mixture is high enough, increasing the pressure beyond a particular value (called the first explosion limit) causes explosion. If the pressure is increased sufficiently above the first limit, a second limit is reached, above which no explosion occurs. This quenching of

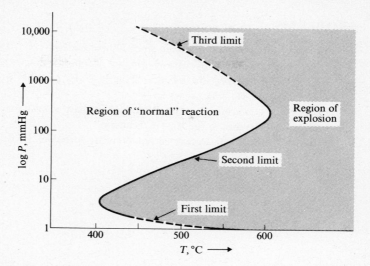

FIGURE 4.3

Plot showing the explosion limits of a stoichiometric mixture of $H_2 + O_2$ in a KCl-coated spherical vessel (7.4-cm diameter). Dashed parts of the curve are extrapolations. (*From S. W. Benson, "Foundations of Chemical Kinetics," McGraw-Hill, New York, 1960.*)

the explosion is probably caused by the following reaction which forms the relatively unreactive HO_2 species:

$$H + O_2 + M \xrightarrow{15} HO_2 + M$$

The third body (M), which can be any other molecule such as O_2 or H_2O, serves to carry off some of the heat of reaction and prevents the HO_2 from falling apart to OH and O. Note that reaction 15 is of higher kinetic order than reactions 11 and 12 and therefore would be expected to predominate at higher pressures. Although HO_2 is a radical, it serves to break the chain reaction by reactions such as

$$H + HO_2 \rightarrow H_2 + O_2$$
$$OH + HO_2 \rightarrow H_2O + O_2$$

Above the second limit, the reaction rate is moderate. Increasing the pressure causes the rate to increase until yet a third limit is reached above which explosion again occurs, probably of the thermal type. The regions of explosion and the various limits are marked in the pressure-temperature plot of Fig. 4.3.

PROBLEMS

4.1 The conversion of p-H_2 (nuclear spins antiparallel) to o-H_2 (nuclear spins parallel) follows the rate law: Rate $= k[p\text{-}H_2][H_2]^{1/2}$. Propose a reaction mechanism.

4.2 Why would you expect the kinetics of the $H_2 + F_2$ reaction to be more complicated than that of the $H_2 + Br_2$ reaction?

4.3 Under what conditions would you expect the chain mechanism to compete with the nonchain mechanism in the $H_2 + I_2$ reaction?

4.4 Which of the following concerted reactions are forbidden by orbital symmetry rules?

$$2NO \rightarrow N_2 + O_2$$
$$N_2H_2 + C_2H_4 \rightarrow N_2 + C_2H_6$$
$$H_2 + N_2 \rightarrow N_2H_2$$
$$SO_2 + F_2 \rightarrow SO_2F_2$$
$$SF_2 + F_2 \rightarrow SF_4$$

4.5 A spark between two wires can set off an explosion in an $H_2 + O_2$ mixture which lies outside the explosion region of Fig. 4.3. Explain.

4.6 Discuss the nuclear fission of ^{235}U in terms of a branching-chain mechanism. Note the analogies between this reaction and the $H_2 + O_2$ reaction.

COMPOUNDS OF HYDROGEN

CLASSIFICATION OF THE HYDRIDES

Hydrides, that is, compounds of hydrogen, represent a wide variety of compounds having many different types of structure and bonding. This variety in structure and bonding is attributable to the unique electronic structure of the hydrogen atom: The atom possesses no core electrons, it has only one valence electron (and thus is *formally* analogous to the alkali-metal atoms), and it is only one electron short of a stable rare-gas electronic configuration (and thus is analogous to halogen atoms). The known hydrides may be roughly categorized as saline hydrides, metallic hydrides, transition-metal hydride complexes, and nonmetal hydrides.

Saline Hydrides

At high temperatures hydrogen reacts with the alkali metals and the alkaline-earth metals heavier than beryllium to form salts, such as NaH and CaH_2, containing the hydride ion H^-. Inasmuch as the ionic radii of H^- and F^- are very similar (1.40 and 1.36 Å, respectively), the physical properties of the saline hydrides are similar to those of the corresponding fluorides. Even the periodic-table trend in the heats of formation of the alkali-metal hydrides resembles that for the alkali-metal fluorides. In each case, the lithium compound is more stable than the cesium compound. (See Table 5.1.)

Striking evidence for the negative charge of the hydrogen in lithium hydride is obtained when molten lithium hydride (mp 691°C) is electrolyzed: lithium metal forms at the cathode, and hydrogen is evolved at the anode. The saline hydrides react vigorously with protonic solvents such as water, alcohol, and liquid ammonia to form molecular hydrogen and the basic anions of the solvents:

$$CaH_2 + 2H_2O \rightarrow Ca^{2+} + 2OH^- + 2H_2$$
$$LiH + C_2H_5OH \rightarrow Li^+ + C_2H_5O^- + H_2$$
$$KH + NH_3 \rightarrow K^+ + NH_2^- + H_2$$

Sodium hydride finds considerable use as a reducing agent in synthetic chemistry. Sodium hydroborate, $NaBH_4$, is prepared by the reaction of sodium hydride with methyl borate at elevated temperatures:

$$4NaH + B(OCH_3)_3 \rightarrow NaBH_4 + 3NaOCH_3$$

Aqueous solutions of sodium hydroborate are kinetically stable at high pH; however, acidic solutions decompose rapidly to boric acid and hydrogen:

$$BH_4^- + H^+ + 3H_2O \rightarrow 4H_2 + H_3BO_3$$

Sodium hydroborate is often used as a reducing agent in both inorganic and

Table 5.1 PROPERTIES OF THE SALINE HYDRIDES†

Formula	Density, g ml^{-1}	$\Delta H°$ of formation, kcal mol^{-1}	Crystal structure analog
LiH	0.778	−21.67	NaCl
NaH	1.36	−13.49	NaCl
KH	1.43	−13.82	NaCl
RbH	2.60	−11.3	NaCl
CsH	3.41	−11.92	NaCl
MgH$_2$	1.42	−17.79	TiO$_2$
CaH$_2$	1.9	−41.65	Distorted PbCl$_2$
SrH$_2$	3.27	−42.3	Distorted PbCl$_2$
BaH$_2$	4.15	−40.9	Distorted PbCl$_2$

† K. M. Mackay, "Hydrogen Compounds of the Metallic Elements," Spon, London, 1966.

organic syntheses. When ether solutions of sodium hydroborate are heated at 100°C with diborane, the octahydrotriborate ion, $B_3H_8^-$, is formed:

$$B_2H_6 + NaBH_4 \rightarrow NaB_3H_8 + H_2$$

By suitable changes in reaction time, temperature, diborane pressure, and solvent, these same reagents will yield various anions of the general formula $B_nH_n^{2-}$ (n ranging from 6 to 12). These ions are remarkably resistant toward hydrolysis, and their boron skeletons remain intact throughout a wide variety of reactions in which the hydrogen atoms are replaced by other groups.

Lithium hydride is used in the preparation of lithium aluminum hydride (lithium hydroaluminate), $LiAlH_4$:

$$4LiH + AlCl_3 \rightarrow LiAlH_4 + 3LiCl$$

Both $LiAlH_4$ and $NaAlH_4$ can be prepared directly from the elements:

$$M + Al + 2H_2 \xrightarrow[\substack{1000-5000 \text{ lb in}^{-2} \\ 120-150°C}]{\text{ether solvent}} MAlH_4$$

Lithium aluminum hydride is an important reagent in synthetic chemistry; it is a much more powerful reducing agent than an alkali-metal hydroborate, and it reacts vigorously with protonic solvents such as alcohols.

Metallic Hydrides

The binary hydrides of the transition metals have metallic properties and are usually obtained as powders or brittle solids which are dark and metallic in appearance, and which have electric conductivities and magnetic properties characteristic of metals. Most of these hydrides have nonstoichiometric compositions; the stoichiometries and properties depend on the purity of the metals used in the preparations. Some typical limiting formulas are ScH_2, LaH_2, SmH_2, ThH_2, PaH_3, UH_3, VH, and CrH. An early theory of the metallic hydrides was implied in their classification, along with carbides, nitrides, borides, etc., as "interstitial." It was believed that the hydrogen entered the interstices of the metal lattice as atoms. This theory, which suggested very little interaction between the interstitial atoms and the metal, soon became untenable in the light of structural studies. In most hydrides their metal atoms are not in the same positions as in the parent metal (although in some hydrides metal-atom positions correspond to a high-temperature modification of the metal). Other facts which must be explained by any theory of metallic hydrides are the following:

1 In several of the hydrides, the bonded hydrogen is very mobile and has only a small activation energy of diffusion.

2 The magnetic susceptibility of palladium falls as hydrogen is added, and the material becomes diamagnetic at a composition of about $PdH_{0.6}$.

3 If a potential is applied across a filament of a hydride such as one of palladium, titanium, or tantalum, the hydrogen migrates toward the negative electrode.

In one theory which is used to describe the metallic hydrides, hydrogen atoms are assumed to have lost their electrons to the *d* orbitals of the metal atoms and to be present in the lattice as protons. Thus the compound is like an alloy. In another theory of metallic hydrides, the hydrogen atoms are assumed to have acquired electrons from the conduction band and to be present as hydride ions. The partially depleted conduction band gives residual metallic bonding in the compound. Thus the compound can be described as a saline hydride with metallic properties. The various properties of metallic hydrides have been explained on the basis of both theories.

Transition-Metal Hydride Complexes

Until the late 1950s, very few hydrides were known in which hydrogen atoms were covalently bonded directly to transition-metal atoms. Since then, numerous compounds of this type have been characterized. Most of them have the general formula MH_xL_y, where M is a transition metal and L is a ligand capable of acting both as a σ donor and π acceptor. Examples are $K_3[CoH(CN)_5]$ (a homogeneous hydrogenation catalyst for the conversion of alkynes to alkenes), *trans*-$PtHCl[P(C_2H_5)_3]_2$ (a compound so stable that it can be sublimed in a vacuum), $OsHCl_2[P(C_4H_9)_2C_6H_5]_3$ (a complex with an odd number of electrons and therefore paramagnetic), and K_2ReH_9 (remarkable because the transition metal is bonded to an unusually large number of hydrogen atoms, and no other ligands).

Some transition-metal ions and complexes react directly with molecular hydrogen to form complexes containing metal-hydrogen bonds. These reactions will be discussed in Chap. 18. In general, transition-metal hydride complexes are prepared by the reduction of transition-metal compounds in the presence of π-bonding ligands. Besides the obvious reducing agents BH_4^- and $LiAlH_4$, reagents such as alcoholic KOH, hydrazine, and hypophosphorous acid have been used. A typical procedure is to treat the anhydrous halide with $NaBH_4$ or $LiAlH_4$ in an ether solvent. For example,

$$C_5H_5Fe(CO)_2Cl \xrightarrow[\text{tetrahydrofuran}]{NaBH_4} C_5H_5Fe(CO)_2H$$

$$\textit{trans-}Pt(PR_3)_2Cl_2 \xrightarrow[\text{tetrahydrofuran}]{LiAlH_4} \textit{trans-}Pt(PR_3)_2(H)Cl$$

Many hydrides and carbonyl hydrides can be prepared by the treatment of anhydrous halides with triphenylphosphine or triphenylarsine in an alcohol. For example,

$$(NH_4)_2OsCl_6 \xrightarrow[\substack{\text{diethyleneglycol} \\ \text{monomethyl ether}}]{PPh_3;\ 165°C} HOsCl(CO)(PPh_3)_3$$

In early research, many of the transition-metal hydride complexes were believed to involve low oxidation states of the metals, and the presence of hydrogen was not suspected. It is now known that the hydrogen in such compounds can be identified by an infrared absorption, due to the M—H stretch, in the range 1600 to 2250 cm^{-1} and a proton magnetic resonance signal at very high field, corresponding to τ values in the range 15 to 50.

Nonmetal Hydrides

Nonmetal hydrides contain hydrogen atoms covalently bonded to the elements of main groups III to VII of the periodic table. The compounds generally consist of discrete molecules having considerable volatility. Properties of some of the volatile nonmetal hydrides are listed in Table 5.2.

A wide variety of synthetic methods have been used in the synthesis of nonmetal hydrides. Some of these methods and typical examples are:

1 The reaction of lithium aluminum hydride with a nonmetal halide,

$$LiAlH_4 + SiCl_4 \xrightarrow{\text{ether}} SiH_4 + LiCl + AlCl_3$$

2 The reduction of a nonmetal oxyacid with borohydride in aqueous solution,

$$3BH_4^- + 4H_3AsO_3 + 3H^+ \xrightarrow{H_2O} 4AsH_3 + 3H_3BO_3 + 3H_2O$$

3 The solvolysis of a binary compound of a nonmetal (e.g., calcium phosphide) in acid solution,

$$Ca_3P_2 + 6H^+ \xrightarrow{H_2O} 3Ca^{2+} + 2PH_3$$

4 The pyrolysis of a nonmetal hydride,

$$B_2H_6 \xrightarrow{\Delta} H_2 + B_4H_{10} + B_5H_{11} + \cdots$$

5 The treatment of a nonmetal hydride, or a mixture of such hydrides, with an electric discharge,

$$GeH_4 + PH_3 \xrightarrow[\text{discharge}]{\text{electric}} GeH_3PH_2 + Ge_2H_6 + P_2H_4 + \cdots$$

6   The coupling of an alkali-metal salt of a nonmetal hydride with a halogen derivative of a nonmetal hydride,

$$KSiH_3 + SiH_3Br \rightarrow KBr + Si_2H_6$$

The study of the hydrides of carbon and their derivatives constitutes the enormous discipline of organic chemistry. Carbon atoms are remarkable for their ability to catenate, i.e., to form chains, as in the aliphatic hydrocarbons, or alkanes. Other elements show this ability to a much lesser extent. Silicon and germanium hydrides analogous to the organic alkanes have been characterized up to $Si_{10}H_{22}$ and $Ge_{10}H_{22}$, that is, up to silicon and germanium chain lengths of 10. The longest-molecular-chain hydrides which have been definitely characterized for other elements are Sn_2H_6, HN_3, P_3H_5, As_3H_5, Sb_2H_4, H_2O_3, and H_2S_6.

Table 5.2   PROPERTIES OF SOME VOLATILE HYDRIDES[†], [‡]

Formula	Melting point, °C	Boiling point, °C	$\Delta H°$ of formation of gas, kcal mol^{-1}
B_2H_6	-165	-90	7.5
B_4H_{10}	-120	16.1	13.8
B_5H_9	-46.1	58.4	15.0
B_5H_{11}	-123.3	63	22.2
B_6H_{10}	-63.2	108	19.6
$B_{10}H_{14}$	100	· · · · · · · ·	4.4
SiH_4	-185	-111.2	7.3
Si_2H_6	-129.3	-14.2	17.1
Si_3H_8	-114.8	53.0	25.9
GeH_4	-165.9	-88.51	21.6
Ge_2H_6	-109	29	38.7
Ge_3H_8	-101.8	110.5	53.6
SiH_3GeH_3	-119.7	7.0	27.8
SnH_4	-150	-51.8	38.9
PbH_4	· · · · · · · ·	~ -13	60
PH_3	-133.8	-87.74	1.3
P_2H_4	-99	63.5	5.0
SiH_3PH_2	< -135	12.7	2
AsH_3	-116.9	-62.5	15.9
SbH_3	-88	-18.4	34.7
BiH_3	· · · · · · · ·	~ 17	66

† M. F. Hawthorne, in E. L. Muetterties (ed.), "The Chemistry of Boron and Its Compounds," pp. 223–323, Wiley, New York, 1967.
‡ W. L. Jolly and A. D. Norman, Hydrides of Groups IV and V, *Prep. Inorg. React.*, **4**, 1 (1968).

HYDRIDIC AND PROTONIC CHARACTER

The hydrides of boron, a relatively electropositive nonmetal, react as if the hydrogen atoms had "hydridic" character (that is, as if they were H^- ions). Thus these hydrides react with water to form hydrogen: $B_2H_6 + 6H_2O \rightarrow 6H_2 + 2H_3BO_3$. Hydrides of the more electronegative elements such as those of groups VI and VII react as if the hydrogen atoms had protonic (H^+) character. Thus these react with bases to form salts or adducts: $HCl + C_5H_5N \rightarrow (C_5H_5NH)Cl$. Hydrides of the elements of intermediate electronegativity, such as those of group IV, have no marked hydridic or protonic character. These hydrides may show, depending on the reagent, either hydridic or protonic character. Thus germane, GeH_4, shows hydridic character in the reaction with hydrogen bromide, $GeH_4 + HBr \rightarrow GeH_3Br + H_2$, and shows protonic character in its reaction with sodium amide, $GeH_4 + NaNH_2 \rightarrow NaGeH_3 + NH_3$.

It is instructive to study the trends in the hydridic and protonic characters of the binary molecular hydrides from a thermodynamic viewpoint. The hydridic character of a hydride H_nX can be measured by the energy of the process:

$$H_nX(g) \rightarrow H^-(g) + H_{n-1}X^+(g) \tag{5.1}$$

The question is: How does this energy change as element X is changed from one place to another in the periodic table? The problem is somewhat simplified if we consider the energy of reaction 5.1 as a combination of energies. The energy can be broken up into the energies of the following separate processes:

$$H_nX(g) \rightarrow H(g) + H_{n-1}X(g) \qquad D(H - XH_{n-1})$$
$$H(g) + e^-(g) \rightarrow H^-(g) \qquad\qquad -A(H)$$
$$H_{n-1}X(g) \rightarrow H_{n-1}X^+(g) + e^-(g) \qquad I(H_{n-1}X)$$

These energies are (1) the energy required to remove a hydrogen atom from the hydride, (2) the negative electron affinity of hydrogen, and (3) the ionization potential of the $H_{n-1}X$ radical. When we go from one hydride to another, the change in hydridic character is the sum of the changes in $D(H - XH_{n-1})$ and $I(H_{n-1}X)$. Changes in these latter energies are fairly easy to predict using concepts learned in Chaps. 1 and 2. Both the bond dissociation energy and the ionization potential would be expected to increase from left to right in a series, or from bottom to top in a family of the periodic table. Hence, we would expect the energy of reaction 5.1 to have the same trend. Hydridic character should be at a maximum for the gaseous molecule FrH and at a minimum for HF. Actual values of the energy of the reaction in Eq. 5.1 for various molecular hydrides are given in Table 5.3; it can be seen that the data are consistent with our expectations.

Table 5.3 VALUES OF ΔH° OF THE REACTION
$H_nX \rightarrow H^- + H_{n-1}X^+$, kcal mol^{-1}†

LiH 165	....	CH$_4$ 312	NH$_3$ 349	H$_2$O 404	HF 521
NaH 149	....	SiH$_4$ ∼260	PH$_3$ ∼290	H$_2$S 315	HCl 386
KH 112	....	......	......	.....	HBr 344
					HI 295

† J. L. Franklin et al., Ionization Potentials, Appearance Potentials, and Heats of Formation of Gaseous Positive Ions, *Natl. Bur. Std. (U.S.)*, *Natl. Std. Ref. Data Ser.*, NSRDS–NBS 26, 1969; R. S. Berry, *Chem. Rev.*, **69**, 533 (1969).

Table 5.4 THERMODYNAMIC DATA FOR MOLECULAR BINARY HYDRIDES, H_nX
Dissociation energies, electron affinities of $H_{n-1}X$, and energies for the reaction $H_nX \rightarrow H^+ + H_{n-1}X^-$, all in kilocalories per mole†,‡

	LiH	CH$_4$	NH$_3$	H$_2$O	HF
D	58	104	110	119	136
A	14	25	17	42	78
ΔH°	358	393	407	391	372
	NaH	SiH$_4$	PH$_3$	H$_2$S	HCl
D	48	80	84	90	103
A	12	33	29	53	83
ΔH°	350	361	369	351	334
	KH	GeH$_4$	AsH$_3$	H$_2$Se	HBr
D	44	∼75	∼73	∼75	88
A	7	40	29	51	78
ΔH°	351	349	358	338	324
					HI
D					71
A					71
ΔH°					314

† J. L. Franklin et al., Ionization Potentials, Appearance Potentials, and Heats of Formation of Gaseous Positive Ions, *Nat. Std. Bur. (U.S.)*, *Natl. Std. Ref. Data Ser.*, NSDR–NBS 26, 1969; R. S. Berry, *Chem. Rev.*, **69**. 533 (1969).

‡ J. I. Brauman et al., *J. Am. Chem. Soc.*, **93**, 6360 (1971); K. C. Smyth and J. I. Brauman, *J. Chem. Phys.*, **56**, 1132 (1972); J. I. Brauman, private communication.

The protonic character of a hydride H_nX can be measured by the energy of the process

$$H_nX(g) \rightarrow H^+(g) + H_{n-1}X^-(g) \tag{5.2}$$

This energy can be broken up into the energies of the following processes:

$$H_nX(g) \rightarrow H(g) + H_{n-1}X(g) \qquad D(H - XH_{n-1})$$
$$H(g) \rightarrow H^+(g) + e^-(g) \qquad I(H)$$
$$e^-(g) + H_{n-1}X(g) \rightarrow H_{n-1}X^-(g) \qquad -A(H_{n-1}X)$$

We have already pointed out that $D(H - XH_{n-1})$ would be expected to increase toward the upper right-hand corner of the periodic table. By recognizing that the electron affinity of a species is the ionization potential of the negative ion, we predict the same sort of trend for $A(H_{n-1}X)$. However, a change in protonic character is equal to the corresponding change in the *difference* $D(H - XH_{n-1}) - A(H_{n-1}X)$, and therefore it is not possible to predict trends in protonic character without quantitative data on the trends in $D(H - XH_{n-1})$ and $A(H_{n-1}X)$. $D(H - XH_{n-1})$, $A(H_{n-1}X)$, and the energy of reaction 5.2 for various molecular hydrides are given in Table 5.4. It can be seen that, within a given family, the dissociation energy changes more rapidly than the electron affinity. Hence the general tendency is for the gas-phase acidity to increase on descending a given family. However, from left to right in the periodic table, the change in dissociation energy almost matches the change in electron affinity. Hence there is not a strong general trend in gas-phase acidity from left to right in the table. Tables 5.3 and 5.4 reveal some interesting facts. For example, the gas-phase acidity of NaH is greater than that of HF! Note, however, that the ionization of NaH to give Na^+ and H^- is favored by 210 kcal mol^{-1} over the ionization to give H^+ and Na^-. On the other hand, the ionization of HF to given H^+ and F^- is favored by 149 kcal mol^{-1} over the ionization to give F^+ and H^-.

HYDROGEN BONDING[1]

In some compounds, individual hydrogen atoms are bound simultaneously to two (or sometimes more) electronegative atoms. The electronegative atoms are said to be "hydrogen-bonded." The first proposal of such bonding was made by Moore and Winmill,[2] who explained the weakness of trimethyl-

---

[1] L. Pauling, "The Nature of the Chemical Bond," 3d ed., pp. 449–504, Cornell University Press, Ithaca, N.Y., 1960; G. C. Pimentel and A. L. McClellan, "The Hydrogen Bond," Freeman, San Francisco, 1959.
[2] T. S. Moore and T. F. Winmill, *J. Chem. Soc.*, **101**, 1635 (1912). Also see W. M. Latimer and W. H. Rodebush, *J. Am. Chem. Soc.*, **42**, 1419 (1920), and M. L. Huggins, *Phys. Rev.*, **18**, 333 (1921); **19**, 346 (1922).

ammonium hydroxide relative to tetramethylammonium hydroxide by postulating the structure

$$CH_3-\underset{\underset{\displaystyle CH_3}{|}}{\overset{\overset{\displaystyle CH_3}{|}}{N}}-H-OH$$

Of course it is now recognized that the hydrogen atom, with only one valence atomic orbital, cannot exceed a total covalency of 1. Hydrogen bonds are generally indicated by dotted lines, as follows:

$$CH_3-\underset{\underset{\displaystyle CH_3}{|}}{\overset{\overset{\displaystyle CH_3}{|}}{N}}-H\cdots OH$$

Evidence for hydrogen bonding is found in a wide variety of experimental data. In the following paragraphs we shall describe some of this evidence.

The data in Table 5.4 show that the gas-phase acidities of water and methane are almost identical. That is, the proton affinities of CH_3^- and OH^- are practically the same. Yet every chemist knows that methane shows no significant ionization in aqueous solution, whereas water itself self-ionizes to an important extent. It has been estimated that the pK_a of methane in water is about 44; this should be compared with the pK_a of water which, when the concentration of water is expressed as molarity, is about 16. The enormous change in the relative acidities upon going to aqueous solution is due to the relatively strong stabilization of the aqueous hydroxide ion by hydrogen bonding. It is believed that the aqueous hydroxide ion is hydrated to form species of the type

$$\begin{bmatrix} H \\ OH\cdots OH \end{bmatrix}^- \qquad \begin{bmatrix} OH \\ H \\ OH \\ H \end{bmatrix}^- \qquad \text{etc.}$$

To some extent, this hydrogen bonding may be looked upon as an interaction of the dipole moment of the O—H bonds in the water molecules with the negative charge of the OH^- ion. Because the OH^- ion is very small, the water dipoles can approach the ion very closely and interact strongly. On the other hand, the methyl anion is relatively large, and the energy of interaction with water molecules is relatively small.

The effect of anion size on gas-phase hydration energy is shown by the data in Table 5.5. Notice that the fluoride ion, which has a size comparable to that of the hydroxide ion, also has very high hydration energies. One would expect that the $-\Delta H^\circ$ for the reaction $X^-(H_2O)_{n-1}(g) + H_2O(g) \rightarrow X^-(H_2O)_n(g)$ would approach the heat of vaporization of water, 10.5 kcal mol^{-1}, as n approached ∞. The data of Table 5.5 show that the $-\Delta H^\circ$ values for iodide have essentially this value even for low values of n. In other words, a water molecule interacts with an iodide ion with about the same energy that it does with other water molecules in liquid water.

Because of the strong hydrogen bonding of the hydroxide ion in water and other hydroxylic solvents such as alcohols, much of the intrinsic basicity of the ion is lost when it is dissolved in such solvents. However, by using a solid alkali-metal hydroxide it is possible to take advantage of the strong basicity of the hydroxide ion. Thus a suspension of potassium hydroxide in a nonhydroxylic solvent such as dimethyl sulfoxide can be used to deprotonate extremely weak acids, such as triphenylmethane (aqueous p$K \approx 30$), phosphine (aqueous p$K \approx 27$), and germane (aqueous p$K \approx 25$).[1]

$$2KOH(s) + HA \rightarrow K^+ + A^- + KOH \cdot H_2O(s)$$

This result is particularly impressive when one remembers that, in aqueous solution, the hydroxide ion is incapable of quantitatively deprotonating any acid with a pK higher than about 14.

The compounds HF, H_2O, and NH_3 are strongly hydrogen-bonded in both the solid and liquid states. In ice each oxygen atom is surrounded by four other oxygen atoms, and hydrogen atoms are located between the oxygen atoms, but not midway. Each oxygen atom is bonded to two hydrogen atoms at a distance of 1.01 Å and is hydrogen-bonded to two hydrogen atoms at a distance

---

[1] W. L. Jolly, *J. Chem. Educ.*, **44**, 304 (1967); *Inorg. Synth.*, **11**, 113 (1968).

Table 5.5 **GAS-PHASE HYDRATION ENERGIES OF ANIONS†**

$-\Delta H^\circ$ in kilocalories per mole for $X^-(H_2O)_{n-1}(g) + H_2O(g) \rightarrow X^-(H_2O)_n(g)$

n	OH$^-$	F$^-$	Cl$^-$	Br$^-$	I$^-$
1	22.5	23.3	13.1	12.6	10.2
2	16.4	16.6	12.7	12.3	9.8
3	15.1	13.7	11.7	11.5	9.4
4	14.2	13.5	11.1	10.9	

† M. Arshadi, R. Yamdagni, and P. Kebarle, *J. Phys. Chem.*, **74**, 1475 (1970); M. Arshadi and P. Kebarle, *J. Phys. Chem.*, **74**, 1483 (1970).

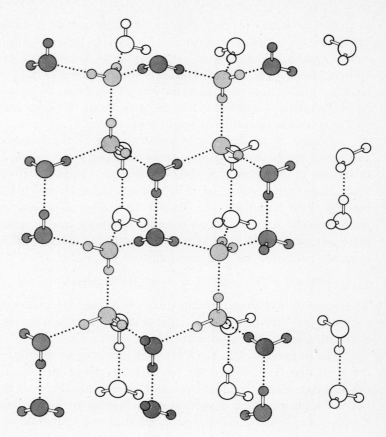

FIGURE 5.1
Structure of ice, showing the hydrogen bonds. Notice the irregular arrangement
of the O—H···O bonds. (*From L. Pauling, "The Nature of the Chemical Bond,"
3d ed., pp. 449–504, Cornell University, Press, Ithaca, N.Y., 1960. Used by permission of Cornell University Press.*)

of 1.75 Å. (This asymmetry is typical of most hydrogen bonds.) Except for the
restriction that there are always two hydrogen bonds and two normal O—H
bonds to each oxygen atom, the distribution of these bonds throughout the
lattice is random, even in ice which has been slowly cooled to the neighbor-
hood of 0 K. This randomness is responsible for a residual entropy of $R \ln \frac{3}{2}$
$= 0.81$ eu in ice at 0 K. The structure of ice is pictured in Fig. 5.1. From
this picture it can be seen that ice has a rather open structure, with many
holes in the lattice. When ice melts, many of the hydrogen bonds are broken,
permitting the structure to collapse and causing an increase in density. Because

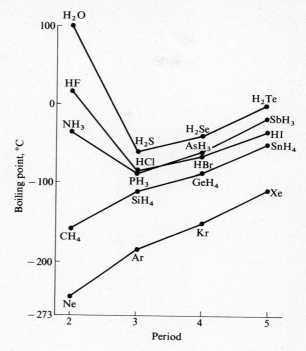

FIGURE 5.2
Boiling points of some molecular hydrides and, for comparison, the noble gases.
(*Reproduced with permission from W. L. Jolly, "The Chemistry of the Non-Metal,"
Prentice-Hall, Englewood Cliffs, N.J., 1966.*)

the processes of melting and vaporization for HF, H_2O, and NH_3 require the
breaking of hydrogen bonds, these hydrides have abnormally high heats of melting
and vaporization as well as high melting points and boiling points. In Fig. 5.2
the boiling points for various nonmetal hydrides are plotted against the periodic-
table positions of the nonmetals. Clearly the points for HF, H_2O, and NH_3 deviate
markedly from the positions expected if these compounds were unassociated in
the liquid state. The point for HCl also shows a significant deviation, probably
attributable to a small amount of hydrogen bonding.

The hydrogen bonding in HF is so strong that it persists even in the gaseous state. At 20°C and 745 torr, 80 percent of the HF molecules are in the
form of $(HF)_6$ polymers. In aqueous solution, HF is a weak acid which interacts
with the fluoride ion to form the bifluoride ion HF_2^-:

$$F^- + HF \rightleftharpoons HF_2^- \qquad K = 3.9$$

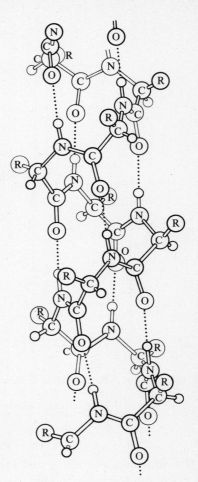

FIGURE 5.3
The α helix of polypeptide chains, showing
the N—H···O hydrogen bonds. (*From L.
Pauling, "The Nature of the Chemical
Bond," 3d ed., pp. 449–504, Cornell University, Press, Ithaca, N.Y., 1960. Used by
permission of Cornell University Press.*)

In salts containing this anion, the proton is located midway between the two
fluoride ions; hence this hydrogen-bonded system has no residual entropy at 0 K.

The open structure of ice is a consequence of the hydrogen bonding. Many
other compounds have structures based on hydrogen bonding. For example, the
helical configurations of polypeptide chains would not be possible if the chains
were not "knit" together by N—H···O hydrogen bonds. (See Fig. 5.3.) Crystalline boric acid contains layers of $B(OH)_3$ molecules held together by hydrogen
bonds, as shown in Fig. 5.4. Crystalline potassium dihydrogen phosphate,

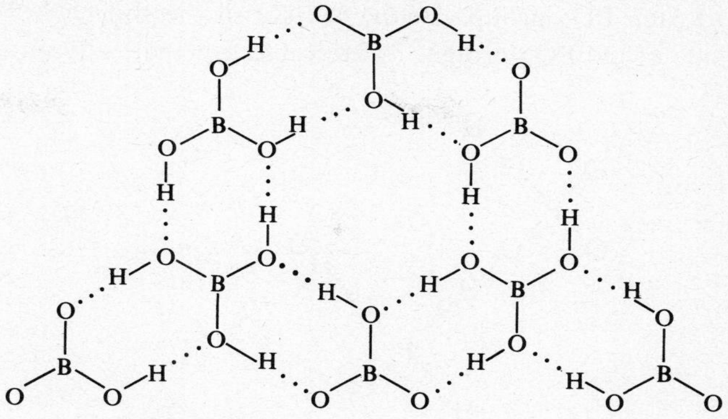

FIGURE 5.4
Arrangement of atoms in a layer of crystalline boric acid.

KH_2PO_4, contains PO_4 tetrahedra, each of which is surrounded tetrahedrally by four other PO_4 groups. A drawing of the structure is shown in Fig. 5.5. A feature not shown by the drawing is that the PO_4 groups are connected by hydrogen bonds, with two short O—H bonds and two long O$\cdots$H bonds

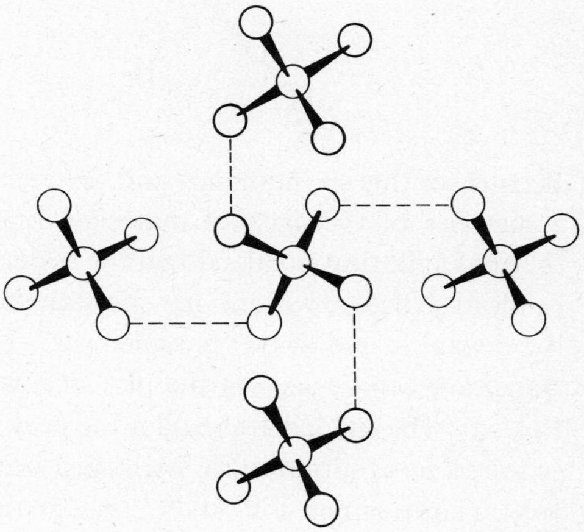

FIGURE 5.5
The environment of an $H_2PO_4^-$ ion in KH_2PO_4.

to each PO_4 group. Below 121 K, all the short O—H bonds are on the same side of the PO_4 groups, as indicated schematically in the following:

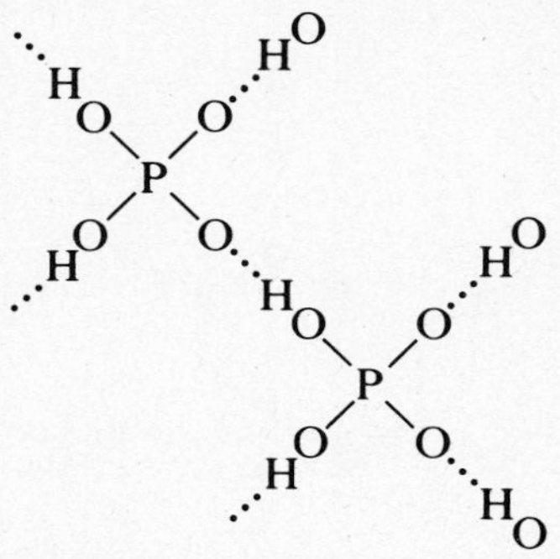

By appropriate application of an electric field to the crystal, the polarization of the hydrogen bonds can be reversed, as indicated in the following:

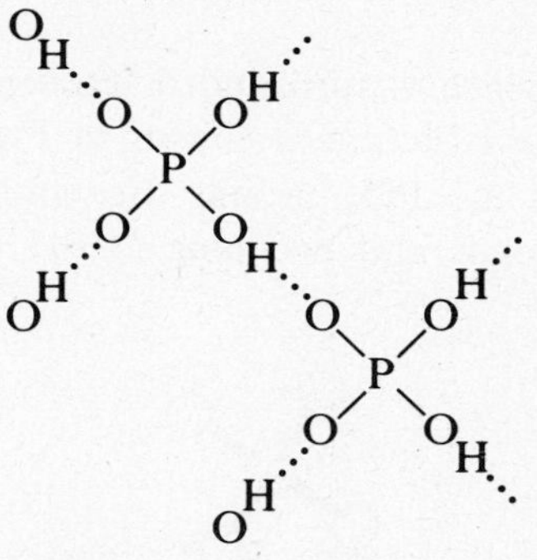

Because of this spontaneous and reversible electric polarization of KH_2PO_4, it is a member of the class of materials known as "ferroelectrics." Above 121 K, thermal agitation breaks down the ordering of the hydrogen bonds to produce a random structure with no net spontaneous polarization. Above this temperature, the crystal is said to be "paraelectric." Evidence for the ferroelectric-paraelectric transition can be seen in the plot of heat capacity versus temperature shown in Fig. 5.6. The shaded area under the peak around 121 K corresponds to the energy required to randomize the hydrogen bonds.

The oxonium ion, H_3O^+, is a pyramidal species found in many solid acid hydrates.[1] Thus $HClO_4 \cdot H_2O$ is really a salt, $H_3O^+ClO_4^-$. Similarly, $HCl \cdot H_2O$ is $H_3O^+Cl^-$ and $HNO_3 \cdot 3H_2O$ is $H_3O^+NO_3^- \cdot 2H_2O$. In aqueous solutions of strong acids, the oxonium ion undoubtedly exists in a highly hydrated, or hy-

---

[1] W. C. Hamilton and J. A. Ibers, "Hydrogen Bonding in Solids," W. A. Benjamin, Inc., New York, 1968.

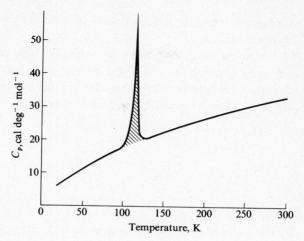

FIGURE 5.6
Plot of heat capacity versus temperature for KH_2PO_4. The peak corresponds to the ferroelectric-paraelectric transition. (*Reproduced with permission from W. C. Hamilton and J. A. Ibers, "Hydrogen Bonding in Solids," W. A. Benjamin, Inc., New York, 1968.*)

drogen-bonded, form. Cations of the general formula $H(H_2O)_n{}^+$ have been identified in the gas phase by mass spectroscopy and in various solid compounds by X-ray and neutron diffraction. For example, the ion $H_5O_2{}^+$ exists in $HCl \cdot 2H_2O$, $HCl \cdot 3H_2O$, and $[Co(en)_2Cl_2]Cl \cdot HCl \cdot 2H_2O$.

Infrared spectroscopy is commonly used to detect, and to measure quantitatively, hydrogen bonding. A change in the frequency, width, and intensity of an X—H stretching band is particularly diagnostic. For example, the stretching frequency band of an isolated O—H bond might occur around 3500 cm^{-1} and have a width of less than 10 cm^{-1}. After strong hydrogen bonding, the frequency might occur around 1700 cm^{-1}, the band width would be several hundred wave numbers, and the band intensity would probably have increased markedly.

The data in Table 5.6 show that the aqueous ionic conductances of the

Table 5.6 IONIC CONDUCTANCES IN WATER AT 25°C

H^+	350	OH^-	192
Na^+	51	Cl^-	76
K^+	74	$NO_3{}^-$	71
Ag^+	64	$C_2H_3O_2{}^-$	41

hydrogen ion and the hydroxide ion are extremely high in comparison with those of other ions.

It is believed that these abnormally high conductances are due to the hydrogen bonding in water and to the fact that the H^+ and OH^- ions are strongly hydrogen-bonded. Consider the following picture of an oxonium ion and some of its associated water molecules:

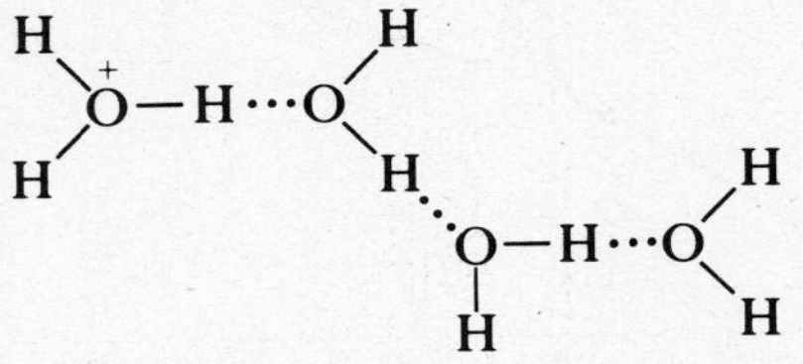

By the slight shift of the protons in three hydrogen bonds, the oxonium ion can in effect move to the other end of the chain of water molecules:

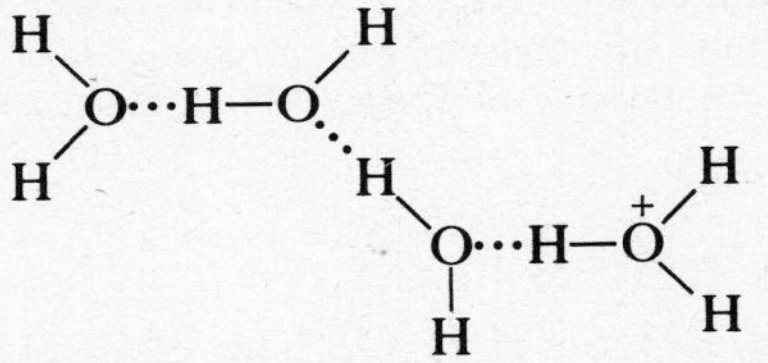

A similar conduction mechanism is possible for the hydroxide ion:

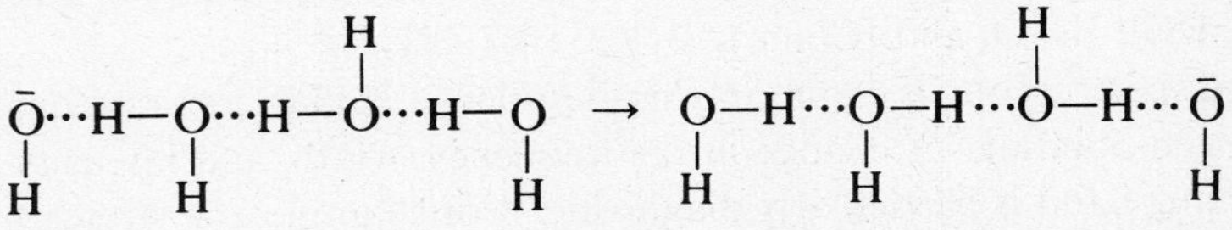

Thus these ions can migrate through an aqueous solution without having to shove aside water molecules and without having to carry along shells of co-ordinated water molecules, as is necessary in the migration of other aqueous ions.

PROTON TRANSFER REACTIONS

The migration of a proton through water involves steps corresponding to the transfer of a proton from one water molecule to another. This process may be written as a chemical reaction:

$$H_2O + H_3O^+ \rightarrow H_3O^+ + H_2O$$

The rate constant for this bimolecular aqueous reaction has been determined by nmr line width measurements[1] to be $k = 6.0 \times 10^{11}e^{-2.4/RT}M^{-1}$ s^{-1}, or $k = 1 \times 10^{10}$ M^{-1} s^{-1} at 25°C. The migration of a hydroxide ion through water likewise corresponds to successive transfers of protons, but between *hydroxide* ions, $H_2O + OH^- \rightarrow OH^- + H_2O$. The nmr-determined[1] rate constant for this reaction is $k = 1.0 \times 10^{11}e^{-2.1/RT}$ M^{-1} s^{-1}, or $k = 3 \times 10^9$ M^{-1} s^{-1} at 25°C. Both of these reactions are among the fastest known reactions in aqueous solution. The fastest aqueous reaction is the transfer of a proton from a water molecule to a hydroxide ion: $H_3O^+ + OH^- \rightarrow H_2O + H_2O$. This reaction, for which $k = 1.4 \times 10^{11}$ M^{-1} s^{-1} at 25°C, is the net reaction which occurs whenever a strong acid is neutralized by a strong base.[2]

We may state as a general rule than any hydrogen-bond proton transfer reaction for which $\Delta H° < 0$ has a very low activation energy and is very fast, with $k \sim 10^{10}$ M^{-1} s^{-1}. Consequently endothermic proton transfer reactions have activation energies slightly greater then $\Delta H°$ and rate constants approximately equal to $K_{eq} \times 10^{10}$, where K_{eq} is the equilibrium constant of the reaction. The forward and reverse rate constants for various reactions of the type

$$\begin{array}{c} H \\ \diagdown \\[-4pt] \overset{+}{O}-H\cdots B^- \\[-4pt] \diagup \\ H \end{array} \rightarrow \begin{array}{c} H \\ \diagdown \\[-4pt] O\cdots H-B \\[-4pt] \diagup \\ H \end{array}$$

and

$$\begin{array}{c} H \\ | \\ {}^-O\cdots H-B \end{array} \rightarrow \begin{array}{c} H \\ | \\ O-H\cdots B^- \end{array}$$

are given in Table 5.7.[2]

A technique based on the measurement of nmr line widths has been used to determine the rates of symmetric proton exchange for ammonia and various amines.[3] Two types of processes were detected: direct exchange,

$$H_3\overset{+}{N}-H\cdots NH_3 \xrightarrow{k_1} H_3N\cdots H-\overset{+}{N}H_3$$

and exchange via an intervening water molecule,

$$\begin{array}{c} H_3\overset{+}{N}-H\cdots O-H\cdots NH_3 \\ | \\ H \end{array} \xrightarrow{k_2} \begin{array}{c} H_3N\cdots H-O\cdots H-\overset{+}{N}H_3 \\ | \\ H \end{array}$$

[1] Z. Luz and S. Meiboom, *J. Am. Chem. Soc.*, **86**, 4768 (1964).
[2] M. Eigen, *Angew. Chem. Int. Ed.*, **3**, 1 (1964).
[3] A. Loewenstein and S. Meiboom, *J. Chem. Phys.*, **27**, 1067 (1957); S. Meiboom, A. Loewenstein, and S. Alexander, *J. Chem. Phys.*, **29**, 969 (1958).

Table 5.7   RATE CONSTANTS FOR AQUEOUS PROTON TRANSFER REACTIONS†

Reaction	$k_f, M^{-1} s^{-1}$	k_r, s^{-1} or $M^{-1} s^{-1}$
$H^+ + OH^- \rightarrow H_2O$	1.4×10^{11}	2.5×10^{-5}
$H^+ + NH_3 \rightarrow NH_4^+$	4.3×10^{10}	24
$H^+ + HS^- \rightarrow H_2S$	7.5×10^{10}	8.3×10^3
$H^+ + OAc^- \rightarrow HOAc$	4.5×10^{10}	8.4×10^5
$H^+ + F^- \rightarrow HF$	1.0×10^{11}	7×10^7
$H^+ + SO_4^{2-} \rightarrow HSO_4^-$	1×10^{11}	1×10^9
$H^+ + H_2O \rightarrow H_3O^+$	1×10^{10}	1×10^{10}
$H^+ + SO_2(NH_2)_2 \rightarrow NH_2SO_2NH_3^+$	3×10^7	$> 10^8$
$OH^- + H^+ \rightarrow H_2O$	1.4×10^{11}	2.5×10^{-5}
$OH^- + HATP^{3-} \rightarrow H_2O + ATP^{4-}$‡	1.2×10^9	38
$OH^- + NH_4^+ \rightarrow H_2O + NH_3$	3.4×10^{10}	6×10^5
$OH^- + HCO_3^- \rightarrow H_2O + CO_3^{2-}$	6×10^9	1.2×10^6
$OH^- + HPO_4^{2-} \rightarrow H_2O + PO_4^{3-}$	2×10^9	4×10^7
$OH^- + SO_2(NH_2)_2 \rightarrow H_2O + NH_2SO_2NH^-$	10^{11}	
$OH^- + H_2O \rightarrow H_2O + OH^-$	3×10^9	3×10^9
$OH^- + SO_3NH_2^- \rightarrow H_2O + SO_3NH^{2-}$	$\sim 10^8$	
$OH^- + CH_3OH \rightarrow H_2O + CH_3O^-$	3×10^6	

† M. Eigen, *Angew. Chem. Int. Ed.*, **3**, 1 (1964).
‡ ATP = adenosine triphosphate.

Table 5.8   RATE CONSTANTS FOR PROTON EXCHANGE REACTIONS†

Reaction	$k_1, M^{-1} s^{-1}$	$k_2, M^{-1} s^{-1}$	k_2/k_1
$NH_4^+ + NH_3$	10.6×10^8	0.9×10^8	0.09
$CH_3NH_3^+ + CH_3NH_2$	2.5×10^8	3.4×10^8	1.4
$(CH_3)_2NH_2^+ + (CH_3)_2NH$	0.4×10^8	5.6×10^8	14
$(CH_3)_3NH^+ + (CH_3)_3N$	$<0.3 \times 10^8$	3.1×10^8	>10

† A. Loewenstein and S. Meiboom, *J. Chem. Phys.*, **27**, 1067 (1957); S. Meiboom, A. Loewenstein, and S. Alexander, *J. Chem. Phys.*, **29**, 969 (1958).

The rate constants k_1 and k_2 are listed in Table 5.8. It can be seen that, as the hydrogen atoms of ammonia are replaced by methyl groups, the rate of the direct process decreases relative to that of the indirect process.

ACIDITIES OF PROTONIC ACIDS

The aqueous pK values (pK = $-\log K_a$) of the binary hydrides of the nonmetals are listed in Table 5.9. These values have been obtained by various methods which we shall briefly describe. The pK values for HF, H_2O, H_2S, H_2Se, and H_2Te were obtained by conventional electrochemical methods, generally involving emf data from cells with pH-sensitive glass electrodes. The values for HCl, HBr, and HI were estimated from thermodynamic data, making the assumption that the standard free energy of dissolution of gaseous HX to form undissociated aqueous HX is zero. Thus, in the case of HBr,

$$HBr(g) \rightarrow H^+(aq) + Br^-(aq) \qquad \Delta G° = -11.85 \text{ kcal mol}^{-1}$$

$$HBr(aq) \rightarrow HBr(g) \qquad \Delta G° \approx 0$$

$$HBr(aq) \rightarrow H^+(aq) + Br^-(aq) \qquad \Delta G° \approx -11.85 \text{ kcal mol}^{-1}$$

Hence

$$pK = -\log K \approx \frac{-11.85}{2.3RT} = \frac{-11.85}{1.364} = -8.7$$

The pK value for PH_3 was estimated from kinetic data on the base-catalyzed exchange of hydrogen and deuterium atoms between PH_3 and D_2O.[1] The rate of reaction was found to be first order in both PH_3 and OD^-, and the mechanism of the reaction was assumed to be

$$PH_3 + OD^- \xrightarrow{\text{slow}} PH_2^- + HOD$$

$$PH_2^- + D_2O \xrightarrow{\text{fast}} PH_2D + OD^-$$

The rate constant for the exchange, and presumably for the slow step of the proposed mechanism, is 0.4 M^{-1} s^{-1} at 25°C. Now, in ordinary aqueous solution, exchange probably proceeds by the same sort of mechanism, i.e.,

$$PH_3 + OH^- \underset{k_2}{\overset{k_1}{\rightleftharpoons}} PH_2^- + H_2O$$

[1] R. E. Weston, Jr., and J. Bigeleisen, *J. Am. Chem. Soc.*, **76**, 3078 (1954).

Table 5.9 AQUEOUS pK VALUES OF THE BINARY HYDRIDES OF THE NONMETALS

CH_4	NH_3	H_2O	HF
~ 44	39	16	3
SiH_4	PH_3	H_2S	HCl
~ 35	27	7	-7
GeH_4	AsH_3	H_2Se	HBr
25	≤ 23	4	-9
		H_2Te	HI
		3	-10

If we neglect isotope effects, then $k_1 = 0.4\ M^{-1}\ s^{-1}$, and, because the reverse reaction is an extremely exothermic proton transfer, we may assume $k_2 \approx 10^{11}$ $M^{-1}\ s^{-1}$. Hence for the latter reaction we can calculate the equilibrium constant, $K' = k_1/k_2 = 4 \times 10^{-12}$. This value is based on the concentrations of all species expressed as molarity. To obtain the value based on the usual convention of unit activity for water, we must divide K' by the molarity of water; hence, $K = (4/55.5) \times 10^{-12} \approx 10^{-13}$. Multiplication by the ionization constant of water yields $K_a = 10^{-27}$ for

$$PH_3 \rightleftharpoons H^+ + PH_2^-$$

The relative pK values in liquid ammonia for PH_3, GeH_4, and AsH_3 were determined from equilibrium measurements.[1] The tabulated values for the aqueous pKs of GeH_4 and AsH_3 are based on the assumption that the relative pK values of the acids are the same in water and liquid ammonia. The pK value for NH_3 is based on the measured self-ionization constant for liquid ammonia and the general observation that pK values of acids are about 10 units lower in ammonia than in water. The value for CH_4 is an estimate based on correlations between base-catalyzed hydrogen exchange rates for hydrocarbons and the corresponding pK values. The value for SiH_4 is a simple interpolation.

From the data in Table 5.9 it is obvious that the trend in the aqueous acidity of binary hydrides within a family or group of elements is the same as the corresponding trend in gas-phase acidity; that is, acidity increases on descending the family. It is also obvious that there is a marked trend toward greater acidity from left to right within a horizontal series of elements. This horizontal trend is in contrast to the absence of a marked trend in the corresponding gas-phase acidities. Undoubtedly the strong horizontal trend in aqueous acidities is due to the increase in anionic hydration energy from left to right (i.e., from larger ions to smaller ions). The interaction energy of a point charge with a dipole increases with decreasing distance between the charge and the dipole; therefore ionic hydration energy is expected to increase with decreasing ionic radius. For most practical purposes, only the hydrides of groups VI and VII show any detectable acidity in aqueous solution. The acids with negative pK values (HCl, HBr, and HI) are essentially completely ionized at all concentrations.

The acidity of a hydride of group IV, V, or VI can be increased by replacing one (or more, when possible) of the hydrogen atoms in the molecule by a relatively electronegative atom or group. The relative effects of various substituents on the aqueous pK values of water, ammonia, and methane can be

---

[1] T. Birchall and W. L. Jolly, *Inorg. Chem.*, **5**, 2177 (1966).

seen in the data of Table 5.10. Notice that substitution of ammonia with the acetyl group, CH_3CO, yields a compound (acetamide) which has a barely detectable acidity in aqueous solution. Methane must be substituted with the much more electronegative nitro group, NO_2, to obtain a compound (nitro-methane) with detectable acidity in aqueous solution. It is significant that the order of the acidifying strength of the various groups is the same for the aquo acids, the ammono acids, and the methane acids and that for a given substituent, X, the acidity always increases in the order $CH_3X < NH_2X < HOX$.

By far the most important family of acids in aqueous solutions are the aquo or *hydroxy* acids, HOX. A very simple method for estimating the pK value of such an acid is based on the formal charge of the atom to which the OH group is attached.[1] When the formal charge on this atom is zero, as in H—O—Cl, the pK value is about 8. When the formal charge is +1, as in phosphorous acid,

$$
\begin{array}{c}
\text{OH} \\
| \\
\text{HO}-\overset{}{\text{P}^+}-\text{H} \\
| \\
\text{O}^-
\end{array}
$$

the pK value is about 3. When the formal charge is +2, as in sulfuric acid,

$$
\begin{array}{c}
\text{O}^- \\
| \\
\text{HO}-\overset{}{\text{S}^{2+}}-\text{OH} \\
| \\
\text{O}^-
\end{array}
$$

the pK is about −2. When the formal charge is +3, as in perchloric acid,

---

[1] J. Ricci, *J. Am. Chem. Soc.*, **70**, 109 (1948); W. M. Latimer, "Oxidation Potentials," 2d ed., Prentice-Hall, Englewood Cliffs, N.J., 1952.

Table 5.10 AQUEOUS pK VALUES OF SOME DERIVATIVES OF WATER, AMMONIA, AND METHANE

X	pK		
	HOX	NH_2X	CH_3X
H	16	39	~44
C_6H_5	10	27	38
CH_3CO	5	15	~20
CN	4	10.5	20
NO_2	−2	7	10

$$
\begin{array}{c}
\text{O}^{-} \\
| \\
\text{HO} - \overset{\displaystyle |}{\underset{\displaystyle |}{\text{Cl}}}{}^{3+} - \text{O}^{-} \\
\text{O}^{-}
\end{array}
$$

the pK is about -7. In the case of di- and tribasic acids (for example, H_3PO_3 and H_3PO_4, respectively) the successive ionization constants decrease by factors of 10^{-5}. The pK values for various hydroxy acids given in Table 5.11 generally agree with the values predicted by these simple rules within ± 1 pK unit.[1] For many purposes rough estimates of this sort are quite adequate. However, the data in Table 5.11 show that the pK values are influenced by more than just the formal charge of the atom bonded to the hydroxyl group. It can be seen that, for acids having analogous structures and the same formal charges, the acidity increases with increasing electronegativity of the atom bonded to the hydroxyl group. When greater predictive accuracy is required, the following equation of Branch and Calvin[2] can be used for estimating the pK of a hydroxy acid.

$$
pK = 16 - \sum I_\alpha \left(\frac{1}{2.8}\right)^i - \sum I_c \left(\frac{1}{2.8}\right)^i - \log \frac{n}{m}
$$

[1] W. L. Jolly, "The Chemistry of Non-Metals," p. 16, Prentice-Hall, Englewood Cliffs, N.J., 1966.
[2] G. E. K. Branch and M. Calvin, "The Theory of Organic Chemistry," Prentice-Hall, Englewood Cliffs, N.J., 1941.

Table 5.11 pK VALUES FOR SOME HYDROXY ACIDS†

Formal charge‡ 0			Formal charge $+1$				Formal charge $+2$			Formal charge $+$	
	pK_1	pK_2		pK_1	pK_2	pK_3		pK_1	pK_2		pK
$HOCH_3$	~16		H_2CO_3§	3.6¶	10.3		HNO_3§	<0		$HClO_4$	≪
HOH	16		HNO_2§	3.2			H_2SO_4	<0	1.9	$HMnO_4$	≪
HOI	11.0		H_2TeO_3	2.7	8.0		H_2SeO_4	<0	2.0		
H_3AsO_3	9.2		H_2SeO_3	2.6	8.3						
H_6TeO_6	8.8		H_3AsO_4	2.3	7.0	13.0					
HOBr	8.7		H_3PO_4	2.2	7.1	12.3					
H_4GeO_4	8.6	12.7	$HClO_2$	2.0							
HOCl	7.2		H_3PO_2	2.0							
			H_2SO_3	1.9	7.2						
			H_3PO_3	1.8	6.2						
			H_5IO_6	1.6	6.0						

† W. L. Jolly, "The Chemistry of Non-Metals," p. 16, Prentice-Hall, Englewood Cliffs, N.J., 1966.
‡ Formal charge of atom bonded to hydroxy group.
§ Formal charge for structure in which π bonds are ignored and in which the central atom lacks an octet.
¶ Corrected for unhydrated CO_2.

Table 5.12 VALUES OF I_z FOR USE IN BRANCH AND
CALVIN'S EQUATION†

H	0	O	4.0	P	1.1
Cl	8.5	S	3.4	As	1.0
Br	7.5	Se	2.7	C	−0.4
I	6.0	N	1.3	C_6H_5	2.0

† G. E. K. Branch and M. Calvin, "The Theory of Organic Chemistry," Prentice-Hall, Englewood Cliffs, N.J., 1941.

The summations are carried out over all the atoms in the group attached to the OH group. The value of I_α is a measure of the electron-attracting ability of the atom; values of I_α for several important elements are given in Table 5.12. The term I_c is 12.3 times the formal charge of the atom, and i is the number of atoms separating the OH group from the atom in question. The last term in the equation accounts for the increase in acidity due to multiple —OH groups in the acid and the decrease in acidity due to multiple —O⁻ groups in the anion; n is the number of —OH groups in the acid, and m is the number of —O⁻ groups in the anion. We shall illustrate the use of this equation in the estimation of the pK of the dihydrogen phosphate ion,

$$\begin{array}{c} \text{O}^- \\ | \\ \text{HO}-\text{P}^+-\text{OH} \\ | \\ \text{O}^- \end{array}$$

$$\sum I_\alpha \left(\frac{1}{2.8}\right)^i = 1.1\left(\frac{1}{2.8}\right)^0 + 3(4)\left(\frac{1}{2.8}\right)^1 = 5.3_8$$

$$\sum I_c \left(\frac{1}{2.8}\right)^i = 12.3\left(\frac{1}{2.8}\right)^0 - 2(12.3)\left(\frac{1}{2.8}\right)^1 = 3.5_2$$

$$\log \frac{n}{m} = \log \frac{2}{3} = -0.1_8$$

$$\text{p}K = 16 - 5.3_8 - 3.5_2 + 0.1_8 = 7.3$$

The calculated value of 7.3 is in good agreement with the experimental value of 7.1.

PROBLEMS

5.1 Write equations for the net reactions which occur in the following cases:
    a   Calcium hydride is added to an aqueous solution of sodium carbonate.
    b   Sodium hydride is heated to 500°C in vacuo.

     c $LiAlH_4$ and $(C_2H_5)_2O \cdot BF_3$ react in ether to form diborane.

     d   Germane is passed into a stirred suspension of finely divided potassium in a high-molecular-weight ether such as diglyme.

     e   Arsine is passed into a liquid ammonia solution of $KGeH_3$.

     f   Excess of water is added to an ether solution of $LiAlH_4$.

5.2   Suggest a sequence of reactions for preparing GeH_3AsH_2 from As_2O_3 and $GeCl_4$.

5.3   Suggest a sequence of reactions for preparing CH_3GeH_2Br from GeH_4 and CH_3I.

5.4   In the synthesis of Ge_2H_6 from GeH_3Br and $KGeH_3$, large amounts of $(GeH_2)_x$ and GeH_4 form as by-products. Explain. [Hint: GeH_2Br^- would be expected to decompose to give $(GeH_2)_x$ and Br^-.]

5.5   The base-catalyzed exhange reaction of PH_3 and D_2O may possibly proceed by a one-step process such as the following:

$$D_2O + PH_3 + OD^- \rightarrow \left[DOD \cdots \overset{\displaystyle H}{\underset{\displaystyle H}{\overset{|}{\underset{|}{P}}}} - H \cdots OD \right]^{\ddagger} \rightarrow DO^- + PH_2D + HOD$$

    Consequently what may one say about the pK of PH_3?

5.6   How do we know that phosphorous acid does not have the following structure?

$$:P \underset{\displaystyle OH}{\overset{\displaystyle OH}{{-}OH}}$$

5.7   Arrange the following aqueous acids in order of increasing acidity:

        HSO_4^- H_3O^+ C_2H_5OH H_4SiO_4 $HSeO_4^-$ CH_3GeH_3 NH_3 AsH_3

        $HClO_4$ HSO_3F

5.8   According to the Branch and Calvin equation for estimating the pK of hydroxy acids, by what amount should the successive pK values of a polybasic acid differ?

5.9   Which of the following aqueous reactions would you expect to be the slowest? The fastest?

$$NH_4^+ \rightarrow NH_3 + H^+$$
$$NH_3 + OH^- \rightarrow H_2O + NH_2^-$$
$$HS^- \rightarrow H^+ + S^{2-}$$
$$HSO_4^- + OH^- \rightarrow H_2O + SO_4^{2-}$$

5.10   Predict the pK of diamidophosphoric acid, $HOPO(NH_2)_2$.

6

ACID-BASE SYSTEMS

THE BRÖNSTED CONCEPT

Under ordinary conditions, the ionization of a protonic acid does not involve the liberation of free protons. In Chap. 5 we have seen that such a process occurring in aqueous solution is best looked upon as the transfer of a proton to a water molecule (or, better, a group of water molecules). In fact, almost all reactions of protonic acids consist of transfers of protons from one base to another. Long ago Brönsted[1] pointed out that acids and bases may be related by the following half-reaction:

$$\text{Acid} \rightleftharpoons H^+ + \text{base}$$

Thus an acid is a species which can act as a proton donor, and a base is a species which can act as a proton acceptor. A typical acid-base reaction may then be written

$$\text{Acid 1} + \text{base 2} \rightleftharpoons \text{acid 2} + \text{base 1}$$

[1] J. N. Brönsted, *Rec. Trav. Chim.*, **42**, 718 (1923).

By removing a proton from acid 1, we obtain base 1. Acid 1 is said to be the conjugate acid of base 1 and base 1 the conjugate base of acid 1. Similar relationships exist between acid 2 and base 2.

PROTONIC SOLVENTS

Certain solvents resemble water in that they self-ionize to give solvated protons and anions and dissolve salts to give conducting solutions. In such solvents the Brönsted concept is quite applicable. For example, in liquid ammonia[1] the following self-ionization, or "autoprotolysis," occurs:

$$2NH_3 \rightleftharpoons NH_4^+ + NH_2^- \qquad K_{-33^\circ C} \approx 10^{-30}$$

In liquid ammonia any species which gives (or reacts to give) ammonium ions is an acid, and any species which gives (or reacts to give) amide ions is a base. Thus the following reactions are acid-base reactions in ammonia.

$$H_2NCONH_2 + NH_2^- \rightarrow H_2NCONH^- + NH_3$$
$$NH_4^+ + KOH(s) \rightarrow K^+ + NH_3 + H_2O$$
$$H_2O + K^+ + NH_2^- \rightarrow KOH(s) + NH_3$$

In anhydrous sulfuric acid,[2] the following autoprotolysis occurs:

$$H_2SO_4 \rightleftharpoons H^+ + HSO_4^-$$

For simplicity, we write H^+ for the solvated proton in H_2SO_4 rather than $H_3SO_4^+$ or $H(H_2SO_4)_n^+$. When water is dissolved in anhydrous sulfuric acid, the following solvolysis takes place:

$$H_2O + H_2SO_4 \rightleftharpoons H_3O^+ + HSO_4^-$$

Hence water is a base in sulfuric acid. Similar reactions occur with many other oxygen-containing compounds. Often dehydration accompanies the basic hydrolysis:

$$HONO_2 + 2H_2SO_4 \rightarrow NO_2^+ + H_3O^+ + 2HSO_4^-$$

In solutions of nitric acid in concentrated sulfuric acid, which are often used in the nitration of aromatic compounds, the active agent is the nitryl ion,

---

[1] W. L. Jolly and C. J. Hallada, in T. C. Waddington (ed.), "Non-aqueous Solvent Systems," pp. 1–45, Academic, London, 1965; J. J. Lagowski and G. A. Moczygemba, in J. J. Lagowski (ed.), "The Chemistry of Non-aqueous Solvents," vol. 2, pp. 320–371, Academic, New York, 1967.

[2] R. J. Gillespie and E. A. Robinson, in T. C. Waddington (ed.), "Non-aqueous Solvent Systems," pp. 117–210, Academic, London, 1965.

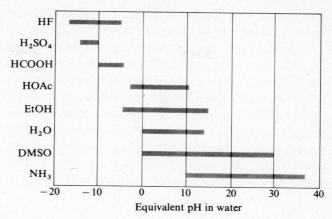

FIGURE 6.1
Effective pH ranges of various protonic solvents. (*Reproduced with permission from W. L. Jolly, "The Synthesis and Characterization of Inorganic Compounds," Prentice-Hall, Englewood Cliffs, N.J., 1970.*)

NO_2^+. This electrophilic species attacks the π-bonded system, with displacement of a proton to the solvent.

$$NO_2^+ + ArH \rightarrow ArNO_2 + H^+$$

Because of the high acidity of sulfuric acid itself, relatively few substances are acids in this solvent. However, SO_3 does dissolve to form disulfuric acid, which is a slightly stronger acid than sulfuric acid.

$$SO_3 + H_2SO_4 \rightarrow H_2S_2O_7$$
$$H_2S_2O_7 \rightleftharpoons H^+ + HS_2O_7^-$$

If one wishes to have a protonic acid and its anion present in comparable amounts at equilibrium in aqueous solution, the pK value of the acid must lie within, or near, the range 0 to 14. If the pK is much less than 0, the acid will be completely ionized even in a strongly acidic solution, and if the pK is much greater than 14, the acid will be completely nonionized even in strongly basic solution. To have a measurable acid-anion equilibrium for p$K \ll 0$, one must use a solvent that is less basic than water, such as sulfuric acid or acetic acid. To have a measurable acid-anion equilibrium for p$K \gg 14$, one must use a solvent that is more basic than water, such as liquid ammonia. In Fig. 6.1, the effective pH ranges corresponding to aqueous solutions that can be achieved in several protonic solvents are shown graphically. It can be seen that in a solvent such as formic acid one can differentiate the acidities of HCl (p$K \approx -7$) and

HBr ($pK \approx -9$). On the other hand, in a solvent such as liquid ammonia, one can differentiate the acidities of $PH_3(pK \approx 27)$ and GeH_4 ($pK \approx 25$).

In aqueous solutions, the hydrogen-ion activity is closely proportional to the concentration of strong acid down to about pH 0 and closely proportional to the reciprocal of the concentration of strong base up to about pH 14. Beyond these pH limits, the hydrogen-ion activity and the effective pH change much more rapidly than one would calculate from changes in the concentration of strong acid or strong base. In such solutions, and even in completely nonaqueous solutions, it is possible to define an "acidity function"[1] which is the effective aqueous pH. The acidity function H_0 is defined in terms of a system buffered with a base B and its cationic conjugate acid BH^+:

$$H_0 = pK_a + \log \frac{C_B}{C_{BH^+}}$$

where pK_a is the ionization constant of BH^+ and C_{BH^+} and C_B are the concentrations of BH^+ and B, respectively. The function H_- is defined in terms of a system buffered with a base B^- and its neutral conjugate acid HB:

$$H_- = pK_a + \log \frac{C_{B^-}}{C_{HB}}$$

[1] L. P. Hammett, "Physical Organic Chemistry," McGraw-Hill, New York, 1940.

Table 6.1 COMPOSITIONS OF SOME SOLUTIONS WITH HIGH $-H_0$ AND H_- VALUES

Solution	$-H_0$†	Solution	H_-‡
7% $SbF_5 \cdot 3SO_3$ in HSO_3F	19.35	95% DMSO§, 5% EtOH; 10^{-2} M KOEt	20.68
10% SbF_5 in HSO_3F	18.94	90% DMSO§, 10% EtOH; 10^{-2} M KOEt	19.68
HSO_3F	15.07	80% DMSO§, 20% EtOH; 10^{-2} M KOEt	18.97
$H_2S_2O_7$	14.14		
100% H_2SO_4	11.93	15 M KOH (aqueous)	18.23
98% H_2SO_4	10.44	10 M KOH (aqueous)	16.90
90% H_2SO_4	8.92	5 M KOH (aqueous)	15.44
60% H_2SO_4	4.46		

† R. L. Gillespie, T. E. Peel, and E. A. Robinson, *J. Am. Chem. Soc.*, **93**, 5083 (1971); R. J. Gillespie and T. E. Peel, *J. Am. Chem. Soc.*, **95**, 5173 (1973); M. J. Jorgenson and D. R. Harrter, *J. Am. Chem. Soc.*, **85**, 878 (1963).
‡ K. Bowden and R. Stewart, *Tetrahedron*, **21**, 261 (1965); G. Yagil, *J. Phys. Chem.*, **71**, 1034 (1967).
§ DMSO = dimethyl solfoxide.

By appropriate choices of solvent and acid or base concentration, it is possible to prepare solutions having overlapping H_0 or H_- ranges that collectively span an effective pH range from -19 to $+21$. The compositions and H_0 or H_- values for some representative solutions are listed in Table 6.1.

APROTIC SOLVENTS

Some solvents do not self-ionize to any significant extent to give solvated protons. This category of "aprotic" solvents includes most hydrocarbons, ethers, and of course solvents containing no hydrogen such as bromine trifluoride and sulfur dioxide. Obviously the Brönsted concept of acids and bases is inapplicable in such solvents. If an aprotic solvent undergoes self-ionization, the cation formed in the self-ionization is considered the acidic species, and the anion formed in the self-ionization is considered the basic species. For example, BrF_2^+ and BrF_4^- are the acidic and basic species, respectively, in bromine trifluoride.

$$2BrF_3 \rightleftharpoons BrF_2^+ + BrF_4^-$$

Thus antimony pentafluoride, which reacts with BrF_3 as follows,

$$SbF_5 + BrF_3 \rightarrow BrF_2^+ + SbF_6^-$$

is an acid in BrF_3. Potassium fluoride, which reacts as follows,

$$KF + BrF_3 \rightarrow K^+ + BrF_4^-$$

is a base in BrF_3. When solutions of these reagents are mixed, a neutralization reaction occurs:

$$BrF_2^+ + BrF_4^- \rightleftharpoons 2BrF_3$$

In the case of aprotic solvents which do not self-ionize, acid-base reactions are best treated in terms of the very general Lewis acid-base theory, which we shall now discuss.

LEWIS ACID-BASE THEORY

The Lewis acid-base theory is completely independent of solvent considerations. Indeed, many Lewis acid-base reactions proceed in the gas phase. In a Lewis acid-base reaction, a pair of electrons from one species is used to form a covalent bond to another species. The species that "donates" the electron pair is the base, and the species that "accepts" the electron pair is the acid. The reaction may be written

$$A + :B \rightarrow A—B$$

Sometimes the acid is referred to as an electron acceptor or an electrophile and the base is referred to as an electron donor or a nucleophile. Many reactions involve the transfer of an acid from one base to another, or vice versa, or both transfers at once. In such cases the net reactions are written

$$B + A—B' \rightarrow B' + A—B$$
$$A + A'—B \rightarrow A' + A—B$$
$$A—B + A'—B' \rightarrow A—B' + A'—B$$

A few examples of some Lewis acids, bases, and the known adducts of these are listed in Table 6.2.

One of the difficulties of systematizing the relative basicities of Lewis bases is the fact that different trends in K, $\Delta G°$, or $\Delta H°$ are obtained when different reference acids are used. For example, the complexing ability of the halide ions toward Al^{3+} increases in the order $I^- < Br^- < Cl^- < F^-$. On the other hand, the order is $F^- < Cl^- < Br^- < I^-$ for Hg^{2+}. A similar reversal can be seen in the heats of reaction of the acids I_2 and C_6H_5OH with the bases $(C_2H_5)_2O$ and $(C_2H_5)_2S$.[1] The heat of reaction of I_2 with $(C_2H_5)_2S$ is greater than that with $(C_2H_5)_2O$, whereas the heat of reaction of C_6H_5OH with $(C_2H_5)_2S$ is less than that with $(C_2H_5)_2O$. To resolve these problems, Pearson proposed the principle of hard and soft acids and bases.[2] A soft acid or base is one in which the valence electrons are easily polarized or removed, and a hard acid or base is one which holds its valence electrons tightly and is not easily distorted. The principle

[1] R. J. Niedzielski, R. S. Drago, and R. L. Middaugh, *J. Am. Chem. Soc.*, **86**, 1694 (1964).
[2] R. G. Pearson, *J. Am. Chem. Soc.*, **85**, 3533 (1963); *Science*, **151**, 172 (1966); *Chem. Br.*, **3**, 103 (1967); *J. Chem. Educ.*, **45**, 581, 643 (1968).

Table 6.2 SOME LEWIS ACIDS, BASES, AND THEIR ADDUCTS

Lewis acids	Lewis bases				
	H^-	OH^-	NH_3	Cl^-	C_6H_6
H^+	H_2	H_2O	NH_4^+	HCl	$[C_6H_7^+]$
$B(CH_3)_3$	$B(CH_3)_3H^-$	$B(CH_3)_3OH^-$	$B(CH_3)_3NH_3$		
SO_3	....	HSO_4^-	NH_3SO_3	SO_3Cl^-	
Ag^+	....	$[AgOH]$	$Ag(NH_3)_2^+$	$AgCl_2^-$	$AgC_6H_6^+$
I_2	....	I_2OH^-	I_2NH_3	I_2Cl^-	$I_2C_6H_6$

is: Hard acids prefer to coordinate to hard bases, and soft acids prefer to coordinate to soft bases. Tables 6.3 and 6.4 list some examples of acids and bases in the hard, soft, and borderline categories. The opposing trends in the reactivity of the halide ions toward Al^{3+} and Hg^{2+} are now easily rationalized. The Al^{3+} ion is a hard acid that prefers to bond to hard bases (such as F^-), and the Hg^{2+} ion is a soft acid that prefers to bond to soft bases (such as I^-). We can similarly explain the thermal data for I_2, C_6H_5OH, $(C_2H_5)_2O$, and $(C_2H_5)_2S$. Iodine and $(C_2H_5)_2S$ interact relatively strongly because they are both soft, and C_6H_5OH and $(C_2H_5)_2O$ interact strongly because they are both hard.

Table 6.3 THE CLASSIFICATION OF SOME LEWIS ACIDS[†]

Hard	Borderline	Soft
H^+, Li^+, Na^+, K^+	Fe^{2+}, Co^{2+}, Ni^{2+}	Cu^+, Ag^+, Au^+, Tl^+, Hg^+
Be^{2+}, Mg^{2+}, Ca^{2+}, Sr^{2+}, Mn^{2+}	Cu^{2+}, Zn^{2+}, Pb^{2+}	Pd^{2+}, Cd^{2+}, Pt^{2+}, Hg^{2+}
Al^{3+}, Sc^{3+}, Ga^{3+}, In^{3+}, La^{3+}	Sn^{2+}, Sb^{3+}, Bi^{3+}, Rh^{3+}	CH_3Hg^+, $Co(CN)_5^{2-}$, Pt^{4+}
N^{3+}, Cl^{3+}, Gd^{3+}, Lu^{3+}, Cr^{3+}	Ir^{3+}, $B(CH_3)_3$, SO_2	Te^{4+}
Co^{3+}, Fe^{3+}, As^{3+}, CH_3Sn^{3+}	NO^+, Ru^{2+}, Os^{2+}	Tl^{3+}, $Tl(CH_3)_3$, BH_3, $Ga(CH_3)_3$
Si^{4+}, Ti^{4+}, Zr^{4+}, Th^{4+}, U^{4+}	R_3C^+, $C_6H_5^+$, GaH_3	$GaCl_3$, GaI_3, $InCl_3$
Pu^{4+}, Ce^{3+}, Hf^{4+}, WO^{4+}, Sn^{4+}		RS^+, RSe^+, RTe^+
UO_2^{2+}, $(CH_3)_2Sn^{2+}$		I^+, Br^+, HO^+, RO^+
VO^{2+}, MoO^{3+}, $BeMe_2$, BF_3		I_2, Br_2, ICN, etc.
$B(OR)_3$, $Al(CH_3)_3$, $AlCl_3$		Trinitrobenzene, etc.
AlH_3, RPO_2^+, $ROPO_2^+$		Chloranil, quinones, etc.
RSO_2^+, $ROSO_2^+$, SO_3		Tetracyanoethylene, etc.
I^{7+}, I^{5+}, Cl^{7+}, Cr^{6+}, RCO^+		O, Cl, Br, I, N, RO, RO_2
CO_2, NC^+		M^0 (metal atoms)
HX (hydrogen-bonding		Bulk metals
molecules)		CH_2, carbenes

[†] R. G. Pearson, *J. Am. Chem. Soc.*, **85**, 3533 (1963); *Science*, **151**, 172 (1966); *Chem. Br.*, **3**, 103 (1967); *J. Chem. Educ.*, **45**, 581, 643 (1968).

Table 6.4 THE CLASSIFICATION OF SOME LEWIS BASES[†]

Hard	Borderline	Soft
H_2O, OH^-, F^-	$C_6H_5NH_2$, C_5H_5N, N_3^-	R_2S, RSH, RS^-
$CH_3CO_2^-$, PO_4^{3-}, SO_4^{2-}	Br^-, NO_2^-, SO_3^{2-}, N_2	I^-, SCN^-, $S_2O_3^{2-}$, R_3P, R_3As
Cl^-, CO_3^{2-}, ClO_4^-, NO_3^-		$(RO)_3P$
ROH, RO^-, R_2O		CN^-, RNC, CO
NH_3, RNH_2, N_2H_4		C_2H_4, C_6H_6
		H^-, R^-

[†] R. G. Pearson, *J. Am. Chem. Soc.*, **85**, 3533 (1963); *Science*, **151**, 172 (1966); *Chem. Br.*, **3**, 103 (1967); *J. Chem. Educ.*, **45**, 581, 643 (1968).

The principle of hard and soft acids and bases has many qualitative applications in inorganic chemistry. Because the hardness of an element usually increases with increasing oxidation state, to stabilize an element in a very high oxidation state, the element should be coordinated to hard bases, such as O^{2-}, OH^-, and F^-. Thus iron(VI), silver(III), and Pt(VI) can be obtained in the compounds K_2FeO_4, AgO, and PtF_6, respectively. To stabilize an element in a low oxidation state, it should be coordinated to soft bases, such as CO and PR_3. Thus cobalt($-I$) and Pt(0) are found in compounds such as $Na[Co(CO)_4]$ and $Pt[P(CH_3)_3]_4$. In the series of compounds $R_3SiI \rightarrow ((R_3Si)_2S \rightarrow R_3SiBr \rightarrow R_3SiNC \rightarrow R_3SiCl \rightarrow R_3SiNCS \rightarrow R_3SiNCO \rightarrow R_3SiF$, a given compound can be converted to any other on its right by refluxing it with the appropriate silver salt.[1] However, a compound cannot be converted in good yield to a compound on its left by this method. For example, the following reaction is irreversible:

$$Et_3SiI + AgBr \rightarrow Et_3SiBr + AgI$$

The soft silver ion prefers to bond to a relatively soft group, and the hard silicon atom prefers to bond to a relatively hard group.

The qualitative concept of hard and soft acids and bases must be used with caution. If one does not take into account the *strengths* of Lewis acids and bases as well as their hardness or softness, wrong conclusions regarding the driving forces of acid-base reactions can be drawn. For example, consider the reaction

$$CHCl_3 \cdot NH_3 + B(CH_3)_3 \cdot C_6H_6 \rightleftharpoons CHCl_3 \cdot C_6H_6 + B(CH_3)_3 \cdot NH_3 \qquad (6.1)$$

According to Tables 6.3 and 6.4, $CHCl_3$ and NH_3 are both hard, $B(CH_3)_3$ is borderline, and C_6H_6 is soft. Therefore one might predict that the reaction would go in the reverse direction. However, the reaction actually goes in the forward direction because of the very strong interaction between $B(CH_3)_3$ and NH_3. One way of quantitatively accounting for strength as well as hardness and softness is by assigning two empirical parameters to each acid and each base and by equating the interaction energy of a given acid-base pair to a suitable algebraic function of the four parameters involved. For example, Drago et al.[2] have shown that the heat of formation of an adduct of an acid A and a base B can be calculated from the equation

$$-\Delta H^\circ = E_A E_B + C_A C_B \qquad \text{kcal mol}^{-1}$$

[1] C. Eaborn, *J. Chem. Soc.*, p. 3077, 1950; A. G. MacDiarmid, *Prep. Inorg. React.*, **1**, 165 (1964).

[2] R. S. Drago and B. B. Wayland, *J. Am. Chem. Soc.*, **87**, 3571 (1965); R. S. Drago, *Chem. Br.*, **3**, 516 (1967); *Struct. Bonding*, **15**, 73 (1973); *J. Chem. Educ.*, **51**, 300 (1974), and references therein.

where E_A, E_B, C_A, and C_B are empirical constants characteristic of the acid A and the base B. Values of these parameters have been tabulated for a wide variety of Lewis acids and bases.[1] These may be used to predict heats of adduct formation with an average error of 0.7 kcal mol^{-1} or less. For example, we may use this method to calculate the heat of reaction 6.1. The E and C parameters are given below.

$CHCl_3$	$E_A = 3.31$	$C_A = 0.150$
$B(CH_3)_3$	$E_A = 6.14$	$C_A = 1.70$
C_6H_6	$E_B = 0.486$	$C_B = 0.707$
NH_3	$E_B = 1.36$	$C_B = 3.46$

From these data we calculate $\Delta H° = -6.7$ kcal mol^{-1} for reaction 6.1 as written.

PROBLEMS

6.1 Because fluorine is more electronegative than chlorine, one might expect the boron atom in BF_3 to be more electron-deficient than that in BCl_3. However, BCl_3 is a stronger acid toward $N(CH_3)_3$ than BF_3. Explain.

6.2 Write equations for the net reactions which occur in the following cases:
  a Aniline is added to a solution of potassium amide in liquid ammonia.
  b Lithium nitride is added to a solution formed by addition of acetic acid to excess liquid ammonia.
  c NO_2Cl is formed by passing HCl into a solution of nitric acid in concentrated sulfuric acid.
  d Boric acid is dissolved in sulfuric acid. (Six moles of dissolved species, including 2 mol of bisulfate ion, are formed per mole of boric acid.)

6.3 Explain how potassium acid phthalate can be a primary standard acid in water, and a primary standard base in anhydrous acetic acid.

6.4 What is the ammonium ion concentration in pure liquid ammonia?

6.5 Explain the fact that KOH is much more soluble in liquid ammonia containing water than in pure liquid ammonia.

6.6 Explain the fact that $FeCl_4^-$ is formed when $FeCl_3$ is dissolved in $OP(OEt)_3$.

6.7 Which of the following reactions have equilibrium constants greater than 1 at ordinary temperatures? (All species may be assumed to be gaseous unless otherwise indicated.)
  a $R_3PBBr_3 + R_3NBF_3 \rightleftharpoons R_3PBF_3 + R_3NBBr_3$
  b $SO_2 + (C_6H_5)_3PHOC(CH_3)_3 \rightleftharpoons (CH_3)_3COH + (C_6H_5)_3PSO_2$
  c $AgCl_2^-(aq) + 2CN^-(aq) \rightleftharpoons Ag(CN)_2^-(aq) + 2Cl^-(aq)$
  d $MeOI + Et(CO)OMe \rightleftharpoons MeOOMe + Et(CO)I$

[1] Ibid.

6.8 $\Delta H^\circ = -3.9$ and -17.0 kcal mol^{-1}, respectively, for the reactions

$$CHCl_3 + C_5H_5N \rightarrow C_5H_5NHCCl_3$$

and

$$B(CH_3)_3 + C_5H_5N \rightarrow C_5H_5NB(CH_3)_3$$

Calculate E_B and C_B values for C_5H_5N.

6.9 The heats of reaction of $B(CH_3)_3$ with NH_3, CH_3NH_2, $(CH_3)_2NH$, and $(CH_3)_3N$ are 13.8, 17.6, 19.3, and 17.6 kcal mol^{-1}, respectively. Can you explain why the value for $(CH_3)_3N$ is out of line?

SOME ASPECTS OF AQUEOUS SOLUTION CHEMISTRY

THE USE OF REDUCTION POTENTIALS[1]

One of the most effective methods for systematizing inorganic chemical reactions is the use of oxidation-reduction (redox) potentials in quantitatively summarizing the driving forces of oxidation-reduction reactions in aqueous solutions. Because the method is an application of thermodynamics, it is important to keep in mind the limitations of thermodynamics. Thermodynamic data give information only regarding the extent to which reactions can go, not regarding the rates at which they go. For example, although both of the following disproportionation reactions have large equilibrium constants, only the first proceeds at an appreciable rate under ordinary conditions:

$$2NO_2 + 2OH^- \rightarrow NO_2^- + NO_3^- + H_2O \qquad \text{(fast)}$$
$$2N_2O \rightarrow N_2 + 2NO \qquad \text{(slow)}$$

The science, or art, of predicting the rates of chemical reactions is in a developmental stage. At present, often the best that can be done is to roughly

[1] W. L. Jolly, *J. Chem. Educ.*, **43**, 198 (1966).

classify certain classes of compounds as fast-reacting and others as slow-reacting. However, sometimes thermodynamic data are useful even in the absence of kinetic information. Thus, when thermodynamics tells us that a reaction cannot go, we are saved the trouble of trying the reaction.[1]

We shall write half-reactions corresponding to *reduction potentials*, thus:

$$\text{Na}^+ + e^- \rightarrow \text{Na} \qquad E° = -2.71 \text{ V}$$

Of course there is no difficulty in applying the same methods, after taking account of the sign change, to *oxidation potentials:*

$$\text{Na} \rightarrow e^- + \text{Na}^+ \qquad E° = +2.71 \text{ V}$$

Obviously, in order to use any potential, it is necessary to know whether it is an oxidation or reduction potential, and whether or not the above sign convention is used. Particular care must be taken when using potentials from different sources. All the potentials cited in this chapter refer to 25°C.

It should be remembered that the potentials are expressed in volts (a difference in electrical potential). Therefore a potential is a constant which does not change when the coefficients of the half-reaction are changed. Thus both of the following half-reactions have the same potential:

$$\tfrac{1}{2}\text{I}_2 + e^- \rightarrow \text{I}^- \qquad E° = 0.536 \text{ V}$$
$$\text{I}_2 + 2e^- \rightarrow 2\text{I}^- \qquad E° = 0.536 \text{ V}$$

For simplicity, we shall hereafter omit the symbol V (for volt) in such expressions involving potentials.

The Use of Tabulated Potentials

Table 7.1 is a table of "half-reactions" and their standard potentials.[2] The reducing agents are listed in order of decreasing strength (and consequently the oxidizing agents are listed in order of increasing strength) from the top to the bottom of the table. An oxidizing agent should be able to oxidize all reducing agents lying above it in the table, and a reducing agent should be able to reduce all oxidizing agents lying below it in the table. However, because the potentials are *standard* potentials, this statement is strictly correct only when all

---

[1] One must be cautious in such applications of thermodynamics. Consider the reaction of a nonvolatile phase (or phases) to yield a volatile product (or products) for which $\Delta G° \gg 0$. Reactions of this type often can be carried to completion by pumping on the reactants or by continuous sweeping of the nonvolatile phase with an inert carrier gas. The reaction of a gas with a nonvolatile phase to yield a volatile product (for which $\Delta G° \gg 0$) often can be carried to completion by sweeping the nonvolatile phase with the reactant gas.

[2] Most of the potentials given in this chapter are from W. M. Latimer, "Oxidation Potentials," 2d ed., Prentice-Hall, Englewood Cliffs, N.J., 1952.

Table 7.1 THE REDUCTION POTENTIALS OF SOME AQUEOUS HALF-REACTIONS

Couple (Acid solutions)	E	Couple (Basic solutions)	E
$Na^+ + e^- \rightarrow Na$	-2.71	$Mg(OH)_2 + 2e^- \rightarrow Mg + 2OH^-$	-2.69
$Mg^{2+} + 2e^- \rightarrow Mg$	-2.37	$H_2AlO_3^- + H_2O + 3e^- \rightarrow Al + 4OH^-$	-2.35
$Al^{3+} + 3e^- \rightarrow Al$	-1.66	$B(OH)_4^- + 2H_2 + 4e^- \rightarrow BH_4^- + 4OH^-$	-1.60
$Zn^{2+} + 2e^- \rightarrow Zn$	-0.76	$HPO_3^{2-} + 2H_2O + 2e^- \rightarrow H_2PO_2^- + 3OH^-$	-1.57
$H_3PO_3 + 2H^+ + 2e^- \rightarrow H_3PO_2 + H_2O$	-0.50	$ZnO_2^{2-} + 2H_2O + 2e^- \rightarrow Zn + 4OH^-$	-1.22
$H_3BO_3 + 7H^+ + 8e^- \rightarrow BH_4^- + 3H_2O$	-0.47	$2SO_3^{2-} + 2H_2O + 2e^- \rightarrow S_2O_4^{2-} + 4OH^-$	-1.12
$Sn^{2+} + 2e^- \rightarrow Sn$	-0.14	$SO_4^{2-} + H_2O + 2e^- \rightarrow SO_3^{2-} + 2OH^-$	-0.93
$2H_2SO_3 + H^+ + 2e^- \rightarrow HS_2O_4^- + 2H_2O$	-0.08	$HSnO_2^- + H_2O + 2e^- \rightarrow Sn + 3OH^-$	-0.91
$2H^+ + 2e^- \rightarrow H_2(g)$	0.00	$2H_2O + 2e^- \rightarrow H_2(g) + 2OH^-$	-0.83
$HCOOH + 2H^+ + 2e^- \rightarrow HCHO + H_2O$	0.06	$AsO_4^{3-} + 3H_2O + 2e^- \rightarrow H_2AsO_3^- + 4OH^-$	-0.67
$S + 2H^+ + 2e^- \rightarrow H_2S$	0.14	$S + 2e^- \rightarrow S^{2-}$	-0.48
$Sn^{4+} + 2e^- \rightarrow Sn^{2+}$	0.15	$CrO_4^{2-} + 4H_2O + 3e^- \rightarrow Cr(OH)_3 + 5OH^-$	-0.13
$SO_4^{2-} + 4H^+ + 2e^- \rightarrow H_2SO_3 + H_2O$	0.17	$O_2(g) + H_2O + 2e^- \rightarrow HO_2^- + OH^-$	-0.08
$Fe(CN)_6^{3-} + e^- \rightarrow Fe(CN)_6^{4-}$	0.36	$PbO_2 + H_2O + 2e^- \rightarrow PbO + 2OH^-$	0.25
$I_3^- + 2e^- \rightarrow 3I^-$	0.54	$O_2(g) + 2H_2O + 4e^- \rightarrow 4OH^-$	0.40
$H_3AsO_4 + 2H^+ + 2e^- \rightarrow H_3AsO_3 + H_2O$	0.56	$IO^- + H_2O + 2e^- \rightarrow I^- + 2OH^-$	0.49
$O_2(g) + 2H^+ + 2e^- \rightarrow H_2O_2$	0.68	$MnO_4^- + 2H_2O + 3e^- \rightarrow MnO_2 + 4OH^-$	0.59
$Fe^{3+} + e^- \rightarrow Fe^{2+}$	0.77	$H_3IO_6^{2-} + 2e^- \rightarrow IO_3^- + 3OH^-$	0.70
$NO_3^- + 4H^+ + 4e^- \rightarrow NO(g) + 2H_2O$	0.96	$BrO^- + H_2O + 2e^- \rightarrow Br^- + 2OH^-$	0.76
$Br_2 + 2e^- \rightarrow 2Br^-$	1.06	$HO_2^- + H_2O + 2e^- \rightarrow 3OH^-$	0.88
$O_2(g) + 4H^+ + 4e^- \rightarrow 2H_2O$	1.23	$ClO^- + H_2O + 2e^- \rightarrow Cl^- + 2OH^-$	0.89
$Cr_2O_7^{2-} + 14H^+ + 6e^- \rightarrow 2Cr^{3+} + 7H_2O$	1.33	$O_3(g) + H_2O + 2e^- \rightarrow O_2(g) + 2OH^-$	1.24
$Cl_2(g) + 2e^- \rightarrow 2Cl^-$	1.36	$S_2O_8^{2-} + 2e^- \rightarrow 2SO_4^{2-}$	2.01
$PbO_2 + 4H^+ + 2e^- \rightarrow Pb^{2+} + 2H_2O$	1.46		
$MnO_4^- + 8H^+ + 5e^- \rightarrow Mn^{2+} + 4H_2O$	1.51		
$H_5IO_6 + H^+ + 2e^- \rightarrow IO_3^- + 3H_2O$	1.60		
$MnO_4^- + 4H^+ + 3e^- \rightarrow MnO_2 + 2H_2O$	1.69		
$H_2O_2 + 2H^+ + 2e^- \rightarrow 2H_2O$	1.77		
$S_2O_8^{2-} + 2e^- \rightarrow 2SO_4^{2-}$	2.01		
$O_3(g) + 2H^+ + 2e^- \rightarrow O_2(g) + H_2O$	2.07		
$F_2(g) + 2e^- \rightarrow 2F^-$	2.87		

the species are at unit activity. As will be shown later, exceptions to this rule can be effected by appropriate adjustment of product and reactant concentrations. Because from any pair of reduction potentials one can calculate the driving force for a complete, unique reaction, it is clear that a table of this sort summarizes an enormous amount of information in a small space. One calculates that a table of 100 reduction potentials would yield the equilibrium constants for 4950 different reactions.

A tabulation of potentials like Table 7.1 is very handy when looking for oxidizing agents or reducing agents having a certain specified strength. However, a chemist often wishes to know whether a particular reaction can go, or what its equilibrium constant is. For such purposes it is more useful to have the half-reactions listed according to the elements whose oxidation states change.[1] Thus, if one wishes to know the driving force for the reaction

$$H_3AsO_3 + I_3^- + H_2O \rightarrow H_3AsO_4 + 3I^- + 2H^+$$

one looks under the arsenic potentials for

$$H_3AsO_4 + 2H^+ + 2e^- \rightarrow H_3AsO_3 + H_2O \qquad E° = 0.559$$

and under iodine for

$$I_3^- + 2e^- \rightarrow 3I^- \qquad E° = 0.536$$

By subtracting the first half-reaction from the second, one gets the desired reaction. Whenever half-reactions are added or subtracted, one should not add or subtract the corresponding $E°$ values, but rather the appropriate number of electronvolts, or $nE°$ values (where n is the number of electrons appearing in a half-reaction). For the reaction under question, $nE° = 2(0.536 - 0.559) = -0.046$. From this one can calculate $\Delta G°$ (in calories per mole) or K, using the relations

$$-\Delta G° = 23,060 \cdot nE° \qquad \text{and} \qquad nE° = 0.05916 \log K$$
$$\Delta G° = 1060 \text{ cal mol}^{-1}$$
$$K = 0.17$$

From these results one must not draw the conclusion that the reaction cannot proceed quantitatively in either direction. Indeed, there are volumetric methods of analysis based both on the quantitative oxidation of arsenious acid by triiodide and on the quantitative reduction of arsenic acid by iodide. These are accomplished by appropriately adjusting the hydrogen-ion concentration of the solution. At pH 7,

$$\frac{[I^-]^3[H_3AsO_4]}{[I_3^-][H_3AsO_3]} = \frac{0.17}{(10^{-7})^2} = 1.7 \times 10^{13}$$

---

[1] Generally there is no ambiguity as to which elements change oxidation state. However, in the case of a half-reaction such as $S_2O_8^{2-} + 2e^- \rightarrow 2SO_4^{2-}$ one might be justified in a double listing.

and arsenious acid can be titrated with triiodide or vice versa. In 6 M HCl,

$$\frac{[I^-]^3[H_3AsO_4]}{[I_3^-][H_3AsO_3]} \approx \frac{0.17}{6^2} \approx 5 \times 10^{-3} \qquad \text{(neglecting activity coefficients)}$$

and consequently arsenic acid can be quantitatively reduced by an excess of iodide. (The liberated triiodide can be titrated with thiosulfate.)

Dependence of Potentials on pH

The general problem of the pH dependence of reduction potentials is of considerable importance. The potential of any half-reaction changes with the concentration of the species involved according to the Nernst equation,

$$E = E^\circ - \frac{0.05916}{n} \log Q$$

where Q has the same form as the equilibrium constant but is a function of the activities of the actual reactants and products and not those of the equilibrium state. If hydrogen ion or hydroxide ion appears in the half-reaction (and the corresponding ionic activity appears in Q), the potential will change with pH. Thus for the hydrogen-ion–hydrogen half-reaction,

$$H^+ + e^- \to \tfrac{1}{2}H_2 \qquad E^\circ = 0$$

we write

$$E = -0.05916 \log \frac{P_{H_2}^{1/2}}{[H^+]}$$

and for the oxygen-water half-reaction,

$$\tfrac{1}{2}O_2 + 2H^+ + 2e^- \to H_2O \qquad E^\circ = 1.23$$

we write

$$E = 1.23 + 0.05916 \log \{[H^+] \cdot P_{O_2}^{1/4}\}$$

The potentials for these half-reactions are plotted versus pH in Fig. 7.1. Theoretically no oxidizing agent whose reduction potential lies above the O_2–H_2O line and no reducing agent whose reduction potential falls below the H^+–H_2 line can exist in aqueous solutions. Actually, for kinetic reasons, these lines can be extended about 0.5 V, and the dotted lines in Fig. 7.1 are more realistic boundaries for the region of stability of oxidizing and reducing agents in aqueous solutions.

One must take care to use reduction potentials only in the pH ranges for which they are valid. For example, the ferrous ion–iron reduction potential

$$Fe^{2+} + 2e^- \to Fe \qquad E^\circ = -0.41$$

has little significance in alkaline solutions because the ferrous ion forms an essentially insoluble hydroxide. In order to calculate the appropriate potential for alkaline solutions, one must know the solubility product for ferrous hydroxide (8×10^{-16}). Note that by adding the Fe^{2+}–Fe half-reaction to the reaction

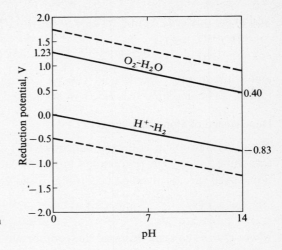

FIGURE 7.1
The H^+–H_2 and O_2–H_2O reduction potentials as a function of pH.

for the dissolution of ferrous hydroxide, we get the $Fe(OH)_2$–Fe half-reaction:

$$
\begin{array}{ll}
 & nE^\circ \\
Fe^{2+} + 2e^- \rightarrow Fe & -0.82 \\
Fe(OH)_2 \rightarrow Fe^{2+} + 2OH^- & -0.89 \\
\hline
Fe(OH)_2 + 2e^- \rightarrow Fe + 2OH^- & -1.71
\end{array}
$$

One adds the electronvolts (not the potentials!) for each step to obtain the electronvolts corresponding to the new half-reaction. The potential for the new half-reaction is then obtained by dividing by n, the number of electrons. Thus, $E^\circ = -1.71/2 = -0.86$.

Reduction-Potential Diagrams

If an element can exist in several oxidation states, it is convenient to display the reduction potentials corresponding to the various half-reactions in diagrammatic form, as shown in the following diagrams for iron and oxygen.

$$Fe \xleftarrow{-0.41} Fe^{2+} \xleftarrow{0.77} Fe^{3+} \xleftarrow{2.20} FeO_4^{2-}$$

$$-0.02$$

$$H_2O \xleftarrow{1.77} H_2O_2 \xleftarrow{0.68} O_2$$

$$1.23$$

One of the most important facts which can be learned from a reduction-potential diagram is which oxidation states, if any, are unstable with respect to disproportionation. If a given oxidation state is a stronger oxidizing agent than the next higher state, disproportionation can occur. (This is the situation when a reduction potential on the right is more negative than one on the left.) It will be noted that both Fe^{2+} and Fe^{3+} are stable with respect to disproportionation and that H_2O_2 is unstable with respect to disproportionation. Thus the following reactions proceed spontaneously:

$$Fe + 2Fe^{3+} \rightarrow 3Fe^{2+}$$

$$2H_2O_2 \rightarrow 2H_2O + O_2$$

The latter reaction is very slow under ordinary conditions but rapid in the presence of certain catalysts.

Often species which are thermodynamically unstable with respect to disproportionation are intermediates in oxidation-reduction reactions and are the cause of slow reactions. Hydrogen peroxide is such a species. Most oxidations by molecular oxygen proceed with intermediate formation of hydrogen peroxide. Thus, although any reducing agent with a reduction potential more negative than that of the O_2–H_2O half-reaction (1.23 V) can theoretically be oxidized by oxygen, those with reduction potentials more positive than that of the O_2–H_2O_2 half-reaction (0.68 V) generally react slowly. In agreement with this rule, bromide ($Br_3{}^-$–Br^-, $E° = 1.05$) is oxidized very slowly, and iodide ($I_3{}^-$–I^-, $E° = 0.54$) is oxidized rapidly.

Sometimes it is not immediately obvious from an abbreviated reduction-potential diagram that a particular species is unstable with respect to disproportionation. Consider the diagram for phosphorus in basic solution (1 M OH^-):

$$PH_3 \xleftarrow{-0.89} P_4 \xleftarrow{-2.05} H_2PO_2{}^- \xleftarrow{-1.57} HPO_3{}^{2-} \xleftarrow{-1.12} PO_4{}^{3-}$$

Hypophosphite ($H_2PO_2{}^-$) is stable with respect to disproportionation into phosphorus and phosphite. But if we consider the phosphine-hypophosphite half-reaction as shown in the more complete diagram,

$$PH_3 \xleftarrow{-0.89} P_4 \xleftarrow{-2.05} H_2PO_2{}^- \xleftarrow{-1.57} HPO_3{}^{2-} \xleftarrow{-1.12} PO_4{}^{3-}$$

we see that hypophosphite is unstable with respect to the formation of phosphine and phosphite. This reaction takes place when solid hypophosphites are heated, but in hot aqueous solution the principal reaction is the reduction of water to give hydrogen:

$$H_2PO_2{}^- + OH^- \rightarrow HPO_3{}^{2-} + H_2$$

It is appropriate at this point to show how one calculates potentials when adding or subtracting two half-reactions to form a third half-reaction. As shown below, the phosphorus-phosphine and hypophosphite-phosphorus half-reactions may be added to give the hypophosphite-phosphine half-reaction

$$
\begin{array}{ll}
 & nE° \\
\frac{1}{4}P_4 + 3H_2O + 3e^- \rightarrow PH_3 + 3OH^- & -2.67 \\
H_2PO_2^- + e^- \rightarrow \frac{1}{4}P_4 + 2OH^- & -2.05 \\
\hline
H_2PO_2^- + 3H_2O + 4e^- \rightarrow PH_3 + 5OH^- & -4.72
\end{array}
$$

We divide the sum of the electronvolts by the number of electrons in the new half-reaction to get the new potential. Thus, $E° = -4.72/4 = -1.18$. It can be seen that the new potential is a weighted average of the other two.

One of the nicest applications of reduction-potential diagrams is the prediction of the products of reactions involving elements having several oxidation states. Let us consider the reactions of iodide with permanganate in acid solution. The pertinent diagrams are shown below.

$$
I^- \xleftarrow{0.54} I_3^- \xleftarrow{1.45} HOI \xleftarrow{1.14} IO_3^- \xleftarrow{\sim 1.7} H_5IO_6
$$

$$
\text{(1.20)}
$$

$$
Mn \xleftarrow{-1.20} Mn^{2+} \xleftarrow{1.5} Mn^{3+} \xleftarrow{1.0} MnO_2 \xleftarrow{2.26} MnO_4^{2-} \xleftarrow{0.56} MnO_4^-
$$

$$
\text{(1.239)} \qquad \text{(1.69)}
$$

In this case we may correctly assume that the reactions are "thermodynamically controlled," that is, that equilibria are fairly rapidly achieved.[1] We notice that there are three species in these diagrams which are unstable toward disproportionation: HOI, Mn^{3+}, and MnO_4^{2-}. These are therefore eliminated from consideration. The Mn^{2+}–Mn half-reaction is of no concern to us, because its reduction potential is far too negative for Mn to have any stability in acid solution. Therefore we may justifiably consider only the simplified diagrams

$$
I^- \xleftarrow{0.54} I_3^- \xleftarrow{1.20} IO_3^- \xleftarrow{\sim 1.7} H_5IO_6
$$

$$
Mn^{2+} \xleftarrow{1.239} MnO_2 \xleftarrow{1.69} MnO_4^-
$$

If the reaction between iodide and permanganate is carried out with iodide in excess (as when permanganate is added dropwise to a hydroiodic acid solution), then the products of the reaction must be compatible with the presence of iodide ion. Thus under these conditions iodate cannot be formed, because iodate would react with excess iodide to form triiodide. Similarly, manganese dioxide cannot

---

[1] The oxidation of water by H_5IO_6 and MnO_4^- is slow under ordinary conditions. The potentials for these oxidizing agents barely fall between the dotted lines of Fig. 7.1.

be formed because it is capable of oxidizing iodide. The observed net reaction is

$$15I^- + 2MnO_4^- + 16H^+ \rightarrow 5I_3^- + 2Mn^{2+} + 8H_2O$$

If the reaction is carried out with permanganate in excess (as when iodide is added dropwise to an acidic permanganate solution), the products of the reaction must be compatible with the presence of permanganate. Thus manganous ion cannot be formed, because it would react with permanganate to form manganese dioxide. The iodide would not be oxidized just to triiodide, because triiodide is capable of reducing permanganate. The fact that the H_5IO_6–IO_3^- and MnO_4^-–MnO_2 half-reactions have potentials of similar magnitude complicates the problem. It turns out that iodide is not cleanly oxidized to either iodate or periodic acid but to a mixture of these products:

$$I^- + 2MnO_4^- + 2H^+ \rightarrow IO_3^- + 2MnO_2 + H_2O$$
$$3I^- + 8MnO_4^- + 11H^+ + 2H_2O \rightarrow 3H_5IO_6 + 8MnO_2$$

Notice that entirely different products are obtained when the reactant in excess is changed.

Reduction-potential diagrams for the more important elements that exist in several oxidation states are given in Table 7.2.

Table 7.2 REDUCTION-POTENTIAL DIAGRAMS†

A. Potentials in acid solution

(continued)

Table 7.2 (*continued*)

A. Potentials in acid solution

$Ru \xleftarrow{0.45} Ru^{2+} \xleftarrow{1.3} RuO_4^{2-} \xleftarrow{1.6} RuO_4^- \xleftarrow{0.9} RuO_4$

$Co \xleftarrow{-0.28} Co^{2+} \xleftarrow{1.82} Co^{3+}$

$Ni \xleftarrow{-0.25} Ni^{2+} \xleftarrow{1.68} NiO_2$

$Pd \xleftarrow{0.99} Pd^{2+} \xleftarrow{1.6} Pd^{4+} \xleftarrow{\sim 2} PdO_3$

$Cu \xleftarrow{0.521} Cu^+ \xleftarrow{0.153} Cu^{2+} \xleftarrow{1.8} CuO^+$

$Ag \xleftarrow{0.799} Ag^+ \xleftarrow{1.98} Ag^{2+} \xleftarrow{2.1} AgO^+$

$Hg \xleftarrow{0.789} Hg_2^{2+} \xleftarrow{0.920} Hg^{2+}$
   └──────── 0.854 ────────┘

$In \xleftarrow{-0.14} In^+ \xleftarrow{-0.40} In^{2+} \xleftarrow{-0.49} In^{3+}$

$Tl \xleftarrow{-0.336} Tl^+ \xleftarrow{1.25} Tl^{3+}$

$Sn \xleftarrow{-0.136} Sn^{2+} \xleftarrow{0.15} Sn^{4+}$

$Pb \xleftarrow{-0.126} Pb^{2+} \xleftarrow{1.46} PbO_2$

$NH_4^+ \xleftarrow{1.28} N_2H_5 \xleftarrow{1.41} NH_3OH^+ \xleftarrow{-1.87} N_2 \xleftarrow{1.77} N_2O \xleftarrow{1.59} NO \xleftarrow{1.00} HNO_2 \xleftarrow{1.07} N_2O_4 \xleftarrow{0.79} NO$
   └─── 1.35 ───┘   └─── 0.50 ───┘   └─── 0.86 ───┘   └─── 0.94 ───┘

$PH_3 \xleftarrow{-0.05} P_4 \xleftarrow{-0.51} H_3PO_2 \xleftarrow{-0.50} H_3PO_3 \xleftarrow{-0.28} H_3PO_4$
   └──── -0.16 ────┘
   └────────── -0.28 ──────────┘

$AsH_3 \xleftarrow{-0.38} As \xleftarrow{0.247} H_3AsO_3 \xleftarrow{0.559} H_3AsO_4$

$SbH_3 \xleftarrow{-0.51} Sb \xleftarrow{0.212} SbO^+ \xleftarrow{0.58} Sb_2O_5$

$Bi \xleftarrow{0.32} BiO^+ \xleftarrow{\sim 1.6} Bi_2O_5$

$H_2O \xleftarrow{1.77} H_2O_2 \xleftarrow{0.68} O_2$
   └──── 1.23 ────┘

$H_2S \xleftarrow{0.14} S \xleftarrow{0.50} S_2O_3^{2-} \xleftarrow{0.08} S_4O_6^{2-} \xleftarrow{0.51} SO_2(aq) \xleftarrow{0.57} S_2O_6^{2-} \xleftarrow{-0.22} SO_4^{2-}$

with branches: H_2SO_4 (0.88 and −0.08); 0.40; 0.17; 0.45

$H_2Se \xleftarrow{-0.32} Se \xleftarrow{0.74} H_2SeO_3 \xleftarrow{1.15} SeO_4^{2-}$

Table 7.2 (*continued*)

A. Potentials in acid solution

$$H_2Te \xleftarrow{-0.44} Te \xleftarrow{0.53} TeO_2(s) \xleftarrow{1.02} H_6TeO_6(s)$$

$$Cl^- \xleftarrow{1.36} Cl_2 \xleftarrow{1.63} HOCl \xleftarrow{1.64} HClO_2 \xleftarrow{1.21} ClO_3^- \xleftarrow{1.19} ClO_4^-$$
$$Cl_2 \xleftarrow{1.47} ClO_3^-$$

$$Br^- \xleftarrow{1.07} Br_2 \xleftarrow{1.59} HOBr \xleftarrow{1.49} BrO_3^- \xleftarrow{1.82} BrO_4^-$$
$$Br_2 \xleftarrow{1.51} BrO_3^-$$

$$I^- \xleftarrow{0.536} I_2 \xleftarrow{1.45} HOI \xleftarrow{1.14} IO_3^- \xleftarrow{1.7} H_5IO_6$$
$$I_2 \xleftarrow{1.06} ICl_2^- \xleftarrow{1.23} IO_3^-$$
$$I_2 \xleftarrow{1.20} IO_3^-$$

$$Xe \xleftarrow{1.8} XeO_3 \xleftarrow{2.3} H_4XeO_6$$

B. Potentials in basic solution

$$Ti \xleftarrow{-1.69} TiO_2(\text{hydrated})$$

$$Cr \xleftarrow{-1.4} Cr(OH)_2 \xleftarrow{-1.1} Cr(OH)_3 \xleftarrow{-0.13} CrO_4^{2-}$$
$$Cr \xleftarrow{-1.2} CrO_2^-$$

$$Mn \xleftarrow{-1.58} Mn(OH)_2 \xleftarrow{-0.2} Mn(OH)_3 \xleftarrow{0.1} MnO_2 \xleftarrow{0.603} MnO_4^{2-} \xleftarrow{0.558} MnO_4^-$$
$$Mn(OH)_2 \xleftarrow{0.03} MnO_2 \qquad MnO_2 \xleftarrow{0.588} MnO_4^-$$

$$Re^- \xleftarrow{-0.4} Re \xleftarrow{-0.58} ReO_2 \xleftarrow{-0.5} ReO_4^{2-} \xleftarrow{-0.7} ReO_4^-$$
$$ReO_2 \xleftarrow{-0.60} ReO_4^-$$

$$Fe \xleftarrow{-0.86} Fe(OH)_2 \xleftarrow{-0.56} Fe(OH)_3 \xleftarrow{0.72} FeO_4^{2-}$$

$$Co \xleftarrow{-0.72} Co(OH)_2 \xleftarrow{0.14} Co(OH)_3$$

$$Ni \xleftarrow{-0.72} Ni(OH)_2 \xleftarrow{0.49} NiO_2$$

$$Pd \xleftarrow{0.07} Pd(OH)_2 \xleftarrow{\sim 0.73} Pd(OH)_4$$

$$Cu \xleftarrow{-0.36} Cu_2O \xleftarrow{-0.08} Cu(OH)_2$$

$$Ag \xleftarrow{0.344} Ag_2O \xleftarrow{0.57} AgO \xleftarrow{0.74} Ag_2O_3$$

$$Hg \xleftarrow{0.098} HgO$$

$$In \xleftarrow{-1.0} In(OH)_3$$

$$Tl \xleftarrow{-0.344} TlOH \xleftarrow{-0.05} Tl(OH)_3$$

(*continued*)

Table 7.2 (*continued*)

B. Potentials in basic solution

$Sn \xleftarrow{-0.91} HSnO_2^- \xleftarrow{-0.90} Sn(OH)_6^{2-}$

$Pb \xleftarrow{-0.54} PbO \xleftarrow{0.28} PbO_2$

$NH_3 \xleftarrow{0.1} N_2H_4 \xleftarrow{0.73} NH_2OH \xleftarrow{-3.04} N_2 \xleftarrow{0.94} N_2O \xleftarrow{0.76} NO \xleftarrow{-0.46} NO_2^- \xleftarrow{0.88} N_2O_4 \xleftarrow{-0.86} N$

$\qquad\qquad 0.42 \qquad\qquad\qquad -0.76 \qquad\qquad\qquad -0.14 \qquad\qquad\qquad 0.01$

$PH_3 \xleftarrow{-0.89} P_4 \xleftarrow{-2.05} H_2PO_2^- \xleftarrow{-1.57} HPO_3^{2-} \xleftarrow{-1.12} PO_4^{3-}$

$\qquad\qquad -1.18$

$\qquad\qquad\qquad -1.31$

$AsH_3 \xleftarrow{-1.21} As \xleftarrow{-0.68} H_2AsO_3^- \xleftarrow{-0.67} AsO_4^{3-}$

$SbH_3 \xleftarrow{-1.34} Sb \xleftarrow{-0.66} SbO_2^- \xleftarrow{-0.4} Sb(OH)_6^-$

$Bi \xleftarrow{-0.46} Bi_2O_3 \xleftarrow{\sim 0.6} Bi_2O_5$

$OH^- \xleftarrow{0.87} HO_2^- \xleftarrow{-0.08} O_2$

$\qquad\qquad 0.401$

$\qquad\qquad\qquad\qquad\qquad -0.04 \qquad\qquad S_2O_4^{2-} \xleftarrow{-1.12}$

$S^{2-} \xleftarrow{-0.48} S \xleftarrow{-0.74} S_2O_3^{2-} \xleftarrow{0.08} S_4O_6^{2-} \xleftarrow{-0.80} SO_3^{2-} \xleftarrow{-0.93} SO_4^{2-}$

$\qquad\qquad\qquad\qquad\qquad -0.58$

$\qquad\qquad\qquad\qquad -0.66$

$Se^{2-} \xleftarrow{-0.84} Se \xleftarrow{-0.37} SeO_3^{2-} \xleftarrow{0.05} SeO_4^{2-}$

$Te^{2-} \xleftarrow{-0.86} Te \xleftarrow{-0.57} TeO_3^{2-} \xleftarrow{0.4} TeO_2(OH)_4^{2-}$

$Cl^- \xleftarrow{1.36} Cl_2 \xleftarrow{0.40} ClO^- \xleftarrow{0.66} ClO_2^- \xleftarrow{0.33} ClO_3^- \xleftarrow{0.36} ClO_4^-$

$\qquad\qquad 0.88 \qquad\qquad\qquad 0.50$

$Br^- \xleftarrow{1.07} Br_2 \xleftarrow{0.45} BrO^- \xleftarrow{0.54} BrO_3^- \xleftarrow{0.99} BrO_4^-$

$\qquad\qquad 0.76$

$I^- \xleftarrow{0.54} I_2 \xleftarrow{0.45} IO^- \xleftarrow{0.14} IO_3^- \xleftarrow{0.7} H_3IO_6^{2-}$

$\qquad\qquad 0.49$

$Xe \xleftarrow{0.9} HXeO_4^- \xleftarrow{0.9} HXeO_6^{3-}$

† Most of the potentials given in this chapter are from W. M. Latimer, "Oxidation Potentials," 2d ed., Prentice-Hall, Englewood Cliffs, N.J., 1952.

MECHANISMS OF OXYANION REACTIONS[1]

We have already pointed out that even if thermodynamic data such as reduction potentials indicate that a reaction has a large driving force, without kinetic information one cannot know whether or not the reaction proceeds at a reasonable rate. Let us consider oxidation-reduction reactions involving oxyanions such as BrO_3^-, SO_4^{2-} and OCl^-. Most of these reactions involve the transfer of an oxygen atom (or oxygen atoms) from one atom to another. In the oxyanions, the oxygen atoms are held very tightly and therefore direct transfer processes such as the following are very slow:

$$NO_2^- + OCl^- \to NO_3^- + Cl^-$$

$$ClO_3^- + I^- \to ClO_2^- + OI^-$$

Such reactions are usually acid-catalyzed. When one or two protons are attached to a coordinated oxide ion, the negative charge of the oxide ion is at least partly canceled and an OH^- ion or H_2O molecule can then leave the central atom of the oxyanion relatively easily. Consider the reaction of chlorate with a halide ion:

$$ClO_3^- + 6X^- + 6H^+ \to Cl^- + 3X_2 + 3H_2O$$

The rate law for this reaction is

$$Rate = k[ClO_3^-][X^-][H^+]^2$$

and the initial steps of the mechanism are believed to be as follows:

$$2H^+ + ClO_3^- \rightleftharpoons H_2OClO_2^+ \qquad \text{(fast)}$$

$$X^- + H_2OClO_2^+ \to XClO_2 + H_2O \qquad \text{(slow)}$$

$$XClO_2 + X^- \to X_2 + ClO_2^- \qquad \text{(fast)}$$

The ClO_2^- ion is then reduced to Cl^- by a sequence of rapid reactions. It is not known whether or not the species $H_2OClO_2^+$ loses a water molecule before reaction with X^-. The rate law would be unchanged if the rate-determining step were

$$X^- + ClO_2^+ \to XClO_2$$

It is believed that dehydrated intermediates of this type are formed in some reactions. For example, the nitration of an aromatic hydrocarbon by a solution of nitric acid in concentrated sulfuric acid probably proceeds as follows:

$$HNO_3 + H^+ \rightleftharpoons NO_2^+ + H_2O \qquad \text{(fast)}$$

$$NO_2^+ + ArH \to ArNO_2 + H^+ \qquad \text{(slow)}$$

[1] J. O. Edwards, "Inorganic Reaction Mechanisms," W. A. Benjamin, Inc., New York, 1964; A. G. Sykes, "Kinetics of Inorganic Reactions," Pergamon, New York, 1966; D. Benson, "Mechanisms of Inorganic Reactions in Solution," McGraw-Hill, London, 1968.

The NO_2^+ ion can be detected by Raman spectroscopy in HNO_3–H_2SO_4–H_2O mixtures. The nitrosation of a secondary amine by nitrous acid is acid-catalyzed and is believed to proceed as follows:

$$HNO_2 + H^+ \rightleftharpoons NO^+ + H_2O \qquad K = 2 \times 10^{-7}$$

$$NO^+ + R_2NH \rightarrow R_2NNO + H^+ \qquad \text{(slow)}$$

The NO^+ can be detected in HNO_2–H_2SO_4–H_2O mixtures by its ultraviolet absorption spectrum. The exchange of ^{18}O between $H_2{}^{18}O$ and $SO_4{}^{2-}$ is acid-catalyzed and may involve the following mechanism:

$$2H^+ + SO_4{}^{2-} \rightleftharpoons H_2SO_4$$

$$H_2SO_4 \rightarrow SO_3 + H_2O \qquad \text{(slow)}$$

$$H_2{}^{18}O + SO_3 \rightarrow 2H^+ + SO_3{}^{18}O^{2-} \qquad \text{(fast)}$$

The charge of the central atom of an oxyanion seems to be very important in determining reactivity: the lower the charge, the higher the reactivity.[1] Thus, rates for reactions of the chlorine oxyanions increase in the order $ClO_4^- < ClO_3^- < ClO_2^- < ClO^-$. The rate of oxygen exchange with water increases in the order $ClO_4^- < SO_4{}^{2-} < HPO_4{}^{2-} < H_2SiO_4{}^{2-}$. The size of the central atom is also important; the larger it is, the higher the reactivity. For example, iodate reactions are fast, chlorate reactions are slow, and bromate reactions are of intermediate rate.

Proof of the transfer of an oxygen atom from an oxidizing agent to a reducing agent has been obtained in the case of the reaction of nitrite with hypochlorous acid.[2] When ^{18}O-labeled HOCl was used, the ^{18}O ended up in the nitrate product:

$$NO_2^- + H^{18}OCl \rightarrow H^+ + NO_2{}^{18}O^- + Cl^-$$

The reaction may be looked upon as the S_N2 attack of nitrite on the oxygen of HOCl, with displacement of chloride.

The Landolt Clock Reaction

The reduction of iodate by bisulfite is a fascinating reaction that is often used as an entertaining lecture demonstration.[3] In acid solution iodate is reduced slowly by bisulfite to iodide.

$$IO_3^- + 3HSO_3^- \rightarrow I^- + 3HSO_4^-$$

---

[1] It should be remembered that, for a series of closely related species, atomic charge is a function of formal charge and of oxidation state.

[2] H. Taube, *Rec. Chem. Prog. Kresge-Hooker Sci. Libr.*, **17**, 25 (1956).

[3] H. Landolt, *Chem. Ber.*, **18**, 249 (1885); **19**, 1317 (1886); **20**, 745 (1887); J. A. Church and S. A. Dreskin, *J. Phys. Chem.*, **72**, 1387 (1968), and references therein.

Iodate also reacts fairly rapidly with iodide to form iodine,

$$IO_3^- + 5I^- + 6H^+ \rightarrow 3I_2 + 3H_2O$$

but the liberated iodine is reduced very rapidly by bisulfite:

$$I_2 + HSO_3^- + H_2O \rightarrow 2I^- + HSO_4^- + 2H^+$$

The last reaction is so rapid that no iodine appears until all the bisulfite has been oxidized. At that point, especially if a little starch is present, the solution suddenly becomes nearly opaque. The time for the iodine to appear is a function of concentrations and temperature, and the reaction may be used as a clock. The time T, in seconds at 23°C, is approximately given by the expression

$$T = \frac{3.7 \times 10^{-3}}{[IO_3^-][HSO_3^-]}$$

where the concentrations are expressed in molarities.

PROBLEMS

7.1 Write equations for the principal net reactions which occur in the following cases:
  a A little potassium iodide is dissolved in a solution of KIO_3 in 6 M HCl.
  b Aqueous triiodide solution is added to excess cold sodium carbonate solution. (Potentials for half-reactions involving I_3^- are practically the same as those for the corresponding half-reactions involving I_2.)
  c Cl_2O is passed into a hot solution of sodium hydroxide.
  d Iodine is added to an excess of aqueous chloric acid.
  e Sodium hypophosphite is added to excess of an acidic $KMnO_4$ solution.
  f Arsine is bubbled into an excess of an acidic solution of $Fe(NO_3)_3$.
  g A suspension of powdered sulfur in an alkaline solution of Na_2SO_3 is boiled.
  h Arsine is bubbled into a solution of H_3AsO_3.
7.2 Calculate the $MnO_4^- - MnO_2$ reduction potential for pH = 7.
7.3 Calculate the $HPO_3^{2-} - P_4$ reduction potential for $(OH^-) = 1$ M.
7.4 Calculate the equilibrium constants for the following reactions:

$$4H^+ + 2Cl^- + MnO_2 \rightarrow Cl_2 + Mn^{2+} + 2H_2O$$

$$O_3(g) + Xe(g) \rightarrow XeO_3(aq)$$

7.5 The decomposition of Caro's acid (H_2SO_5) to oxygen and bisulfate follows the rate law: Rate = $k[HSO_5^-][SO_5^{2-}]$. Two conceivable "mechanisms" for the reaction are indicate below. Explain how one of these could be eliminated by the use of ^{18}O labeling.

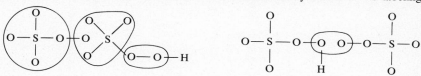

7.6   The rate of exchange of oxygen atoms between water and phosphate passes through a maximum at pH 5. Propose a plausible mechanism involving a reactive intermediate.

7.7   Number the following aqueous reactions in order of increasing rate in $1\ M\ H^+$ at room temperature.

a $IO_3^- + 8I^- + 6H^+ \rightarrow 3I_3^- + 3H_2O$

b $BrO_3^- + 5Br^- + 6H^+ \rightarrow 3Br_2 + 3H_2O$

c $ClO_4^- + 12I^- + 8H^+ \rightarrow 4I_3^- + Cl^- + 4H_2O$

d $HOI + 2I^- + H^+ \rightarrow I_3^- + H_2O$

SOLVATED ELECTRONS

THE HYDRATED ELECTRON[1]

When an aqueous solution is irradiated with X rays, gamma rays, or accelerated electrons, water molecules are ionized as follows:

$$H_2O \xrightarrow[\text{radiation}]{\text{high energy}} H_2O^+ + e^-$$

An ejected electron generally has a high energy which is rapidly dissipated by collisions in which electrons are knocked out of other water molecules. In about 10^{-11} s the ejected electron reaches thermal equilibrium and becomes hydrated; we represent the hydrated species by e^-_{aq}. In the meantime H_2O^+ is transformed into OH by the reaction

$$H_2O^+ + H_2O \rightarrow H_3O^+ + OH$$

Hence the net ionization reaction is

$$2H_2O \rightarrow H_3O^+ + OH + e^-_{aq}$$

---

[1] E. J. Hart, *Acc. Chem. Res.*, **2**, 161 (1969); E. J. Hart and M. Anbar, "The Hydrated Electron," Wiley-Interscience, New York, 1970; M. Anbar and P. Neta, *Int. J. Appl. Radiat. Isot.*, **18**, 493 (1967).

The radiation promotes some water molecules into highly excited eletronic states. These excited water molecules decompose into H and OH radicals:

$$H_2O^* \rightarrow H + OH$$

Thus the irradiation of an aqueous solution gives H, OH, and e^-_{aq} as the major primary reactive species. These species may react with water, with themselves, or with other reagents in the solution to give various products. For example, the hydrated electron is known to react with H_2O_2, O_2, H^+, and NO_2^- as shown in the following reactions:

$$e^-_{aq} + H_2O_2 \rightarrow OH + OH^- \tag{8.1}$$

$$e^-_{aq} + O_2 \rightarrow O_2^- \tag{8.2}$$

$$e^-_{aq} + H^+ \rightarrow H \tag{8.3}$$

$$e^-_{aq} + NO_2^- \rightarrow NO_2^{2-} \tag{8.4}$$

Before 1960, many reduction reactions of this type were mistakenly ascribed to atomic hydrogen. In 1962 Czapski and Schwarz[1] reported data which proved that, in each of the four reactions above, the primary reducing species has a -1 charge and therefore probably is e^-_{aq}. They had no facilities for measuring the absolute rate constants for these reactions, but they were able to measure the ratios of rate constants by determining the relative yields of final products in systems in which two of the reactions were competing. They determined rate-constant ratios as a function of ionic strength and used the following Brönsted-Bjerrum equation[2] to evaluate the charge of the reducing species:

$$\log \frac{k}{k_0} = \frac{1.02 Z_a Z_b \mu^{1/2}}{1 + \mu^{1/2}}$$

In this equation k is the rate constant at ionic strength μ, k_0 is the rate constant at infinite dilution, and Z_a and Z_b are the charges of the species involved in the second-order reaction. In Fig. 8.1, $\log[(k/k_0)_n/(k/k_0)_1]$ is plotted versus $\mu^{1/2}/(1 + \mu^{1/2})$ for $n = 2$, 3, and 4 (that is for the rate constants of reactions 8.2, 8.3, and 8.4 relative to that for reaction 8.1). The points for reaction 8.2 form a line of zero slope corresponding to $Z_a Z_b = 0$. Because the charge of the O_2 molecule is zero, this result is expected. The points for reaction 8.3 form a line of -1 slope, as expected for Z_a and Z_b values of opposite sign. The points for reaction 8.4 form a line of $+1$ slope, as expected for $Z_a = Z_b = -1$. Thus the variation of the rate constants with ionic strength is consistent with a -1 charge for the reducing species in these reactions.

[1] G. Czapski and H. A. Schwarz, *J. Phys. Chem.*, **66**, 471 (1962).
[2] A. A. Frost and R. G. Pearson, "Kinetics and Mechanism," 2d ed., pp. 150–153, Wiley, New York, 1961.

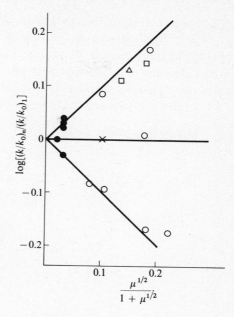

FIGURE 8.1

Plot of $\log[(k/k_0)_n/(k/k_0)_1]$ versus $\mu^{1/2}/(1 + \mu^{1/2})$. Upper curve, $n = 4$; middle curve, $n = 2$; lower curve, $n = 3$. The ionic strength was varied with $LiClO_4$, $\bigcirc$; $KClO_4$, $\square$; $NaClO_4$, $\triangle$; and $MgSO_4$, $\times$. The closed circles represent no added salt other than reactants. [*Reproduced with permission from G. Czapski and H. A. Schwarz, J. Phys. Chem.,* **66,** 471 (1962). *Copyright © by the American Chemical Society.*]

Although the hydrated electron decays very rapidly in aqueous solutions, it is possible to obtain its absorption spectrum by fast-scan spectrophotometry of solutions which have been exposed to an intense pulse of radiation. The absorption spectrum is shown in Fig. 8.2. Notice that e^-_{aq} has a broad absorption band

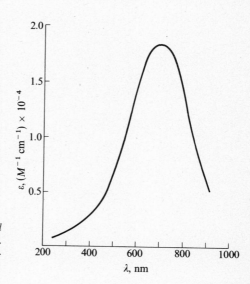

FIGURE 8.2

Absorption spectrum of e^-_{aq} (*Adapted with permission from E. J. Hart and M. Anbar, "The Hydrated Electron," Wiley-Interscience, New York, 1970.*)

at 715 nm, corresponding to a blue solution. By following the absorbance of e^-_{aq} as a function of time (during an interval of approximately 1 ms), it is possible to study directly the kinetics of the reaction of e^-_{aq} with various species in the solution. The second-order rate constants for the reaction of e^-_{aq} with some representative species are given in Tables 8.1 and 8.2.

In alkaline solutions, hydrogen atoms are converted into hydrated electrons.

$$H + OH^- \rightarrow e^-_{aq} + H_2O$$

Table 8.1   RATE CONSTANTS OF SOME e^- REACTIONS[†]

Reaction	Rate constant, $M^{-1}\,s^{-1}$
$e^-_{aq} + H_2O \rightarrow H + OH^-$	16
$e^-_{aq} + e^-_{aq} \xrightarrow{2H_2O} H_2 + 2OH^-$	6.0×10^9
$e^-_{aq} + H \xrightarrow{H_2O} H_2 + OH^-$	2.5×10^{10}
$e^-_{aq} + OH \rightarrow OH^-$	3.0×10^{10}
$e^-_{aq} + H^+ \rightarrow H$	2.1×10^{10}
$e^-_{aq} + H_2O_2 \rightarrow OH + OH^-$	1.2×10^{10}

[†] J. Hart, *Acc. Chem. Res.*, **2**, 161 (1969); E. J. Hart and M. Anbar, "The Hydrated Electron," Wiley-Interscience, New York, 1970; M. Anbar and P. Neta, *Int. J. Appl. Radiat. Isot.*, **18**, 494 (1967).

Table 8.2   RATE CONSTANTS OF e^-_{aq} AND H-ATOM REACTIONS[†]

Reduced species	k, $M^{-1}\,s^{-1}$	
	e^-_{aq}	H atom
Ag^+	3.2×10^{10}	1.1×10^{10}
Zn^{2+}	1.4×10^9	$<1.0 \times 10^5$
Cd^{2+}	5.2×10^{10}	$<1.0 \times 10^5$
O_2	1.9×10^{10}	1.9×10^{10}
N_2O	5.6×10^9	$\sim 10^5$
MnO_4^-	3.0×10^{10}	2.6×10^{10}
Gd^{3+}	5.5×10^8	
Eu^{3+}	6.1×10^{10}	
Cl^-	$<10^5$	
ClO^-	7.2×10^9	
C_6H_6	1.4×10^7	1.1×10^9
C_6H_{10}	$<1.0 \times 10^6$	3×10^9

[†] E. J. Hart, *Acc. Chem. Res.*, **2**, 161 (1969); E. J. Hart and M. Anbar, "The Hydrated Electron," Wiley-Interscience, New York, 1970; M. Anbar and P. Neta, *Int. J. Appl. Radiat. Isot.*, **18**, 494 (1967).

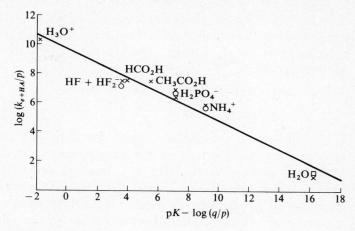

FIGURE 8.3

Plot of log k for e^-_{aq} + HX reactions versus pK_a for HX. The k and k_a values have been slightly corrected for statistical effects. The crosses refer to radiation data; the circles to photochemical data; the square to a study of H_2O in ethylenediamine. [*Reproduced with permission from J. Rabani, Adv. Chem. Ser.*, **50**, 242 (1965). *Copyright © by the American Chemical Society.*]

From the measured rate constants for this reaction ($2.0 \times 10^7\ M^{-1}\ s^{-1}$) and the reverse reaction ($16\ M^{-1}\ s^{-1}$) we can calculate the equilibrium constant

$$\frac{(e^-_{aq})}{(H)(OH^-)} = K = \frac{k_f}{k_r} = \frac{2.0 \times 10^7}{16 \times 55.5} = 2.25 \times 10^4$$

By combining this with the ionization constant of water, we obtain $K_a = 2.25 \times 10^{-10}$ for

$$H \rightleftharpoons H^+ + e^-_{aq}$$

Thus we have a measure of the acidity of the simplest possible aqueous acid!

The reactions of e^-_{aq} with H^+ and H_2O to form atomic H are examples of a general reaction of e^-_{aq} with protonic acids:

$$e^-_{aq} + HX \rightarrow H + X^-$$

As one might expect, the rate constants for such reactions can be correlated with the acidities of the acids. Figure 8.3 shows a plot of log k versus pK for various aqueous acids; it can be seen that these quantities are linearly related.[1] Obviously the relatively low reactivity of e^-_{aq} with the water molecule (see Table 8.1) is related to the weak acidity of the water molecule.

[1] J. Rabani, *Adv. Chem. Ser.*, **50**, 242 (1965).

In Table 8.2, rate constants for $e^-{}_{aq}$ reactions can be compared with rate constants for the analogous atomic H reactions. Although atomic H is a powerful reducing agent, $e^-{}_{aq}$ is stronger. This can be seen from the reduction potentials for these species.[1]

$$e^- + H^+ \rightarrow H \qquad E^\circ = -2.3 \text{ V}$$
$$e^- \rightarrow e^-{}_{aq} \qquad E^\circ = -2.8 \text{ V}$$

In general, the $e^-{}_{aq}$ rate constants are greater than the corresponding atomic H rate constants, except in the case of reacting species which have no stable vacant orbitals to accept electrons. The $e^-{}_{aq}$ reactions probably always involve the formation of intermediate species which correspond to the addition of a negative charge to the reacting species. High rate constants seem to be associated with the formation of stable intermediates, and relatively low rate constants with the formation of unstable intermediates. Thus the reaction with Zn^{2+} is relatively slow because of the instability of Zn^+, and Eu^{3+} reacts much more rapidly than Gd^{3+} because of the stability of Eu^{2+} compared to Gd^{2+}.

METAL-AMMONIA SOLUTIONS[2]

Physical Characteristics

In contrast to water, which reacts rapidly with alkali metals to form hydrogen and the alkali-metal hydroxides, liquid ammonia reversibly dissolves the alkali metals and other electropositive metals to form metastable solutions containing ammoniated electrons. Metal-ammonia solutions have been extensively studied because of their unusual composition and because of their usefulness in inorganic and organic synthesis. We shall first discuss some of the physical characteristics of these solutions which give us insight to the microscopic structure of the solutions.

It is significant that the metals which are soluble in liquid ammonia are the electropositive metals with aqueous reduction potentials more negative than -2.5 V. Obviously, the factors which cause a metal to have a very negative reduction potential (high solvation energy for the ion, low ionization potential, and low sublimation energy) are the same as those which cause it to have a high solubility in ammonia. This observation is consistent with the concept that

---

[1] It should be remembered that in half-reactions of this type the symbol e^- stands for $[\frac{1}{2}H_2 - H^+]$ and does not represent an electron of any type.

[2] W. L. Jolly, *Prog. Inorg. Chem.*, **1**, 235 (1959); Solvated Electron, *Adv. Chem. Ser.*, **50**, 1–304 (1965); J. J. Lagowski and M. J. Sienko (eds.), "Metal-Ammonia Solutions," Butterworth, London, 1970, Supplement to *Pure and Applied Chemistry*; W. L. Jolly, "Metal-Ammonia Solutions," Dowden, Hutchinson, and Ross, Stroudsburg, Pa., 1972.

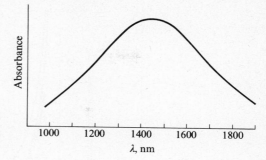

FIGURE 8.4
Absorption spectrum of a dilute ($\sim 10^{-4}$ M) solution of a metal in liquid ammonia at $-70°C$.

metal-ammonia solutions are electrolytic, that is, that they contain ammoniated metal ions and electrons.

Dilute solutions of metals in ammonia are bright blue; the color is due to the short wavelength tail of a very broad and intense absorption band with a peak at approximately 1500 nm (1.5 μm). This absorption spectrum is shown in Fig. 8.4. The fact that the spectra of the various metals are essentially identical proves that the absorption is due to a common species and strongly suggests that this species is the ammoniated electron. Further indication that the blue color is due to the ammoniated electron is found in the electrolytic behavior of metal-salt solutions. When a solution of a salt such as sodium bromide in liquid ammonia is electrolyzed with direct current using inert electrodes (e.g., platinum electrodes), a blue color appears in the region of the cathode. The electrode reaction may be written $e^- \rightarrow e^-_{am}$, where e^-_{am} stands for the ammoniated electron.

Metal-ammonia solutions are extremely good conductors of electricity; at all concentrations the equivalent conductances are greater than those found for any other known electrolyte in any known solvent. In Fig. 8.5 the equivalent conductances for Li, Na, and K are plotted against the negative logarithm of the molarity of the metal in the ammonia. In the very dilute region, the equivalent conductances approach a limiting value characteristic of the ions M^+_{am} and e^-_{am}. As the solutions are made more concentrated, the equivalent conductances decrease, as in the case of most electrolytic solutions, because of ionic association. However, at approximately 0.05 M the conductances reach a minimum value. Further increase in concentration causes the conductances to increase rapidly until, in the saturated solutions, conductances comparable to those of metals are reached. These concentrated, highly conducting solutions are not blue, like the relatively dilute, electrolytic solutions, but rather bronze-colored, like a molten metal. Indeed, these metallike solutions have been likened to molten "expanded metals" in which ammoniated metal ions are held together by electrons. Further evidence for such a description is found in the fact that the metallike concen-

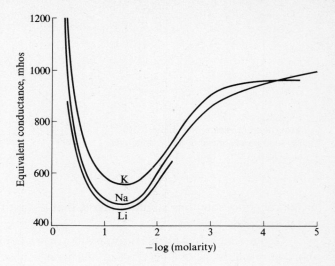

FIGURE 8.5

Equivalent conductances of metal-ammonia solutions at $-33.5°C$ versus the logarithm of dilution. (*Reproduced with permission from D. M. Yost and H. Russell, Jr., "Systematic Inorganic Chemistry," Prentice-Hall, Englewood Cliffs, N.J., 1946.*)

tration regions correspond to NH_3/metal ratios which are reasonable coordination numbers for the metal ions (e.g., in the case of Na, the metallike region ranges from $NH_3/Na \approx 5.5$ to $NH_3/Na \approx 10$).

The magnetic susceptibility of a very dilute metal-ammonia solution corresponds to that expected for a solution of ammoniated valence electrons with independent unpaired spins. In more concentrated solutions, the molar susceptibility decreases, as shown by the data in Table 8.3. Obviously spin pairing occurs in the more concentrated solutions.

Table 8.3 MOLAR MAGNETIC SUSCEPTIBILITY OF POTASSIUM SOLUTIONS† AT $-33°C$

M	$\chi \times 10^6$
0.00341	1268
0.00406	1240
0.00812	974
0.0318	402
0.482	29.9

† S. Freed and N. Sugarman, *J. Chem. Phys.*, **11**, 354 (1943).

To explain the various physical phenomena, a model has been proposed for metal-ammonia solutions. It is believed that ionic aggregation equilibria of the following type are involved:

$$M^+_{am} + e^-_{am} \rightleftharpoons (M^+_{am})(e^-_{am})$$
$$2M^+_{am} + 2e^-_{am} \rightleftharpoons (M^+_{am})_2(e^-_{am})_2$$

The species $(M^+_{am})(e^-_{am})$ and $(M^+_{am})_2(e^-_{am})_2$ are neutral and nonconducting; their formation would account for the decrease in equivalent conductance with increasing concentration in the concentration region around 10^{-2} M. The latter species is believed to be diamagnetic; its formation would account for the decrease in magnetic susceptibility with increasing concentration in the same concentration region. In very concentrated solutions (> 1 M), there is not enough ammonia to adequately solvate the electrons; the solution may then be described as $(M^+_{am})e^-$, that is, as a molten expanded metal.

Reactions

Metal-ammonia solutions are unstable with respect to decomposition to hydrogen and the metal amide:

$$e^-_{am} + NH_3 \rightarrow \tfrac{1}{2}H_2 + NH_2^-$$

This decomposition is very slow in clean, cold solutions, but it is fast when suitable transition-metal compounds are added.

The following reactions are examples of simple electron addition without bond cleavage:

$$MnO_4^- + e^-_{am} \rightarrow MnO_4^{2-}$$
$$NO_2^- + e^-_{am} \rightarrow NO_2^{2-}$$
$$O_2 + e^-_{am} \rightarrow O_2^-$$
$$Ni(CN)_4^{2-} + 2e^-_{am} \rightarrow Ni(CN)_4^{4-}$$

The following reactions involve the cleavage of a bond and the addition of one electron:

$$NH_4^+ + e^-_{am} \rightarrow \tfrac{1}{2}H_2 + NH_3$$
$$PH_3 + e^-_{am} \rightarrow \tfrac{1}{2}H_2 + PH_2^-$$
$$Et_3SnBr + e^-_{am} \rightarrow Et_3Sn\cdot + Br^-$$

The addition of two electrons to a molecule generally causes bond cleavage and the formation of two ions:

$$Mn_2(CO)_{10} + 2e^-_{am} \rightarrow 2Mn(CO)_5^-$$
$$C_6H_5NHNH_2 + 2e^-_{am} \rightarrow C_6H_5NH^- + NH_2^-$$
$$R_3GeGeR_3 + 2e^-_{am} \rightarrow 2GeR_3^-$$

Frequently one of the anions formed in such a reaction undergoes ammonolysis, with formation of amide ion:

$$N_2O + 2e^-_{am} + NH_3 \rightarrow N_2 + OH^- + NH_2^-$$
$$NEt_4^+ + 2e^-_{am} + NH_3 \rightarrow NEt_3 + C_2H_6 + NH_2^-$$
$$RBr + 2e^-_{am} + NH_3 \rightarrow Br^- + RH + NH_2^-$$

The amide ion formed in the latter reactions reacts with the starting material to give by-products, as shown in the following side reactions:

$$N_2O + 2NH_2^- \rightarrow N_3^- + OH^- + NH_3$$
$$NEt_4^+ + NH_2^- \rightarrow NEt_3 + C_2H_4 + NH_3$$
$$RBr + NH_2^- \rightarrow RNH_2 + Br^-$$

PROBLEMS

8.1 Hydrated electrons can be generated by ultraviolet irradiation of the aqueous I^- ion: $I^- \xrightarrow{h\nu} I + e^-_{aq}$. What is the overall net reaction for the ultraviolet irradiation of an aqueous HI solution?

8.2 Which species would you expect to be more reactive toward e^-_{aq}: ClO_4^- or ClO^-? Cu^{2+} or Ni^{2+}? $Fe(CN)_6^{4-}$ or $Fe(CN)_6^{3-}$?

8.3 If a sodium-ammonia solution were electrolyzed by using direct current and inert electrodes, what would be the electrode reactions?

8.4 A solvated electron in H_2O or NH_3 is believed to occupy a cavity in the solvent, in which the protons of adjacent solvent molecules are oriented toward the electron in the cavity. Thus a solvated electron is situated in a potential well created by the polarization of solvent molecules. The absorption spectra are analogous to the $1s \rightarrow 2p$ transition of an H atom. What can you say about the relative sizes of the electron cavities in H_2O and NH_3?

8.5 Write equations for the net reactions which occur when the following materials are added to a sodium-ammonia solution.

a Methylgermane.

b Iodine.

c Diethyl sulfide.

BORON HYDRIDES AND THEIR DERIVATIVES

SYNTHESES

Boron hydrides were first isolated and characterized by Alfred Stock and his coworkers in Germany. In the period from 1912 to 1930 they used vacuum-line techniques (most of which they devised themselves) to prepare B_2H_6, B_4H_{10}, B_5H_9, B_5H_{11}, B_6H_{10}, $B_{10}H_{14}$, and many derivatives of these compounds.[1] Further progress in the field was rather slow until, during World War II, Schlesinger's group at the University of Chicago developed easy methods for preparing $NaBH_4$ (sodium tetrahydroborate, usually called sodium borohydride), which is useful for the preparation of B_2H_6 (diborane).[2] Physical properties of some of the many boron hydrides which are now known are listed in Table 9.1.

The Synthesis of Boron Hydrides[3,4]

Diborane can be prepared by the reaction of a boron trihalide with a strong hydriding agent such as sodium tetrahydroborate or lithium aluminum hydride

[1] A. Stock, "Hydrides of Boron and Silicon," Cornell University Press, Ithaca, N. Y., 1933.
[2] H. I. Schlesinger and H. C. Brown, *J. Am. Chem. Soc.*, **75**, 219 (1953).
[3] E. L. Muetterties (ed.), "The Chemistry of Boron and Its Compounds," Wiley, New York, 1967; E. L. Muetterties and W. H. Knoth, "Polyhedral Boranes," Marcel Dekker, Inc., New York, 1968.
[4] R. W. Parry and M. K. Walter, *Prep. Inorg. React.*, **5**, 45 (1968).

in an aprotic solvent. This synthesis is usually carried out with BF_3 and $NaBH_4$ in diglyme (the dimethyl ether of diethylene glycol).

$$3NaBH_4 + BF_3 \rightarrow 2B_2H_6 + 3NaF$$

A convenient laboratory synthesis involves the careful addition of $NaBH_4$ to concentrated H_2SO_4 or H_3PO_4.

$$2NaBH_4 + 2H_2SO_4 \rightarrow 2Na^+ + 2HSO_4^- + B_2H_6 + 2H_2$$

A synthetic method of potential industrial significance is the reduction of boric oxide with aluminum and hydrogen at high pressure in the presence of aluminum chloride catalyst:

$$B_2O_3 + 2Al + 3H_2 \xrightarrow{Al_2Cl_6} B_2H_6 + Al_2O_3$$

Several of the higher boron hydrides, B_4H_{10}, B_5H_9, B_5H_{11}, B_6H_{10}, and $B_{10}H_{14}$, can be prepared by the thermal decomposition of diborane. It is believed that the initial stages of the thermal decomposition involve the formation at low concentrations of high-energy intermediates such as BH_3, B_3H_9, and B_3H_7. For example, the following mechanism has been proposed for the formation of tetraborane:

$$B_2H_6 \rightleftharpoons 2BH_3$$
$$B_2H_6 + BH_3 \rightleftharpoons B_3H_9$$
$$B_3H_9 \rightleftharpoons B_3H_7 + H_2$$
$$B_3H_7 + B_2H_6 \rightarrow B_4H_{10} + BH_3$$

Table 9.1   SOME PHYSICAL PROPERTIES OF THE BORON HYDRIDES

	mp, °C	bp, °C	ΔH (25°C, kcal mol^{-1} for gases)
B_2H_6	−164.86	−92.84	7.5
B_4H_{10}	−120.8	16.1	13.8
B_5H_9	−46.75	58.4	15.0
B_5H_{11}	−123.2	65	22.2
B_6H_{10}	−63.2	108	19.6
B_6H_{12}	−83	80–90	
B_8H_{12}			
B_8H_{18}			
B_9H_{15}	2.6	0.8 mm at 28°	
$B_{10}H_{14}$	99.5	~213	2.8
$B_{10}H_{16}$			
$B_{18}H_{22}$	178.5		
Iso-$B_{18}H_{22}$			
$B_{20}H_{16}$	199		

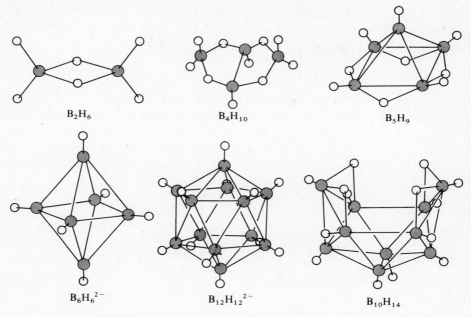

FIGURE 9.1

Structures of some boron hydrides and borane anions.

The same higher hydrides, as well as B_8H_{12}, B_9H_{15}, and $B_{10}H_{16}$, can be prepared by the appropriate treatment of lower hydrides with an electric discharge. For other preparative procedures, the review by Parry and Walter[1] should be consulted. The structures of several boron hydrides are illustrated in Fig. 9.1. It should be noted that the boron-atom frameworks of B_4H_{10}, B_5H_9, and $B_{10}H_{14}$ are fragments of regular polyhedra. For example, the framework of $B_{10}H_{14}$ is that of an icosahedron (a 20-faced polyhedron) which lacks two adjacent vertices.

The Synthesis of Borane Anions

In Chap. 5 we briefly discussed the synthesis of the tetrahydroborate ion, BH_4^-, and the octahydrotriborate ion, $B_3H_8^-$. A large number of higher borane anions are known; these have been prepared mainly by two general methods. The first method, the "BH condensation method," consists of the effective addition of BH groups to borane anions by treatment with diborane or other sources of BH groups. The general mechanism is believed to involve a sequence of steps in

[1] Ibid.

which BH_3 groups are added and H_2 molecules are lost. This sort of process is undoubtedly involved in the synthesis of $B_3H_8^-$ from BH_4^- and diborane:

$$BH_4^- + [BH_3] \rightleftharpoons B_2H_7^-$$

$$B_2H_7^- + [BH_3] \rightarrow B_3H_8^- + H_2$$

Similar mechanisms have been postulated for the synthesis of the anions $B_nH_n^{2-}$, where n ranges from 6 to 12. The second method for preparing higher borane anions, specifically the $B_nH_n^{2-}$ ions, is the pyrolysis of salts of lower borane anions. The particular product obtained is highly dependent on the temperature, the cation, and the solvent (if any). Some useful conversions are indicated in the following:

$$(CH_3)_4NB_3H_8 \xrightarrow{\Delta} (CH_3)_3NBH_3 + [(CH_3)_4N]_2B_{10}H_{10}$$
$$+ [(CH_3)_4N]_2B_{12}H_{12}$$

$$CsB_3H_8 \xrightarrow{\Delta} Cs_2B_9H_9 + Cs_2B_{10}H_{10} + Cs_2B_{12}H_{12}$$

$$CsB_3H_8 \xrightarrow[\Delta]{\text{traces of ether}} Cs_2B_{12}H_{12}$$

$$(C_2H_5)_4NBH_4 \xrightarrow{\Delta} [(C_2H_5)_4N]_2B_{10}H_{10}$$

All the $B_nH_n^{2-}$ anions have boron frameworks corresponding to highly symmetric polyhedra. The octahedral and icosahedral structures of $B_6H_6^{2-}$ and $B_{12}H_{12}^{2-}$, respectively, are shown in Fig. 9.1.

The Synthesis of Carboranes

A boron-hydrogen compound in which carbon atoms occupy structural sites similar to those occupied by boron atoms is called a "carborane." The most important carborane synthesis is that of $1,2\text{-}B_{10}C_2H_{12}$, which is isoelectronic with the $B_{12}H_{12}^{2-}$ ion. The reactions involved in the synthesis follow:

$$B_{10}H_{12} + 2Et_2S \xrightarrow{n\text{-}Pr_2O} B_{10}H_{12}(Et_2S)_2 + H_2$$

$$B_{10}H_{12}(Et_2S)_2 + C_2H_2 \xrightarrow{n\text{-}Pr_2O} B_{10}C_2H_{12} + H_2 + 2Et_2S$$

The second step involves the insertion of the two carbon atoms of acetylene into the $B_{10}H_{12}$ framework, yielding the icosahedral $1,2\text{-}B_{10}C_2H_{12}$ framework, with the carbon atoms occupying adjacent positions on the icosahedron. When this compound is heated, rearrangement occurs to the 1,7 and 1,12 isomers, in which the carbon atoms occupy positions separated by one and two boron atoms, respectively.

STRUCTURE AND BONDING[1, 2]

Three-Center Bonding

The boron hydrides, borane anions, and carboranes have extraordinary structures. Some of these structures are shown in Fig. 9.1. It should be noted that the framework atoms of the larger molecules form the vertices of either a regular polyhedron or a polyhedral fragment. Compounds of this type have fascinated chemists not only because of their unusual structures and reactions but also because simple valence bond theory cannot account for their bonding. It is easy to understand why BH_3 is a reactive species; there are only six valence electrons in the compound, and it is impossible for the boron atom to achieve a complete valence octet. However, even the stable dimer, diborane, lacks enough electrons to permit conventional bonding as in ethane. The key to the bonding of diborane is found in the fact that two of the hydrogen atoms form bridges between the two boron atoms. These B—H—B bridges in diborane are examples of "three-center bonds," in which a bonding pair of electrons holds together three atoms. In our previous discussions of valence bond theory, we have considered only "two-center bonds," in which bonding electrons hold together two atoms. By including the three-center bond concept in valence bond theory, we can write octet-satisfying structures for the boron hydrides and their derivatives.

The B—H—B bridge is usually represented by the structure $\overset{\displaystyle H}{\underset{\displaystyle B \qquad B}{\frown}}$. The only other type of three-center bond which is encountered in boron hydrides is the B—B—B bond, usually represented by the structure $\overset{\displaystyle B}{\underset{\displaystyle B \qquad B}{\triangle}}$. In carboranes one must consider analogous B—B—C or B—C—C bonds. To satisfactorily describe the bonding of a boron compound, the following criteria must be fulfilled: (1) The total number of valence electrons must equal twice the number of bonds, including both two- and three-center bonds. (2) Each boron atom must use four orbitals and must achieve a complete octet of valence electrons. (3) The bonding must be consistent with the observed structure of the molecule and with an approximately tetrahedral disposition of bonds to each boron atom. It is useful to express the first two of these rules in the form of topological equations. For the case of a species of formula $B_b H_h{}^q$, where q is the net charge, let α represent the number of two-center bonds and β the number of three-center bonds. Rule 1 corresponds to

$$\alpha + \beta = \tfrac{1}{2}(3b + h - q)$$

---

[1] See footnote 3 on page 175.
[2] W. N. Lipscomb, "Boron Hydrides," W. A. Benjamin, Inc., New York, 1963.

Rule 2 corresponds to equating the total number of two- and three-center connections to the number of atomic orbitals, or

$$2\alpha + 3\beta = 4b + h$$

Solving for α and β, we obtain

$$\alpha = \tfrac{1}{2}(b + h - 3q) \tag{9.1}$$
$$\beta = b + q \tag{9.2}$$

We shall now describe, for several boron compounds, bonding assignments which are consistent with the preceding rules.

First let us consider the tetrahydroborate ion, BH_4^-. The bonding in this simple species can be described by using classic valence bond theory because, according to Eq. 9.2, $\beta = 1 - 1 = 0$. Clearly rule 3 is satisfied because the species is tetrahedral and the boron uses four sp^3 hybrid orbitals.

Next we shall consider diborane, B_2H_6. From Eqs. 9.1 and 9.2 we obtain $\alpha = 4$ and $\beta = 2$, corresponding to the bonding

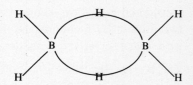

This bonding requires that the bridging hydrogens lie in a plane perpendicular to the plane of the other atoms, as shown in Fig. 9.1.

Now let us consider tetraborane, B_4H_{10}, for which we calculate $\alpha = 7$ and $\beta = 4$. The bonding shown below fits the observed structure of the molecule.

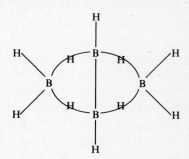

In the case of pentaborane-9, B_5H_9, $\alpha = 7$ and $\beta = 5$. A bonding pattern which is consistent with these parameters and with the structure shown in Fig. 9.1 is as follows:

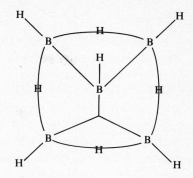

This molecule has the shape of a square pyramid. Inasmuch as the four basal boron atoms are actually equivalent, the bonding must be described as a resonance hybrid of the four bonding structures corresponding to permutation of the two B—B bonds and the B—B—B bond. Pentaborane-9 is the simplest boron hydride which uses all the possible types of two- and three-center bonds.

Finally we shall discuss the bonding in the octahedral anion $B_6H_6^{2-}$ in terms of two- and three-center bonds. For this species, we calculate $\alpha = 9$ and $\beta = 4$. From Fig. 9.1 we see that each boron atom is bonded to a terminal hydrogen atom by a two-center B—H bond; hence the octahedral B_6 framework must be held together by three two-center bonds and four three-center bonds. One of many possible permutations of these framework bonds, consistent with tetrahedral bonding to the boron atoms, is shown in the following sketch:

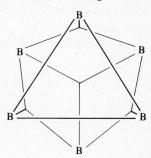

Obviously there are many resonance structures which one can write for this ion.

One of the intriguing aspects of the polyhedral $B_nH_n^{2-}$ ions is the fact that they all have a -2 charge. The simple valence bond method, using two- and three-center bonds, is incapable of rationalizing this fact. For example, the method gives apparently adequate explanations of species such as B_nH_n and $B_nH_n^{4-}$. To obtain an explanation of the stability of the -2 charge of these species we must go to simple MO theory.

MO Theory[1]

We shall apply simple MO theory to the B_6 framework of the $B_6H_6{}^{2-}$ ion. This framework may be looked upon as a square bipyramid. We assume that each of the two apical boron atoms uses an sp hybrid orbital in its B—H bond. The opposite sp hybrid orbital, directed toward the center of the bipyramid, and the two p orbitals perpendicular to the bipyramidal axis are available for interaction with orbitals of the four equatorial boron atoms. We assume that each of the equatorial boron atoms is bonded to a hydrogen atom and to two adjacent equatorial boron atoms by hybrid orbitals lying in the equatorial plane. The p orbital perpendicular to the equatorial plane is available for interaction with orbitals of the axial boron atoms. Twelve of the 26 valence electrons in the ion are used to form the B—H bonds, and eight are used to form the B—B bonds, leaving six electrons for the remaining framework bonding. The remaining framework orbitals and the corresponding group orbitals are pictured in Fig. 9.2. The σ_u group of equatorial orbitals can combine with the σ_u group of apical orbitals, and the π_g group of equatorial orbitals can combine with the π_g group of apical orbitals. The δ_u, σ_g, and π_u group orbitals are nonbonding. Thus three bonding MOs are obtained, as shown in the energy level diagram of Fig. 9.2. The greatest stability is achieved by filling these bonding MOs with the six available valence electrons. It should be noted that a total of 14 electrons are used in the B_6 framework of the $B_6H_6{}^{2-}$ ion.

Similar MO treatment of the other $B_nH_n{}^{2-}$ ions yields similar results; in each case, filling the bonding MOs gives a -2 ion. The general rule for polyhedral species appears to be that stability is achieved when $2n + 2$ electrons are available for a polyhedral framework of n atoms.[2]

[1] Lipscomb, op. cit.
[2] K. Wade, *Chem. Commun.*, p. 792, 1971; R. W. Rudolph and W. R. Pretzer, *Inorg. Chem.*, **11**, 1974 (1972); C. J. Jones, W. J. Evans, and M. F. Hawthorne, *J. Chem. Soc. Chem. Commun.*, p. 543, 1973.

FIGURE 9.2

A molecular orbital treatment of the framework bonding in $B_6H_6{}^{2-}$. Lines have been drawn corresponding to the six B—H bonds and four equatorial B—B bonds. The remaining atomic orbitals may be combined as indicated to form three bonding MOs. The various group orbitals are indicated by the signs of the appropriate atomic orbital wave functions. For the equatorial group orbitals, the signs are given for the p orbitals on one side of the plane; the opposite signs would apply to the other side of the plane. For the apical σ group orbitals, the signs are given for the two sp hybrid orbitals on the apical borons. For the apical π group orbitals, each set of four lobes corresponds to the two $p\pi$ orbitals of an apical boron. Each vertical pair of such sets corresponds to a particular group orbital. (*Adapted from W. N. Lipscomb, "Boron Hydrides," W. A. Benjamin, Inc., New York*, 1963.)

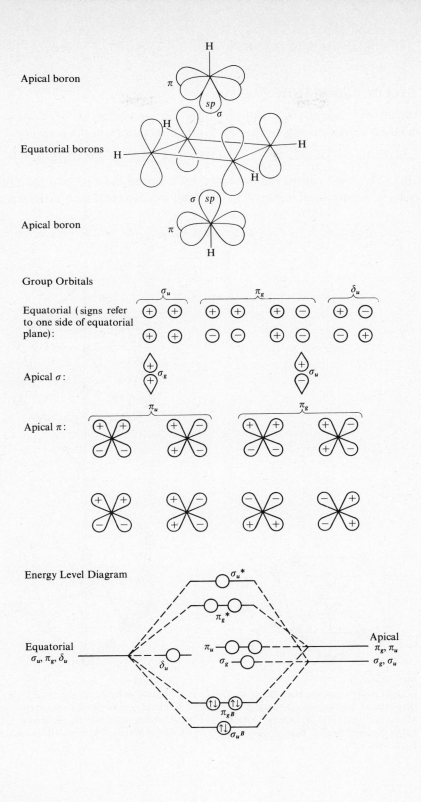

Apical boron

Equatorial borons

Apical boron

Group Orbitals

Equatorial (signs refer to one side of equatorial plane):

Apical σ:

Apical π:

Energy Level Diagram

Equatorial
$\sigma_u, \pi_g, \delta_u$

Apical
π_g, π_u
σ_g, σ_u

METALLOCARBORANES[1,2]

Treatment of $1,2\text{-}B_{10}C_2H_{12}$ with ethoxide ion in ethanol at $70°C$ causes a degradation corresponding to the deletion of a B^+ ion from the molecule:

$$B_{10}C_2H_{12} + OC_2H_5^- + 2C_2H_5OH \rightarrow B_9C_2H_{12}^- + B(OC_2H_5)_3 + H_2$$

The $B_9C_2H_{12}^-$ ion can be deprotonated by a strong base to give the "open" $B_9C_2H_{11}^{2-}$ ion pictured in Fig. 9.3. The latter ion can react with various transi-

[1] See footnote 3 on page 175.
[2] M. F. Hawthorne and G. B. Dunks, *Science*, **178**, 462 (1972).

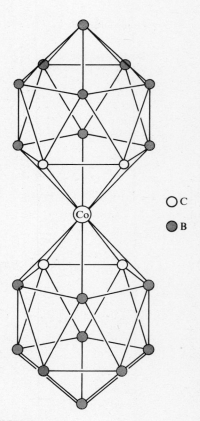

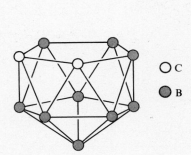

FIGURE 9.3
Skeletal structure of the $B_9C_2H_{11}^{2-}$ ion. (*Reproduced with permission from E. L. Muetterties and W. H. Knoth, "Polyhedral Boranes," Marcel Dekker, Inc., New York, 1968.*)

FIGURE 9.4
Skeletal structure of the $Co(B_9C_2H_{11})_2^-$ ion. (*Reproduced with permission from E. L. Muetterties and W. H. Knoth, "Polyhedral Boranes," Marcel Dekker, Inc., New York, 1968.*)

tion-metal ions or organo transition-metal ions to form complexes in which the metal atom completes the icosahedron. For example,

$$4B_9C_2H_{11}{}^{2-} + 3Co^{2+} \rightarrow Co + 2Co(B_9C_2H_{11})_2{}^-$$

The structure of the $Co(B_9C_2H_{11})_2{}^-$ ion is shown in Fig. 9.4. Each atom in each of the two icosahedra contributes three orbitals to the icosahedral framework bonding. Each icosahedron receives 2 electrons from the Co^- group, 18 electrons from nine BH groups, and 6 electrons from two CH groups. Thus 26 electrons are available per icosahedron, in agreement with the $2n + 2$ rule. The $2n + 2$ rule has been used as a guide in the discovery of a variety of unusual metallocarboranes.[1]

PROBLEMS

9.1 Draw three possible structures for the $B_3H_8{}^-$ ion, showing any three-center bonds clearly. Which structure do you think is more plausible?

9.2 Draw a plausible structure for $B_3C_2H_5$. How many bonds of each type (B—H, B—B, C—H, B—B—B, B—H—B, B—B—C, B—C—C) are there?

9.3 In a polyhedral anion of formula $B_nH_n{}^{2-}$, how many B—B bonds and how many B—B—B bonds are there?

9.4 At least two different mechanisms have been proposed for the isomerization of $1,2\text{-}B_{10}C_2H_{12}$. One involves the rotation of triangular faces of the icosahedron (e.g., a B_2C face). A second involves the conversion of two adjacent triangular faces into a square face, followed by conversion into two triangular faces:

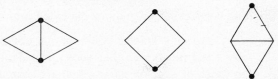

A cube octahedron intermediate, or activated complex, has been proposed for the latter mechanism:

Show that the first mechanism can explain the formation of the 1,7 and 1,12 isomers, but that the second mechanism cannot account for the formation of the 1,12 isomer.

9.5 Construct a molecular orbital energy level diagram for B_2H_6.

[1] See footnote 2 on page 182 and footnote 2 on page 184.

10
THE SOLID STATE

CLASSIFICATION OF BONDS AND CRYSTALS[1]

Bonding between atoms may be classified into the following types:

1 Covalent. This is the type of bonding which we have principally discussed thus far in this book. Covalent bonding is generally described in terms of electron pairs which are shared by the bonded atoms.

2 Ionic. This is the bonding caused by the electrostatic (coulombic) attraction between oppositely charged atoms or ions. In the limit of pure ionic bonding between atoms, each ion would be spherically symmetric and there would be no shared electrons.

3 Van der Waals. This bonding, sometimes referred to as London dispersion forces, corresponds to the relatively weak interaction which occurs between atoms and molecules caused by instantaneous dipole-induced dipole interactions. Even in species which have no permanent dipole moment, instantaneous dipoles arise because of momentary asymmetry in electron distribution. These instantaneous dipoles induce dipoles in adjacent species. In other words, the electrons in separate species tend

---

[1] A. F. Wells, "Structural Inorganic Chemistry," 3d ed., Oxford University Press, Fair Lawn, N.J., 1962.

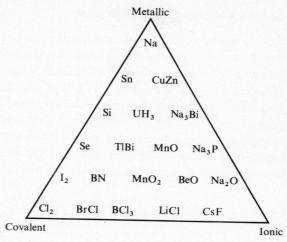

FIGURE 10.1

Classification of compounds according to the degree of covalent, ionic, and metallic character in the bonding.

to synchronize their movements to minimize electron-electron repulsion and to maximize electron-nucleus attraction. Van der Waals forces increase rapidly with molecular volume and with the number of polarizable electrons. They are the principal forces holding together the atoms in solid argon and are the principal *inter*molecular forces in solid CCl_4 and I_2.

4 Metallic. This type of bonding holds together regular lattices of electron-deficient atoms, or groups of atoms, such as in metals and alloys. The characteristic feature of metallic bonding is that the bonding electrons are delocalized over the entire crystal.

It might appear that the best way of classifying solids would be in terms of the types of bonds between the atoms, i.e., covalent, ionic, van der Waals, and metallic. However, bonds which are purely of one type or another are rare, and in most crystals there are bonds of more than one type. Figure 10.1 illustrates how covalent, ionic, and metallic bonding mix together even in simple compounds and shows that the bonding in most compounds cannot be cleanly classified in any one of these groups. It is more useful to classify solids in terms of their structural features, and we shall now show how it is possible to roughly classify crystals in the following four categories: (1) crystals containing finite, discrete complexes, (2) crystals containing infinite one-dimensional complexes, (3) crystals containing infinite two-dimensional complexes, and (4) crystals consisting of infinite three-dimensional complexes.

Crystals Containing Finite Complexes

This category includes most solid compounds of the nonmetals in which the molecules are held together by van der Waals or dipole-dipole forces. In the case of atoms (for example, Ne, Ar, Kr, and Xe) and molecules with approximately spherical shapes (for example, HCl, H_2S, and SiF_4), the so-called close-packed structures and the body-centered cubic structure are generally found. The close-packed structures may be understood by reference to Fig. 10.2. When identical spheres are packed together as closely as possible on a plane surface, they are arranged as shown, with each sphere touching six others. An exactly similar layer of spheres can be laid on top of this layer, so that the spheres of the

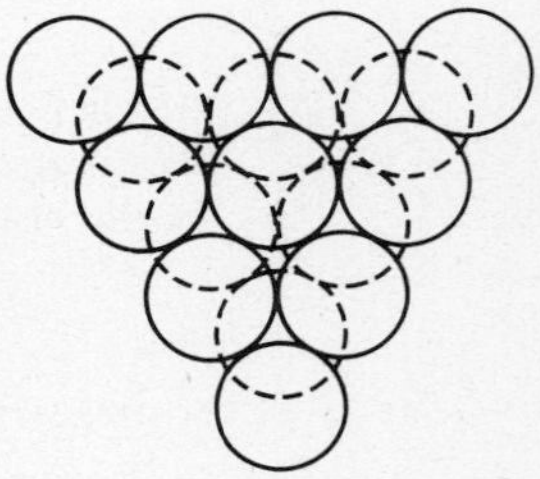

FIGURE 10.2
Two adjacent layers of close-packed spheres. Notice that there are two ways of placing a third layer. If the spheres in the third layer are directly over those of the first layer, the packing is hexagonal close packing. Otherwise the packing is cubic close packing.

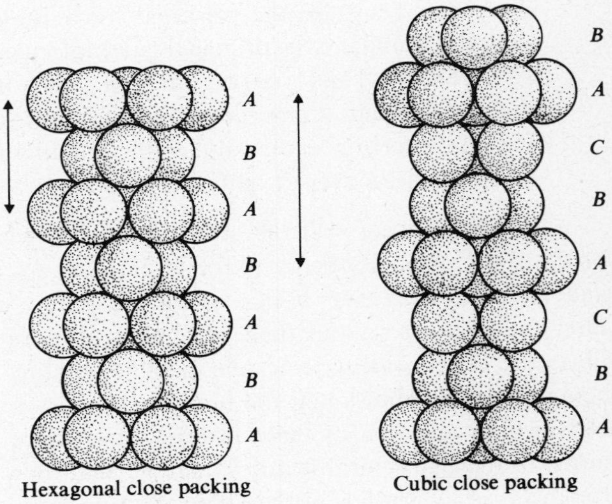

Hexagonal close packing

Cubic close packing

FIGURE 10.3
Arrangement of spheres in hexagonal close packing (left) and cubic close packing (right). (*Adapted with permission from L. Pauling, "The Nature of the Chemical Bond," 3d ed., Cornell University Press, Ithaca, N. Y., 1960.*)

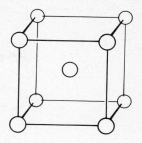

FIGURE 10.4
Body-centered cubic structure. This type of packing is found for the atoms in various metals and for the molecules in SiF_4, $N_4(CH_2)_6$, and $MoAl_{12}$.

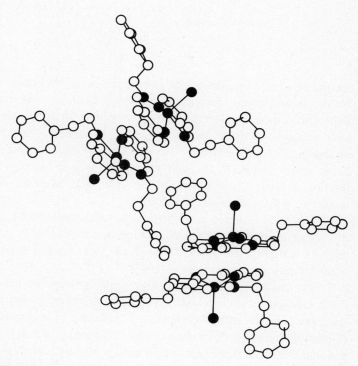

FIGURE 10.5
Packing diagram of $Fe(SANE)_2Cl$. [SANE = N-(2-phenylethyl)salicylaldimine.] The asymmetric units are linked into loose "dimers." Carbon atoms are represented by open circles; other atoms by filled circles. [*Reproduced with permission from J. A. Bertrand et al., Inorg. Chem.,* **13**, 125 (1974). *Copyright © by the American Chemical Society.*]

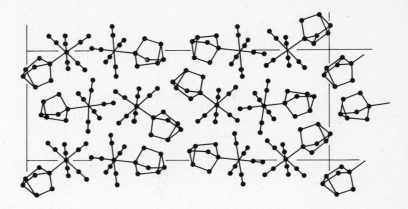

FIGURE 10.6
Projection of the crystal packing of $Mo(CO)_5P_4S_3$. [*Reproduced with permission from A. W. Cordes et al., Inorg. Chem., 13, 132 (1974). Copyright © by the American Chemical Society.*]

second layer rest in depressions of the first layer. When a third layer is placed on the second layer, there are two choices: The spheres may be placed in the depressions directly over the first-layer spheres or in the depressions which lie over depressions in the first layer. The two simplest sequences of layers may be indicated $ABAB \cdots$ (called hexagonal close packing) or $ABCABC \cdots$ (called cubic close packing). In each case each sphere touches 12 others, and the fraction of space occupied by spheres is 0.7405. Sketches of spheres packed in these two ways are shown in Fig. 10.3.

The body-centered cubic structure is illustrated in Fig. 10.4. In the case of molecular crystals, this type of packing is seldom found, probably because the packing is slightly less efficient than in the close-packed structures. (The fraction of space occupied by spheres is 0.68.)

The packing of nonspherical or dipolar molecules generally deviates from the undistorted close-packed and body-centered cubic structures. The packing diagrams of two rather complicated molecules [an iron(III) complex of *N*-(2-phenylethyl)salicylaldimine and chloride,[1] and pentacarbonyl(tetraphosphorus trisulfide)molybdenum[2]] are shown in Figs. 10.5 and 10.6, respectively. Inasmuch as these molecules are not spherically symmetric, it is necessary to specify their angular orientations as well as their positions when describing the packing.

[1] J. A. Bertrand, J. L. Breece, and P. G. Eller, *Inorg. Chem.,* **13,** 125 (1974).
[2] A. W. Cordes, R. D. Joyner, R. D. Shores, and E. D. Dill, *Inorg. Chem.,* **13,** 132 (1974).

Crystals Containing Infinite One-dimensional Complexes

The simplest examples of one-dimensional complexes in the solid state consist of infinite linear molecules held together in lattices by van der Waals bonds. The SiS_2 structure consists of SiS_4 tetrahedra which share edges to form long chains, as shown in Fig. 10.7. In palladium(II) chloride, $PdCl_2$, infinite chains are built of planar $PdCl_4$ groups sharing opposite edges, as shown in Fig. 10.8. In elemental tellurium, the atoms are joined in the form of infinite helical chains. Similar helical chains exist in the "metallic" form of selenium and the fibrous form of sulfur. The fibrous sulfur structure is illustrated in Fig. 10.9.

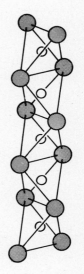

FIGURE 10.7
Structure of SiS_2. Small circles represent silicon atoms; large circles sulfur atoms.

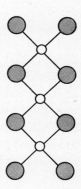

FIGURE 10.8
Structure of $PdCl_2$.

FIGURE 10.9
Structure of fibrous sulfur. (*Adapted with permission from L. Pauling, "The Nature of the Chemical Bond," 3d ed., Cornell University Press, Ithaca, N.Y., 1960.*)

Many crystals contain anionic chains held together by cations. For example, the complex $KCu(CN)_2$ has a spiral polymeric anion in which each Cu(I) atom is bound to two carbon atoms and one nitrogen atom:

Silicate chain ions are found in the two large classes of minerals, the *pyroxenes*, and the *amphiboles*. The former [including enstatite, $MgSiO_3$; diopside, $CaMg(SiO_3)_2$; jadeite, $NaAl(SiO_3)_2$; and spodumene, $LiAl(SiO_3)_2$] contain simple chains of tetrahedral SiO_4 groups connected by shared oxygen atoms, as shown in Fig. 10.10*a*. The latter [including tremolite, $Ca_2Mg_5(OH)_2(Si_4O_{11})_2$, and arfvedsonite, $Na_3Mg_4Al(OH)_2(Si_4O_{11})_2$] contain double chains in which one-half of the SiO_4 groups are joined to two adjacent groups by shared oxygen atoms, and one-half are joined to three adjacent groups by shared oxygen atoms, as shown in Fig. 10.10*b*.

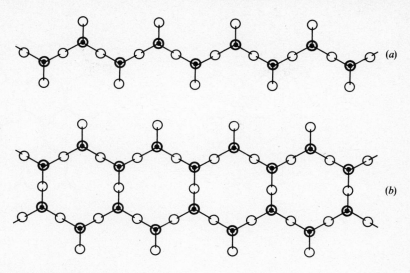

FIGURE 10.10
Silicon-oxygen anionic chains in pyroxenes (*a*) and amphiboles (*b*). (*Reproduced with permission from A. F. Wells, "Structural Inorganic Chemistry," 3d ed., Clarendon Press, Oxford, 1962.*)

Crystals Containing Infinite Two-dimensional Complexes

Structures in which layers are held together by van der Waals bonds are found in elemental arsenic and graphite (Fig. 10.11) and in $CdCl_2$, MoS_2, $CrCl_3$, and HgI_2 (Fig. 10.12). Structures with layers which are held together by hydrogen bonds are found in various hydroxy compounds, such as aluminum

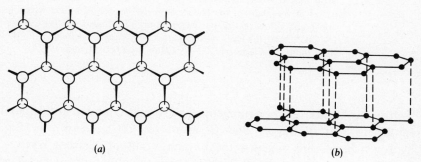

FIGURE 10.11
Layer structures of (*a*) arsenic and (*b*) graphite.

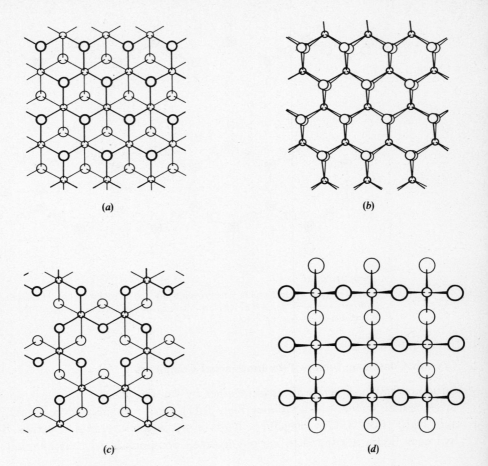

FIGURE 10.12
Structures of some simple MX_2 and MX_3 layers. Small circles represent metal atoms in the plane of the paper; large circles, the X atoms which lie in planes above and below that of the metal atoms. The layers are (a) $CdCl_2$, (b) MoS_2, (c) $CrCl_3$, and (d) HgI_2. (*Reproduced with permission from A. F. Wells, "Structural Inorganic Chemistry,"* 3d ed., *Clarendon Press, Oxford, 1962.*)

hydroxide, $Al(OH)_3$. The arrangement of Al and O atoms in this compound is similar to that of the Cr and Cl atoms in $CrCl_3$, shown in Fig. 10.12c; hydrogen bonds link together the oxygen atoms of adjacent layers. In boric acid (Fig. 5.4) hydrogen bonds hold together the atoms *within* each layer.

Examples of anionic layers held together by cations are found in certain aluminosilicates such as the hexagonal form of $CaAl_2Si_2O_8$. This compound

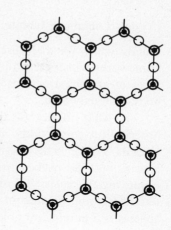

FIGURE 10.13
Structure of a silicon-aluminum-oxygen sheet. Oxygen atoms lying above the Si(Al) atoms (small black circles) are drawn more heavily. (*Reproduced with permission from A. F. Wells, "Structural Inorganic Chemistry," 3d ed., Clarendon Press, Oxford, 1962.*)

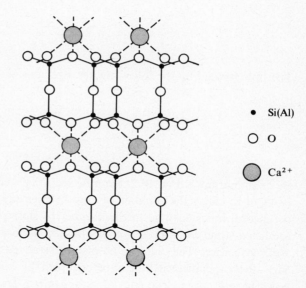

- Si(Al)
○ O
◉ Ca²⁺

FIGURE 10.14
Side view of the structure of hexagonal $CaAl_2Si_2O_8$, showing the double layers, interleaved with Ca^{2+} ions. (*Reproduced with permission from A. F. Wells, "Structural Inorganic Chemistry," 3d ed., Clarendon Press, Oxford, 1962.*)

contains silicon-aluminum-oxygen sheets of the type shown in Fig. 10.13. Such sheets are joined together in pairs by oxygen atoms to form double layers of empirical composition $(AlSiO_4)_n{}^{n-}$. These double layers are held together by calcium ions as shown in Fig. 10.14. The micas have related structures.

Crystals Containing Infinite Three-dimensional Complexes

Metals, which we shall discuss separately in Chap. 11, have three-dimensional frameworks which generally can be described in terms of one of the following types of atomic structures: hexagonal close packing, cubic close packing, and body-centered cubic.

A wide variety of three-dimensional framework structures have been observed for compounds. In the case of simple compounds having the general formula $M_m X_x$, where X is a relatively large atom, most observed structures consist of close-packed arrangements of the X atoms, with the M atoms occupying the holes. There are two types of holes in a close-packed structure: tetrahedral and octahedral. The hole formed by the following arrangement of spheres is a tetrahedral hole:

The hole formed by the following arrangement of spheres is an octahedral hole:

By examination of Fig. 10.2 it can be seen that the number of tetrahedral holes is equal to twice the number of close-packed spheres and that the number of octahedral holes is equal to the number of close-packed spheres. In Table 10.1 are listed some common structures based on close packing. Let us consider the NaCl structure. The chlorine atoms are arranged in a cubic close-packed array, and all the octahedral holes are occupied by sodium atoms. The octahedral holes have a cubic close-packed arrangement, and so the NaCl lattice may be considered as two interpenetrating cubic close-packed lattices. The conventional representation of the NaCl lattice is shown in Fig. 10.15a; however, reference to Fig. 10.15b and c will help make obvious the close-packed arrangements of the two sets of equivalent atoms. Five other structures based on close-packed X atoms (fluorite, rutile, nickel arsenide, wurtzite, and zinc blende) are illustrated in Fig. 10.16. It should be noted that Table 10.1 includes several structures which do not fit strictly into the category of infinite three-dimensional networks. For example, Al_2Br_6 consists essentially of a lattice of close-packed bromine atoms, with one-sixth of the tetrahedral holes occupied by aluminum atoms in such a way that adjacent pairs of tetrahedral holes are occupied. Thus, pairs of bromine atoms on tetrahedral edges are shared, and the structure can be looked upon as a packing of discrete molecules of the following type:

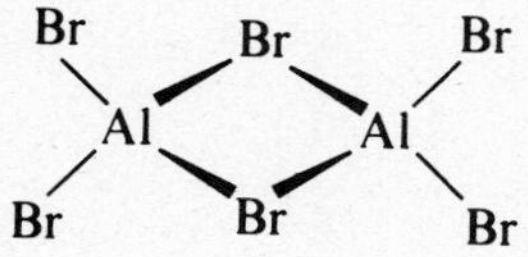

Table 10.1 STRUCTURES BASED ON CLOSE PACKING OF X ATOMS

Formula	Fraction of holes occupied by M atoms		Type of close packing for X atoms		Coordination number	
	Tetrahedral	Octahedral	hcp	ccp	M	X
M_2X	1	0		F_2Ca (fluorite)	4	8
M_3X_2	$\frac{3}{4}$	0		Zn_3P_2 O_3Mn_2	4	6
MX	0	1	NiAs	NaCl	6	6
	$\frac{1}{2}$	0	ZnS (wurtzite)	ZnS (zinc blende)	4	4
M_2X_3	0	$\frac{2}{3}$	α-Al_2O_3 (corundum)		6	4
	$\frac{1}{3}$	0	β-Ga_2S_2	γ-Ga_2S_3	4	
MX_2	0	$\frac{1}{2}$	CdI_2	$CdCl_2$	6	3
			TiO_2 (rutile)	TiO_2 (anatase)		
	$\frac{1}{4}$	0	β-$ZnCl_2$	HgI_2 γ-$ZnCl_2$ SiS_2 OCu_2 α-$ZnCl_2$	4	2
MX_3	0	$\frac{1}{3}$	BiI_3	$CrCl_3$	6	2
	$\frac{1}{6}$	0	Al_2Br_6		4	2
MX_4	$\frac{1}{8}$	0	SnI_4		4	1
MX_6	0	$\frac{1}{6}$		α-WCl_6 UCl_6	6	1

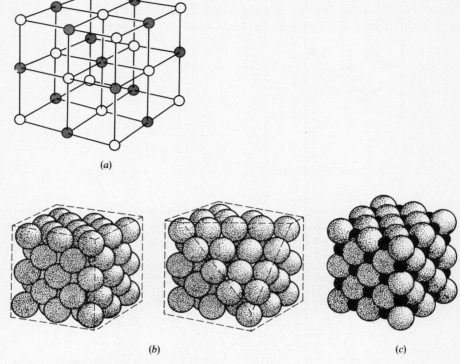

FIGURE 10.15
Conventional NaCl structure (*a*), cubic close packing of spheres (*b*), and another
representation of the NaCl structure (*c*). [*From W. Barlow, Z. Krist*, **29**, 433
(1898).]

An important structure for compounds of the type MX, which does not
correspond to a close packing for either the M or X atoms, is the cesium
chloride structure, shown in Fig. 10.17. Each atom has eight nearest neighbors,
as in the body-centered cubic structure.

There are many ionic compounds of the type $M_m X_x$ in which M, X, or
both M and X are replaced by complex ions. When the complex ions are
approximately spherical, the spatial arrangement of the ions is often the same as
in one of the symmetric structures which we have discussed. For example,
$[Co(NH_3)_6]I_2$ and $[Mg(NH_3)_6]I_2$ adopt the fluorite (CaF_2) structure, and
K_2PtCl_6 and K_2SnCl_6 adopt the anti-fluorite structure (that is, a fluorite-like
structure in which anions occupy the Ca^{2+} positions and cations occupy the F^-
positions). The structure of $[Ni(H_2O)_6][SnCl_6]$ is a slightly deformed version of
that of CsCl. Polyatomic ions can attain effective spherical symmetry by rapid

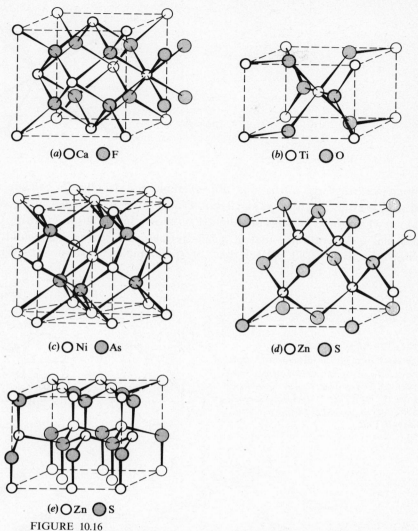

(a) ○ Ca ● F

(b) ○ Ti ● O

(c) ○ Ni ● As

(d) ○ Zn ● S

(e) ○ Zn ● S

FIGURE 10.16

Structures of (a) fluorite (CaF_2), (b) rutile (TiO_2), (c) nickel arsenide (NiAs), (d) zinc blende (ZnS), and (e) wurtzite (ZnS). (*Reproduced with permission from A. F. Wells, "Structural Inorganic Chemistry," 3d ed., Clarendon Press, Oxford, 1962.*)

free rotation or random orientation. Such behavior is often found in salts containing the ions NH_4^+, NO_3^-, CN^-, and SH^-.

Salts containing polyatomic ions sometimes adopt structures similar to the corresponding simple M_mX_x compounds, but of lower symmetry because of the

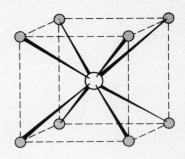

FIGURE 10.17
The CsCl structure.

nonspherical shapes of the ions. For example, the structure of calcium carbide, CaC_2, is essentially that of NaCl, but the structure is distended in the direction along which the C_2^{2-} ions are aligned, as shown in Fig. 10.18. Calcite, $CaCO_3$, also has a structure like that of NaCl, but elongated along a threefold axis, as shown in Fig. 10.19. When one of the complex ions is very large and

FIGURE 10.18
Sections through the structures of NaCl and CaC_2, showing the relationship of the packing. (*Reproduced with permission from A. F. Wells, "Structural Inorganic Chemistry," 3d ed., Clarendon Press, Oxford, 1962.*)

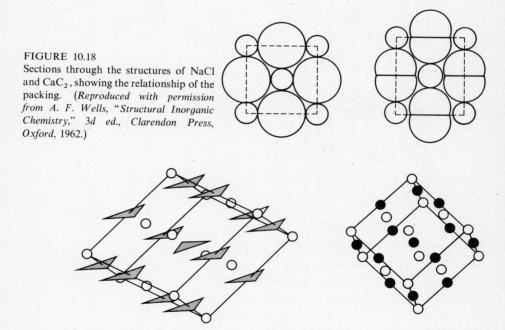

FIGURE 10.19
The $CaCO_3$ and NaCl structures similarly oriented. (*Reproduced with permission from A. F. Wells, "Structural Inorganic Chemistry," 3d ed., Clarendon Press, Oxford, 1962.*)

irregularly shaped, and the counter-ion is small, the structure is usually determined by the packing of the large ions, the counter-ions fitting in the interstices.

THE EFFECT OF RADIUS RATIO AND CHARGE ON STRUCTURE

There is no sharp dividing line between ionic and covalent bonding. The ionic character of a bond increases with increasing electronegativity difference between the atoms. Thus the bonding in solid NaCl is principally ionic, and that in diamond is completely covalent. However, in compounds such as ZnO and AgBr, the bonding is intermediate in character. An abrupt change in a physical property in a series of compounds has often been mistakenly ascribed to an abrupt change in the ionic character of the bonding. For example, consider the melting points of the fluorides of the second-row elements:

NaF	MgF_2	AlF_3	SiF_4	PF_5	SF_6
988°C	1266°C	1291°C (sublimes)	−90°C	−94°C	−50°C

The enormous change in melting points between AlF_3 and SiF_4 is not due to a marked change in ionic character (the electronegativities of Al and Si are 1.5 and 1.8, respectively) but rather to a change from an infinite lattice structure in which each aluminum is coordinated to six fluorine atoms to a lattice of discrete SiF_4 molecules held together by van der Waals bonds.

In most saline halides (that is, halides with infinite three-dimensional ionic lattices), the number of halide ions coordinated to each metal ion is greater than the number of halide ions that could be bound to the metal ion in any conceivable discrete molecule. Generally a halide will exist in the form of molecules only when, in a given molecule, the number of halogen atoms bonded to the electropositive atom equals the maximum coordination number of the electropositive atom for that halogen. Obviously the radius ratio of the atoms is the critical factor. By simple geometric considerations, we may calculate the minimum possible values of the ratio r_M/r_X for various types of coordination of X atoms around M atoms. These results are given in Table 10.2. In principle, these data, when combined with atomic or ionic radii, should be useful for predicting coordination numbers. However, such predictions are quite approximate because atomic and ionic radii are not well-defined, constant quantities. However, by using, for example, a consistent set of ionic radii such as those deduced by Pauling from crystal structure data (Table 10.3), it is possible to observe a correlation of the calculated radius ratios with the observed coordination numbers in crystals. The r_M/r_X values for metals in various oxide lattices and in various halides are listed with the corresponding observed coordination numbers for the metals in these compounds in Table 10.4. It can be seen that r_M/r_X values at the border lines between various coordination numbers are in approximate agreement with the theoretical r_M/r_X values of Table 10.2.

The radius ratio of two atoms is related to the electronegativity difference between the atoms. In general, the greater the electronegativity difference between M and X, the greater is the value of r_M/r_X. Thus it is not surprising that Mooser and Pearson[1] observed that, if the average principal quantum number of the valence shells of M and X is held constant, a transition from tetrahedral coordination to octahedral coordination occurs when the electronegativity difference exceeds a particular value. In Fig. 10.20, the average principal quantum number is plotted against the electronegativity difference for a large number of compounds of type MX. It can be seen that a fairly sharp boundary separates the tetrahedral structures (open circles) from the octahedral structures (solid circles).

Pauling[2] proposed several rules for predicting the types of anion-cation coordination in stable ionic lattices containing more than two kinds of ions. His first rule says, in effect, that the charge on a given anion should be canceled by its share of the charges on the cations to which it is coordinated. Thus an oxide ion will be stable if it is coordinated to any of the following combinations of cations:

---

[1] E. Mooser and W. B. Pearson, *Acta Cryst.*, **12**, 1015 (1959).

[2] L. Pauling, "The Nature of the Chemical Bond," 3d ed., Cornell University Press, Ithaca, N.Y., 1960.

Table 10.2 ALLOWED VALUES OF r_M/r_X FOR VARIOUS TYPES OF COORDINATION

Coordination number	Type of coordination	Minimum values of r_M/r_X
2	Linear	0
3	Triangular planar	0.155
4	Tetrahedral	0.225
4	Square planar	0.414
6	Octahedral	
8	Square antiprismatic	0.645
8	CsCl structure	0.732
12	Cubooctahedral	1.000

Table 10.3 PAULING IONIC RADII (IN ANGSTROMS)†

Ag^+	1.26	Fe^{2+}	0.76	Pb^{4+}	0.84
Al^{3+}	0.50	Fe^{3+}	0.64	P^{3-}	2.12
As^{3-}	2.22	Ga^+	1.13	P^{5+}	0.34
As^{5+}	0.47	Ga^{3+}	0.62	Pd^{2+}	0.86
Au^+	1.37	Ge^{2+}	0.93	Ra^{2+}	1.40
B^{3+}	0.20	Ge^{4+}	0.53	Rb^+	1.48
Ba^{2+}	1.35	H^-	1.40	S^{2-}	1.84
Be^{2+}	0.31	Hf^{4+}	0.81	Sb^{3-}	2.45
Br^-	1.95	Hg^{2+}	1.10	Sc^{3+}	0.81
C^{4-}	2.60	I^-	2.16	Se^{2-}	1.98
C^{4+}	0.15	In^+	1.32	Sr^{2+}	1.13
Ca^{2+}	0.99	In^{3+}	0.81	Sn^{2+}	1.12
Cd^{2+}	0.97	K^+	1.33	Sn^{4+}	0.71
Ce^{3+}	1.11	La^{3+}	1.15	Te^{2-}	2.21
Ce^{4+}	1.01	Li^+	0.60	Ti^{2+}	0.90
Cl^-	1.81	Lu^{3+}	0.93	Ti^{3+}	0.76
Co^{2+}	0.74	Mg^{2+}	0.65	Ti^{4+}	0.68
Co^{3+}	0.63	Mn^{2+}	0.80	Tl^+	1.40
Cr^{2+}	0.84	Mn^{3+}	0.66	Tl^{3+}	0.95
Cr^{3+}	0.69	Mo^{6+}	0.62	U^{3+}	1.11
Cr^{6+}	0.52	N^{3-}	1.71	U^{4+}	0.97
Cs^+	1.69	N^{5+}	0.11	V^{2+}	0.88
Cu^+	0.96	Na^+	0.95	V^{3+}	0.74
Cu^{2+}	0.70	NH_4^+	1.48	V^{4+}	0.60
Eu^{2+}	1.12	Ni^{2+}	0.72	Y^{3+}	0.93
Eu^{3+}	1.03	O^{2-}	1.40	Zn^{2+}	0.74
F^-	1.36	Pb^{2+}	1.20		

† L. Pauling, "The Nature of the Chemical Bond," 3d ed., Cornell University Press, Ithaca, N.Y., 1960.

Table 10.4 RADIUS RATIOS AND COORDINATION NUMBERS IN OXIDES AND HALIDES

Oxide lattices

Cation	r_M/r_O	Commonly observed coordination nos.	Salt	r_M/r_X	Coordination no.
			Halides		
B^{3+}	0.14	3,4	MgI_2	0.30	6
Be^{2+}	0.22	4	$MgBr_2$	0.33	6
Si^{4+}	0.29	4	$MgCl_2$	0.36	6
Al^{3+}	0.36	4,6	CaI_2	0.46	6
Ge^{4+}	0.38	4,6	MgF_2	0.48	6
Li^+	0.43	4	$CaBr_2$	0.51	6
Mg^{2+}	0.46	6	SrI_2	0.52	>6
Ti^{4+}	0.48	6	$CaCl_2$	0.55	6
Zr^{4+}	0.57	6	$SrBr_2$	0.58	>6
Sc^{3+}	0.58	6	BaI_2	0.62	>6
Na^+	0.68	6	$SrCl_2$	0.62	8
Ca^{2+}	0.71	8	$BaBr_2$	0.69	>6
Ce^{4+}	0.72	8	CaF_2	0.73	8
K^+	0.95	9	$BaCl_2$	0.75	>6
Cs^+	1.21	12	SrF_2	0.83	8
			BaF_2	0.99	8

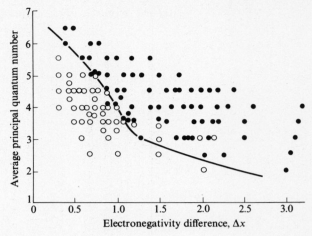

FIGURE 10.20

Average principal quantum number of valence shells versus electronegativity difference for MX compounds. Open circles correspond to tetrahedral structures; solid circles correspond to octahedral structures. [*From E. Mooser and W. B. Pearson, Acta Cryst.,* **12**, 1015 (1959).]

Two tetrahedral Si^{4+} ions, as in many silicates. $(2 = \frac{4}{4} + \frac{4}{4})$.

One tetrahedral Si^{4+} ion and two octahedral Al^{3+} ions, as in topaz, $Al_2SiO_4F_2$. $(2 = \frac{4}{4} + \frac{3}{6} + \frac{3}{6})$.

One tetrahedral Si^{4+} ion and three octahedral Mg^{2+} ions, as in olivine, Mg_2SiO_4. $(2 = \frac{4}{4} + \frac{2}{6} + \frac{2}{6} + \frac{2}{6})$.

One tetrahedral Si^{4+} ion and two tetrahedral Be^{2+} ions, as in phenacite, Be_2SiO_4. $(2 = \frac{4}{4} + \frac{2}{4} + \frac{2}{4})$.

Pauling's second rule is concerned with the number of anions simultaneously coordinated to two different cations, i.e., with the sharing of polyhedron vertices, edges, and faces. The rule states that shared edges, and especially shared faces, decrease the stability of a structure. The effect is greatest for cations of high charge and low coordination number and is caused by the increased electrostatic repulsion between cations. In agreement with this rule, SiO_4 tetrahedra tend to share only vertices with one another and with other polyhedra. However, in compounds in which the bonding is relatively covalent and in which the atomic charges are lower, exceptions to this second rule occur. Thus we have seen, in Fig. 10.7, how SiS_4 tetrahedra share edges in SiS_2 to form infinite chains,

Pauling's third rule, which also has an electrostatic basis, states that in a crystal containing different cations, those with high charge and low coordination number tend not to occupy polyhedra which share vertices or edges with each other. The highly charged cations tend to be as far apart from one another as possible.

LATTICE ENERGY

The energy required to break up a crystal into infinitely separated ions is called the lattice energy U:

$$M_m X_x(s) \rightarrow mM^{z_M}(g) + xX^{z_x}(g)$$

If the bonding in the crystal were completely ionic, the lattice energy could be calculated from the ionic charges and the geometric arrangement of the ions in the lattice. It is convenient to express this theoretical electrostatic lattice energy U_e as a function of the stoichiometric numbers m and x, the ionic charges z_M and z_X, the closest interionic distance r, and a geometric factor called the reduced Madelung constant, M':

$$U_e = -M' \frac{z_M z_X (m + x)e^2}{2r}$$

In the case of a salt for which $m = x = 1$, the quantity $(M' - 1)$ is essentially a measure of the stabilization in the crystal lattice relative to that in a simple ion pair. ($M' = 1$ for an ion pair.) Values of M' for various crystal structures are given in Table 10.5. Values for many other structures are given in the literature.[1] It can be seen that M' has essentially the same magnitude (that is,

[1] Q. C. Johnson and D. H. Templeton, *J. Chem. Phys.*, **34**, 2004 (1961).

Table 10.5 VALUES OF MADELUNG CONSTANTS FOR SEVERAL CRYSTAL STRUCTURES

Crystal structure	Reduced Madelung constant, M'	Conventional Madelung constants	
		$\mathcal{M} = \left(\frac{m+x}{2}\right)M'$	$M = \frac{z_M z_X(m+x)}{2}M'$
NaCl	1.7476	1.7476	1.7476
CsCl	1.7627	1.7627	1.7627
ZnS (zinc blende)	1.6381	1.6381	6.5522
ZnS (wurtzite)	1.6413	1.6413	6.5653
CaF$_2$ (fluorite)	1.6796	2.5194	5.0388
TiO$_2$ (rutile)	1.6053	2.4080	19.264
Al$_2$O$_3$ (corundum)	1.6688	4.172	25.031

~ 1.7) for all the crystal structures listed, showing that the lattice stabilization is of comparable magnitude for these structures.[1] Most texts quote the Madelung constants $\mathscr{M}$ or M, which are related to M' by the equations given in Table 10.5. When these Madelung constants are used, the theoretical electrostatic lattice energy must be calculated by the following equations:

$$U_e = -\mathscr{M} \frac{z_M z_X e^2}{r}$$

$$U_e = -M \frac{e^2}{r}$$

Inasmuch as a Madelung constant of type M contains the $z_M z_X$ factor, it can be used only for crystals having a particular charge type. The Madelung constants M' and $\mathscr{M}$ are more generally useful because they can be applied to crystals of any charge type. Thus $M' = \mathscr{M} = 1.7476$ for any compound of type $M^{n+}X^{n-}$ which has the sodium chloride structure (for example, MnO), whereas $M = 1.7476$ when $n = 1$ and $M = 6.9902$ when $n = 2$.

There is no experimental method known for directly measuring lattice energies. However, it is possible to obtain "experimental" values of lattice energies from appropriate thermodynamic data. Thus the lattice energy of a salt such as NaCl can be calculated as the sum of the energies of the following reactions, whose sum corresponds to the dissociation of the crystal into gaseous ions.

$$NaCl(s) \rightarrow Na(s) + \tfrac{1}{2}Cl_2(g) \qquad -\Delta H_f^{\circ}(NaCl) = 98.2 \text{ kcal mol}^{-1}$$
$$Na(s) \rightarrow Na(g) \qquad S(Na) = 25.9 \text{ kcal mol}^{-1}$$
$$\tfrac{1}{2}Cl_2(g) \rightarrow Cl(g) \qquad \tfrac{1}{2}D(Cl_2) = 29.0 \text{ kcal mol}^{-1}$$
$$Na(g) \rightarrow Na^+(g) + e^-(g) \qquad I(Na) = 118.5 \text{ kcal mol}^{-1}$$
$$Cl(g) + e^-(g) \rightarrow Cl^-(g) \qquad -A(Cl) = -83.5 \text{ kcal mol}^{-1}$$
$$U(NaCl) = -\Delta H_f^{\circ}(NaCl) + S(Na) + \tfrac{1}{2}D(Cl_2) + I(Na) - A(Cl)$$
$$= 188.1 \text{ kcal mol}^{-1}$$

Each of the energy terms involved in the summation has been evaluated by a separate physical chemical method; hence we may consider the sum 188.1 kcal mol^{-1} as the experimental lattice energy of sodium chloride. (Similar calculations can be carried out for other salts using the heats of formation of solids given in Table 10.6, the heats of formation of gaseous atoms in Table 2.12, the ionization potentials given in Table 1.4, and the electron affinities in Table 10.7. It is interesting to compare the experimental value of U for NaCl with the theoretical

[1] D. H. Templeton, *J. Chem. Phys.*, **23**, 1826 (1955).

Table 10.6 HEATS OF FORMATION (AT 25°C),† CRYSTAL STRUCTURES, AND INTERIONIC DISTANCES‡ OF VARIOUS SOLIDS

Compound	$-\Delta H_f^\circ$, kcal mol^{-1}	Crystal structure	$r_{\text{M}-\text{X}}$, Å
LiF	147.1	NaCl	2.009
NaF	137.5	NaCl	2.307
KF	136.2	NaCl	2.664
RbF	133.2	NaCl	2.815
CsF	131.7	NaCl	3.005
LiCl	97.6	NaCl	2.566
NaCl	98.2	NaCl	2.814
KCl	104.4	NaCl	3.139
RbCl	103.3	NaCl	3.285
CsCl	106.8	CsCl	3.47
LiBr	83.9	NaCl	2.747
NaBr	86.3	NaCl	2.981
KBr	94.1	NaCl	3.293
RbBr	93.6	NaCl	3.434
CsBr	97.9	CsCl	3.62
LiI	64.6	NaCl	3.025
NaI	68.8	NaCl	3.231
KI	78.4	NaCl	3.526
RbI	78.8	NaCl	3.663
CsI	83.7	CsCl	3.83
NH$_4$F	110.9	Wurtzite	4.39
NH$_4$Cl	75.2	CsCl	3.347
NH$_4$Br	64.7	CsCl	3.503
NH$_4$I	48.1	NaCl	3.622
BeF$_2$	242	β-cristobalite(SiO$_2$)	
MgF$_2$	268	Rutile	2.02
CaF$_2$	290.2	Fluorite	2.36
SrF$_2$	290.4	Fluorite	2.50
BaF$_2$	287.0	Fluorite	2.68
BeI$_2$	38	?	
MgI$_2$	86	CdI$_2$	
CaI$_2$	127.8	CdI$_2$	
SrI$_2$	136.0	?	
BaI$_2$	144.0	PbCl$_2$	
Li$_2$O	142.6	Anti-fluorite	2.00
Na$_2$O	99.4	Anti-fluorite	2.40
K$_2$O	86.4	Anti-fluorite	2.79
Rb$_2$O	80	Anti-fluorite	2.92
Cs$_2$O	82	Anti-CdCl$_2$	
BeO	143.1	Wurtzite	
MgO	143.8	NaCl	2.10
CaO	151.8	NaCl	2.40
SrO	141.1	NaCl	2.54
BaO	133.5	NaCl	2.75

† Selected Values of Chemical Thermodynamic Constants, *Natl. Bur. Std. (U.S.) Circ.* 500, 1952; *Natl. Bur. Std. Tech. Notes* 270–1 and 270–2, 1965, 1966; L. Brewer, private communication.

‡ L. Pauling, "The Nature of the Chemical Bond," 3d ed., Cornell University Press, Ithaca, N.Y., 1960.

electrostatic value. From the observed interionic distance in crystalline NaCl (2.814 Å) we calculate[1]

$$U_e = \frac{1.7476 \times (4.803 \times 10^{-10})^2 \times 1.4394 \times 10^{13}}{2.814 \times 10^{-8}} = 206.2 \text{ kcal mol}^{-1}$$

The discrepancy of 18.1 kcal mol^{-1} may be attributed to repulsions between the electron clouds of adjacent ions in the lattice. These repulsions may be approximately accounted for by the use of the relation

$$U = U_e\left(1 - \frac{\rho}{r}\right) \tag{10.1}$$

where ρ is a constant, approximately 0.31 Å. In the case of the alkali-metal halides, lattice energies calculated by this equation are accurate to about ± 2 percent.

In Table 10.8, experimental values of lattice energies and lattice energies calculated from Eq. 10.1 are listed for several alkaline-earth halides. It can be seen that the agreement is fairly good for the fluorides and poor for the iodides (especially for CaI_2, for which the experimental value is 13.2 percent greater

---

[1] The factor for converting ergs per molecule to kilocalories per mole is 1.4394×10^{13}.

Table 10.7 SOME ELECTRON AFFINI-TIES† (KILOCALORIES PER MOLE)

H	17.4	O_2	10.6
F	78.6	O_3	66
Cl	83.5	SO_2	64
Br	77.8	NO_2	92
I	70.9	OH	42
		NH_2	33

† H. O. Pritchard, *Chem. Rev.*, **52**, 529 (1953); R. S. Berry, *Chem. Rev.*, **69**, 533 (1969).

Table 10.8 COMPARISON OF EXPERIMENTAL AND CALCULATED LATTICE ENERGIES

Salt	U(exptl.), kcal mol^{-1}	U(calcd.), kcal mol^{-1}†	Δ
CaF_2	628.0	615.9	12.1
CaI_2	494.2	436.5	57.7
BaF_2	561.0	551.8	9.2
BaI_2	446.6	423.5	23.1

† T. E. Brackett and E. B. Brackett, *J. Phys. Chem.*, **69**, 3611 (1965).

than the calculated[1] value). The reason for the poor agreement is undoubtedly the neglect, in the calculation, of the polarizabilities of the ions. The soft iodide ion, when coordinated to the small, doubly charged calcium ion, is appreciably polarized, and the interaction of the induced moment with the charge of the calcium ion increases the bonding energy beyond that expected for spherically symmetric ions. The explanation of the enhanced stability of CaI_2 in terms of polarization is essentially equivalent to ascribing a large covalent contribution to the bonding. Thus we see that the estimation of lattice energies by Eq. 10.1 for relatively covalent compounds such as ZnS and InSb would be almost hopeless.

From Table 10.6 we see that ΔH_f° for the alkali-metal fluorides *decreases* from CsF to LiF, whereas ΔH_f° for the alkali-metal iodides *increases* from CsI to LiI. To understand the reason for these different trends, it is helpful to consider the relation

$$\Delta H_f^\circ = S(M) + \tfrac{1}{2}D(X_2) + I(M) - A(X) - U(MX)$$

The only terms on the right side of this equation that change on going from CsX to LiX are $S(M)$, $I(M)$, and $U(MX)$, each of which increases. Inasmuch as the trend in ΔH_f° is determined by the trend in $S(M) + I(M) - U(MX)$, whether or not ΔH_f° increases or decreases is determined by the relative rates of change of $S(M) + I(M)$ and $U(MX)$. Obviously the trend in ΔH_f° for the fluorides is established by the trend in $U(MF)$, which exceeds the trend in $S(M) + I(M)$, and the trend in ΔH_f° for the iodides is established by the trend in $S(M) + I(M)$, which exceeds the trend in $U(MI)$. The reason for the relatively small trend in $U(MI)$ is the fact that the interionic distance r is principally determined by the larger iodide ion. Hence the fractional change in r on going from CsI to LiI is relatively small, and the lattice energy does not increase markedly. On going from CsF to LiF, the fractional change in r is relatively great; consequently the lattice energy increases markedly.

APPLICATION OF THE ISOELECTRONIC PRINCIPLE

Compounds of elements which have similar electronegativities and polarizabilities often have the same type of structure if the average number of valence electrons per atom is the same. A large class of isostructural compounds, known as "Grimm-Sommerfeld" compounds, have diamondlike structures (wurtzite or zinc blende). These compounds have an average of four valence electrons per atom.

---

[1] T. E. Brackett and E. B. Brackett, *J. Phys. Chem.*, **69**, 3611 (1965).

Examples of such compounds containing group III–group V combinations are BN, AlP, GaAs, and InSb. Group II–group VI combinations are ZnSe and CdTe. Group I–group VII combinations are CuBr and AgI. Even some ternary compounds, such as $CuInTe_2$ and $ZnGeAs_2$, have diamondlike structures which can be rationalized on this basis.

In a binary compound of elements whose electronegativities are quite different, the electropositive (metallic) element generally transfers its electrons to the more electronegative (nonmetallic) element. If there are not enough electrons transferred to give each nonmetal atom a noble-gas configuration, the nonmetal anions may share electrons with each other in the same way as in the corresponding isoelectronic nonmetal. For example, in MgB_2, the B^- ions form planar sheets having the graphite structure. In LiGa and $CaGa_2$, the gallium atoms form an anionic lattice having the germanium structure. In $CaSi_2$, the silicon atoms form anionic puckered layers as in elemental arsenic. (Note, however, that in CaC_2 the carbon atoms form diatomic anions, $C_2{}^{2-}$, isoelectronic with N_2.) And in LiAs the arsenic atoms form anionic spiral chains as in elemental selenium.

PROBLEMS

10.1   From the heat of formation of the ammonium halides and other information, estimate the gas-phase proton affinity of ammonia.

10.2   From the heat of formation of BaO and other information, estimate the heat of the reaction $O(g) + 2e^-(g) \rightarrow O^{2-}(g)$. How reliable do you think your estimate is?

10.3   Estimate the heat of formation of $CaCl(s)$. Do you think $CaCl(s)$, if prepared, would be a stable compound? Explain.

10.4   In the compound MX_3, if all the M atoms are equivalent and all the X atoms are equivalent, what are the *possible* coordination numbers for each atom?

10.5   In solid $BeCl_2$, all the Be atoms and all the Cl atoms are equivalent. Each Be atom is coordinated tetrahedrally to four Cl atoms. Describe a plausible structure.

10.6   Calculate the minimum radius ratio for trigonal prismatic sixfold coordination, as in MoS_2 (see Fig. 10.12b).

10.7   Give the formula of a beryllium nitride that would be expected to have a diamondlike crystal structure.

10.8   The second ionization potential of Mg is approximately twice the first, and the conversion of O^- to O^{2-} is endothermic whereas the conversion of O to O^- is exothermic. Nevertheless we formulate MgO as $Mg^{2+}O^{2-}$ rather than as Mg^+O^-. Why? What simple experiment would show the latter formulation to be unrealistic?

10.9   Some salts can be prepared with either a CsCl structure or an NaCl structure. Which structure would you expect to be preferred at high pressures? Why?

10.10 The cation-to-anion distances for several compounds having the NaCl structure are listed below.

MgO	2.10 Å	MgS	2.60 Å	MgSe	2.73 Å
MnO	2.24 Å	MnS	2.59 Å	MnSe	2.73 Å

How can you explain these data? Calculate the radius of the S^{2-} ion from these data.

11
METALS

BONDING ENERGIES AND STRUCTURES

Metals have very characteristic physical properties, including the following: (1) high reflectivity (often called metallic luster); (2) high electrical conductivity, which decreases with increasing temperature; (3) high thermal conductivity; and (4) malleability and ductility. Many of a metal's properties, such as chemical reactivity, hardness, strength, melting point, and boiling point, can be correlated with the strength with which the atoms of the metal are held together. This bonding strength is most simply measured by the energy required to break up the metal into gaseous atoms, i.e., the atomization energy. As one might expect, metals with low atomization energies generally are soft and have low melting points, and metals with high atomization energies generally are hard and have high melting points. In Table 11.1 we have listed the heats of atomization and melting points of the metals in a periodic-table arrangement. In the case of the nontransition metals (groups I to III), the atomization energies increase from left to right. This trend continues for two or three positions into the transition series, in periods 3, 4, and 5. The trend is strong evidence that metallic bonding energy is directly related to the number of valence electrons. Sodium, with a single $3s$ electron, can form only one electron-pair bond per atom. This bond is equally distributed in fractional bonds to the neighboring atoms in the lattice, and we can crudely represent the metal as a resonance hybrid of structures such as the following:

```
Na   Na         —Na   Na—        Na⁻—Na
|    |                |
Na   Na         Na—Na           Na    Na⁺        etc.
```

The magnesium atom, with the $3s^2$ electron configuration, can be "prepared" for covalent bonding by promotion to the $3s3p$ configuration, with two unpaired electrons. Thus a magnesium atom can distribute two bonds to its neighbors in the lattice. The energetics of the situation may be illustrated as follows:

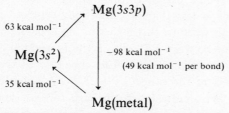

$$Mg(3s3p)$$

63 kcal mol^{-1}

$$Mg(3s^2)$$

-98 kcal mol^{-1}
(49 kcal mol^{-1} per bond)

35 kcal mol^{-1}

$$Mg(metal)$$

The aluminum atom, with the $3s^23p$ ground state, can be promoted to the $3s3p^2$ state, and it can form three bonds per atom in the metal lattice. Further increase in the number of bonds per atom occurs as we move from a group III metal

Table 11.1 HEATS OF ATOMIZATION† (kcal mol^{-1} AT 25°C, EXCEPT AT mp FOR Hg) AND MELTING POINTS (°C) OF THE METALS

Li	Be										
38.6	77.5										
180°	1283°										
Na	Mg	Al									
25.9	35	78									
97.5°	650°	660°									
K	Ca	Sc	Ti	V	Cr	Mn	Fe	Co	Ni	Cu	Zn
21.5	42.5	90	112	123	95	68	99.3	102.4	102.8	81.1	31.2
63.4°	850°	1539°	1725°	1730°	1900°	1247°	1535°	1493°	1455°	1083°	420°
Rb	Sr	Y	Zr	Nb	Mo	Tc	Ru	Rh	Pd	Ag	Cd
19.5	39	101.5	145.5	172	157	158	153	133	91	68	26.7
38.8°	770°	1509°	1852°	2487°	2610°		2400°	1960°	1550°	961°	321°
Cs	Ba	La	Hf	Ta	W	Re	Os	Ir	Pt	Au	Hg
18.7	42.5	103	148	187	203	187	188	160	135	88	15.3
28.7°	704°	920°	2300°	2997°	3380°	3150°	2700°	2454°	1769°	1063°	−38.9°

† L. Brewer, *Science*, **161**, 115 (1968).

such as scandium into the transition-metal series, where d electrons are involved in the bonding. However, because of the pairing of d electrons and the increased promotion energies, the number of bonds per atom eventually drops off. At this point we may state the first rule of Engel and Brewer:[1] *The bonding energy of a metal or alloy depends on the average number of unpaired electrons per atom available for bonding.* Low-lying excited electronic configurations with more unpaired electrons than the ground-state configuration may be important if the bonding energy from the additional electron-pair bonds compensates for the promotion energy.

The second rule of Engel and Brewer is concerned with the crystal structures of metals: *The crystal structure depends on the average number of s and p orbitals per atom involved in bonding,* i.e., upon the average number of unpaired s and p electrons in the atoms in their "prepared-for-bonding" state. When the number of bonding s, p electrons is less than 1.5, the body-centered cubic (bcc) structure is observed. We shall refer to this structure as type I. When the number of bonding s, p electrons is between 1.7 and 2.1, the hexagonal close-packed (hcp) structure is observed. We shall refer to this structure as type II. When the number of s, p electrons is in the range 2.5 to 3.2, the cubic close-packed (ccp) structure is observed. We shall refer to this structure as type III. Of course, when the number of s, p electrons is near 4, a nonmetallic diamond-like structure is found.

In Table 11.2 we have listed the observed crystal structures of the metals. As predicted by the rule, the alkali metals all have the type I (bcc) structure. However, of the alkaline-earth metals, only beryllium and magnesium rigorously obey the rule by having only the type II (hcp) structure. The type I structure shown by calcium, strontium, and barium can be explained by the general rule that the relative importance of d orbitals increases with atomic number. Thus, in these cases, excitation to an $(n-1)d\,ns$ state is energetically favored over excitation to an $ns\,np$ state. The same sort of explanation can account for the type I and type II structures shown by scandium, yttrium, and lanthanum, in contrast to the normal type III structure of aluminum. In fact, the type I structure shown by the first four members of each transition series can be explained by assuming that each atom is prepared for bonding by achieving a $d^{z-1}s$ valence electron configuration (where z is the number of valence electrons).

As we move immediately to the right of chromium, molybdenum, and tungsten, the added electrons are put into p orbitals to maintain the maximum possible amount of d-orbital bonding. However, with continued addition of electrons, d electrons are paired and eventually the bonding is due only to s, p

[1] L. Brewer, *Science,* **161,** 115 (1968); in P. A. Beck, (ed.), "Electronic Structure and Alloy Chemistry of the Transition Elements," pp. 221–235, Interscience, New York, 1963.

orbitals. In copper, silver, and gold, the "prepared" electronic state is d^8sp^2, corresponding to structure type III and five bonds per atom. When we reach zinc and cadmium, the d electrons are held too tightly to be easily promoted, and the "prepared" electronic state is $d^{10}sp$, corresponding to structure type II and two bonds per atom. It can be seen from Table 11.1 that the atomization energies are qualitatively in accord with these bonding descriptions.

The Engel-Brewer rules can be readily applied to alloys.[1] The metals of the first half of a transition series which have fewer than five d electrons in their prepared-for-bonding states can act as sinks for electrons until the d^5 configuration is achieved. According to the second rule, a transition-metal alloy will have a type I (bcc) structure if the average number of valence electrons is less than 6.5, corresponding to the prepared electron configuration $d^5s^1p^{0.5}$. On this basis we predict the maximum solubilities of metals such as Re, Os, Ir, and Pt in the bcc phase of tungsten to be that mole fraction corresponding to 6.5 valence electrons per atom. The predicted and observed mole fractions are given in Table 11.3. Although the agreement is far from perfect, it shows that qualitative and even semiquantitative predictions can be made by using the Engel-Brewer rules.

[1] L. Brewer, *Science*, **161**, 115 (1968); *Acta Metal.*, **15**, 553 (1967); L. Brewer and P. R. Wengert, *Metal. Trans.*, **4**, 83 (1973).

Table 11.2   CRYSTAL STRUCTURES OF METALS†
(I represents bcc; II hcp; III ccp. The structures are listed in order of temperature stability, with the room-temperature structure lowest.)

Li	Be										
I	II										
Na	Mg	Al									
I	II	III									
K	Ca	Sc	Ti	V	Cr	Mn	Fe	Co	Ni	Cu	Zn
I	I	I	I	I	I	I	I	III	III	III	II
	III	II	II			III	III	II			
						β	I				
						χ					
Rb	Sr	Y	Zr	Nb	Mo	Tc	Ru	Rh	Pd	Ag	Cd
I	I	I	I	I	I	II	II	III	III	III	II
	II	II	II								
	III										
Cs	Ba	La	Hf	Ta	W	Re	Os	Ir	Pt	Au	
I	I	I	I	I	I	II	II	III	III	III	
		III	II								
		II									

† L. Brewer, *Science*, **161**, 115 (1968).

Another, related, application of the rules is the prediction of the effect of small additions of alloying metals upon the relative stability of two crystal structures.[1] From Table 11.2 we see that Ti, Zr, and Hf can exist either in structure type I (corresponding to configuration d^3s) or structure type II (corresponding to configuration d^2sp). Addition of d-electron-rich metals to the right of these metals favors d-orbital bonding and thus stabilizes structure type I. Addition of metals containing no d electrons stabilizes structure type II because that structure will suffer less by a reduction in the amount of d-orbital bonding. All the experimental data available confirm these predictions of the effects of alloying on the relative stabilities of the bcc and hcp structures.

Brewer has shown that the alloying of d-electron-poor metals such as Zr, Nb, Ta, and Hf with d-electron-rich metals such as Re, Ru, Rh, Ir, Pt, and Au yields extremely stable alloys.[1] For example, if platinum is heated with ZrC (one of the most stable carbides known), the alloy $ZrPt_3$ and graphite are formed.

$$ZrC + 3Pt \xrightarrow{\Delta} ZrPt_3 + C$$

These interactions of "electron-deficient" and "electron-rich" transition metals may be considered an extension of Lewis acid-base reactions.

Hume-Rothery Compounds[2]

The importance of the average number of s, p valence electrons in the correlation of alloy composition with structure was recognized long ago by Hume-Rothery. The phase diagrams of a large class of binary alloys, for example, those of copper and silver with the zinc family of metals or the aluminum family of metals, show a wide variety of phases. Three of these phases, called Hume-Rothery compounds, have structures which may be associated with certain electron/atom

[1] Brewer and Wengert, op. cit.
[2] H. J. Emeléus and J. S. Anderson, "Modern Aspects of Inorganic Chemistry," 3d ed., pp. 507–510, Van Nostrand, New York, 1960.

Table 11.3 MAXIMUM SOLUBILITIES (MOLE FRACTIONS) OF SOME METALS IN THE BODY-CENTERED CUBIC PHASE OF TUNGSTEN

Metal	Predicted solubility	Observed solubility
Re	50	35–43
Os	25	10–20
Ir	16.7	10–15
Pt	12.5	4–10

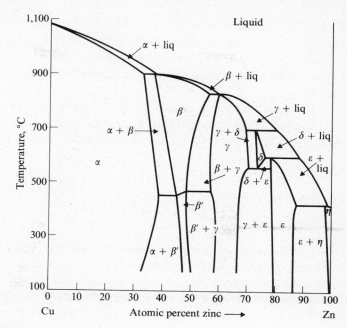

FIGURE 11.1
The Cu–Zn phase diagram. (*From L. V. Azároff, "Introduction to Solids," pp. 294–297, McGraw-Hill, New York, 1960.*)

ratios. The first phase, having the β-brass structure (a type I, bcc structure), corresponds to an s, p electron/atom ratio of 3:2. Examples of such phases are $CuZn$, $AgCd$, Cu_3Al, Cu_5Sn, and $NiAl$. (In these and the following examples, Cu and Ag must be considered as $d^{10}s$ metals, and Ni must be considered as a d^{10} metal.) The second phase, having the γ-brass structure (a complex cubic structure), corresponds to an s, p electron/atom ratio of 21:13. Examples are Cu_5Zn_8, Ag_5Hg_8, Cu_9Al_4, $Cu_{31}Sn_8$, and Ni_5Zn_{21}. The third phase, having the ε-brass structure (a type II, hcp structure), corresponds to an s, p electron/atom ratio of 7:4. Examples are $CuZn_3$, $AgCd_3$, and Cu_3Sn.

It can be seen that the β-brass type of alloys and the ε-brass type of alloys are in accord with the Engel-Brewer rule relating s, p electrons with structure. However, the reader may question the apparently ad hoc assignment of one s, p electron to Cu, Ag, and Au, and no s, p electrons to transition elements such as Ni, Pt, Co, and Fe. Each of these metals, in the pure state, has a prepared-for-bonding electron configuration of d^5sp^2, d^6sp^2, d^7sp^2, or d^8sp^2, corresponding to the type III structure. However, when any one of these metals is diluted sufficiently with a metal such as zinc (with the prepared-for-bonding configuration

$d^{10}sp$), the number of d orbitals available for bonding is reduced so much that there is no advantage in promotion to these excited states, and the metals then have, in effect, prepared-for-bonding electron configurations with more of the d orbitals filled. The limiting electron configurations for such metals are d^8, d^9, d^{10}, and $d^{10}s$.

To show the typical complexity of the phase diagrams of such systems, the Cu–Zn phase diagram is given in Fig. 11.1. As can be seen from this diagram, there are other phases besides the Hume-Rothery phases in this system.

BAND THEORY[1]

Electrical Conductivity

When two lithium atoms are brought together, the 1s atomic orbital levels are split into two levels, corresponding to the $2s\sigma$ bonding molecular orbital and the $2s\sigma^*$ antibonding molecular orbital, as shown in Fig. 11.2. If six lithium atoms were brought together to form a cluster, the atomic orbitals would be split into six MOs, ranging in character from completely bonding to completely anti-bonding, as shown in Fig. 11.3. When a very large number (say Avogadro's number) of lithium atoms are brought together to form a regular array, as in the normal bcc metal, the resulting energy levels are so closely spaced that they form an essentially continuous energy band, as shown in Fig. 11.4. The bottom of the band is a bonding level, and the top of the band is an anti-bonding level. The band is composed of as many levels as there are atoms, and each level can hold two electrons of opposite spin. Inasmuch as each lithium atom furnishes one valence electron, the band is exactly half-filled. Another way of depicting energy bands is in the form of a density-of-states diagram, as shown in Fig. 11.5. Here the number of electrons that can be accommodated in a narrow range of energy is plotted as a function of energy. Shading is used to indicate the filled portion of a band.

The individual levels of a band correspond to MOs which extend throughout the metal lattice. The valence electrons can be thought of as moving throughout the crystal as waves. Of course, in any isolated piece of metal, there are as many electrons moving in one direction as in another. For example, in a horizontal wire stretched from left to right, the number of electrons moving to the left is equal to the number of electrons moving to the right. This situation is depicted by the double density-of-states diagram of Fig. 11.6a, in which points

[1] C. A. Wert and R. M. Thomson, "Physics of Solids," McGraw-Hill, New York, 1964.

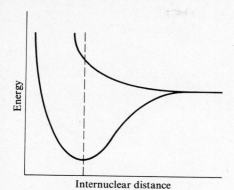

FIGURE 11.2
Interaction of the 2s orbitals of two lithium atoms. The dashed line indicates the internuclear distance in the Li_2 molecule.

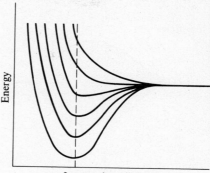

FIGURE 11.3
The molecular orbitals resulting from the interaction of the 2s orbitals of six lithium atoms. The relative positions of the six levels would depend on the arrangement of the atoms in the cluster.

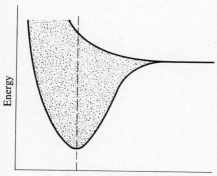

FIGURE 11.4
The valence band in lithium metal, shown as a function of the internuclear distance. The dashed line indicates the equilibrium internuclear distance in the actual metal.

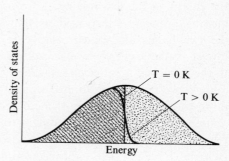

FIGURE 11.5
Density of states as a function of energy for the valence band of a metal such as lithium. The vertical line, which marks the boundary between the filled and empty parts of the band at 0 K, corresponds to the Fermi energy. At a finite temperature, this boundary is fuzzy, as shown.

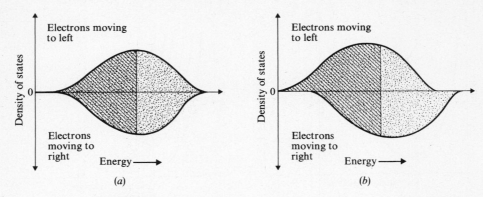

FIGURE 11.6
Density-of-states diagrams for electrons moving to the left (upper curves) and to the right (lower curves). (a) The diagram for a metal wire in the absence of an electric field. (b) The diagram for a metal wire subjected to an electric field, + at the left end, and − at the right end.

above the horizontal axis refer to left-moving electrons, and points below the horizontal axis refer to right-moving electrons. If we apply an electric field to the wire, so that the left side is positively charged relative to the right side, the energy of the left-moving electrons is reduced, and the energy of the right-moving electrons is increased, as shown by the diagram of Fig. 11.6b, in which the two bands are shifted relative to one another. The electrons fill up the lowest available energy levels, resulting in more electrons moving to the left than to the right and a net electric current.

When the temperature of a metal is increased, lattice irregularities become more pronounced because of increased atomic vibrations. These irregularities scatter the electrons, thus reducing the electrical conductivity. This effect of temperature on conductivity may also be explained in terms of MO theory. In a highly regular metal, the electron orbitals extend for great distances through the lattice, and the electrons are highly mobile because of the delocalization. The introduction of irregularities increases the amount of localized bonding and hence decreases the conductivity.

The electrical conductivity of a filled band is zero because, in such a band, the number of electrons moving in one direction must always be equal to the number moving in the opposite direction. For this reason one might expect an alkaline-earth metal such as calcium, which has a filled valence s shell, to be a nonconductor. However, at the interatomic distances found in this metal, the valence s and p bands overlap, as shown in Fig. 11.7. Consequently the density-of-states diagram is as shown in Fig. 11.8, and the metal exhibits typical metallic electrical conductivity.

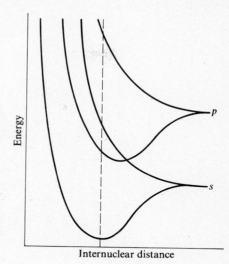

FIGURE 11.7
The valence band of an alkaline-earth metal such as calcium. The dashed line indicates the equilibrium internuclear distance in the actual metal.

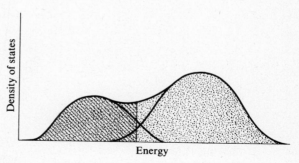

FIGURE 11.8
Density of states as a function of energy for an alkaline-earth metal such as calcium.

The density-of-states diagram for an insulator looks like that in Fig. 11.9, in which the high-energy empty band corresponds to nonbonding or anti-bonding levels. Sometimes it is possible to convert such a material into a metallic conductor by the application of high pressure (for example, 100 kbars). From Fig. 11.10 it can be seen that, if the interatomic distance is sufficiently reduced by increased pressure, the bands will overlap, yielding a density-of-states diagram similar to that of calcium (Fig. 11.8). In Table 11.4 the electrical resistivities of the elements of the first long row of the periodic table are listed. The metals with partially filled d bands generally have higher resistivities than other metals

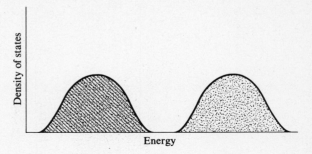

FIGURE 11.9
Density of states as a function of energy for an insulator.

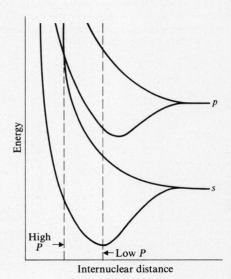

FIGURE 11.10
The valence bands of an insulator which can be converted into a metal by high pressure.

such as K, Ca, Cu, and Zn. This greater resistivity is probably due to the fact that the d orbitals are somewhat "inner" orbitals and do not overlap as effectively as s and p orbitals. Consequently the valence electrons are relatively "localized" in narrow bands and do not hop from one atom to another as rapidly as they would if they occupied broader s or p bands. The very high resistivities of Ge and Se are expected in view of the nonresonating covalent structures of these elements. However, the low resistivity of As is remarkable in view of its structure (Fig. 10.11a). Apparently the bonding valence band and a higher antibonding band are so broad that they overlap and permit metallic conductivity.

Ferromagnetism

Some substances have a permanent magnetic moment even in the absence of an applied magnetic field and are called "ferromagnetic." Only a few elements are ferromagnetic: Fe, Co, Ni, and several lanthanides. A density-of-states diagram for the 3d and 4s shells of Fe, Co, or Ni would look something like that shown in Fig. 11.11. Here the upper diagram corresponds to electrons with their spins oriented one way (say "up"), and the lower diagram corresponds to electrons with their spins oriented the opposite way (say "down"). The remarkable thing about these metals is that, at temperatures below the Curie temperature, the two bands are spontaneously and permanently displaced so that they are unequally filled. Thus more electrons are aligned one way than in the opposite way, and the metal has a permanent magnetic moment. The thermal energy of the crystal tends to misalign the electronic spins and to keep the density-of-states bands together. Obviously energy is required to displace the bands relative to one another as shown in Fig. 11.11. It is believed that the source of this energy is the exchange interaction between neighboring aligned spins. (See Chap. 1.) At temperatures above the Curie temperature (1043, 1404, and 631°K for Fe, Co, and Ni, respectively) the thermal energy of the crystal tending to misalign the spins exceeds the exchange energy, and the metal is no longer ferromagnetic.

Table 11.4 **ELECTRICAL RESISTIVITIES†
OF THE ELEMENTS OF THE
FIRST LONG ROW OF THE
PERIODIC TABLE (22°C)**

Element	Resistivity, $\mu\Omega$ cm
K	7.19
Ca	3.35
Sc	46.8
Ti	43.1
V	19.9
Cr	12.9
Mn	136
Fe	9.8
Co	5.80
Ni	7.04
Cu	1.70
Zn	5.92
Ga	14.8
Ge	$\sim 10^8$
As	29
Se	$\sim 10^{11}$

† D. E. Gray (ed.), "American Institute of Physics Handbook," 3d ed., McGraw-Hill, New York, 1972.

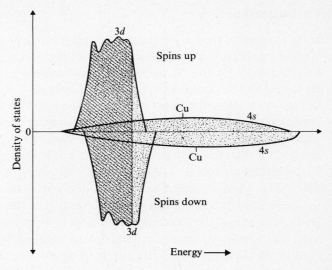

FIGURE 11.11
Density-of-states diagrams for a ferromagnetic transition metal. The upper bands correspond to electrons with spins "up"; the lower bands correspond to electrons with spins "down." The points marked Cu indicate the Fermi energy of copper, for which there is no relative displacement of the bands.

The metals which are ferromagnetic have high densities of states at the Fermi energy (that is, the energy corresponding to the highest filled level in the band). Thus many electrons change their spin direction (and yield exchange energy) for a small displacement of the bands. For ordinary metals, the density of states at the Fermi energy is lower and the energy required to displace the bands is not compensated by the exchange energy. For example, in copper metal the levels are filled to a point where the density of states is very low (see Fig. 11.11); consequently this metal is not ferromagnetic. The reason for the high density of states in Fe, Co, and Ni is the fact that the $3d$ atomic orbitals are somewhat interior orbitals which do not overlap strongly with those on other atoms in the lattice; thus the $3d$ bands for these metals are narrow.

Magnetic data for alloys of the metals in the vicinity of Fe, Co, and Ni in the periodic table show that the number of aligned spins per atom, N_B, is a continuous function of the average number of electrons per atom.[1] In Fig. 11.12 it can be seen that the maximum number of aligned spins per atom is 2.4, a value achieved by an Fe–Co alloy. For some unknown reason, when the constituent metals of the alloy differ by more than two electrons, the points deviate markedly from the curve.

[1] Ibid.

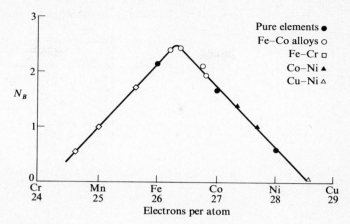

FIGURE 11.12
The number of aligned spins per atom as a function of the average number of electrons per atom. (*From C. A. Wert and R. M. Thomson, "Physics of Solids," McGraw-Hill, New York, 1964.*)

Metallic Compounds

There are numerous binary compounds of transition metals and relatively electronegative elements which show metallic character, at least by virtue of their high electrical conductivity with a negative temperature coefficient. These include some very hard, high-melting, and chemically inert borides, carbides, nitrides, oxides, and sulfides. These compounds may be looked upon as three-dimensional cationic metal lattices, held together by metallic bonding, with interspersed anions. In most cases the compounds are markedly variable in composition. For example, in the case of niobium nitride,[1] four different phases have been identified, having N/Nb ratios of 0.4 to 0.5, 0.8 to 0.9, 0.95, and 1.00. One of the highest-melting substances known is the reaction product of a 4:1 mixture of TaC and ZrC, which melts at 4215K. Niobium carbide, NbC, becomes superconducting at 10.1 K, a temperature at which no pure metal is super-conducting.

The tungsten bronzes are a group of nonstoichiometric bronzelike compounds of general formula M_xWO_3, where $0 < x < 1$ and where M can be an alkali metal, barium, lead, thallium, copper, silver, or a lanthanide.[2] Sodium tungsten bronzes, Na_xWO_3, $(0.32 < x < 0.93)$ can be made by the high-temperature reduction of sodium tungstate, Na_2WO_4, with a wide variety of

[1] Emeléus and Anderson, loc. cit.

[2] E. Banks and A. Wold, *Prep. Inorg. React.*, **4**, 237 (1968).

Table 11.5   PROPERTIES OF Na_xWO_3 CRYSTALS†

x	Color	Phase	Lattice parameter,‡ a_0, Å
0.37	Blue	Tetragonal	
0.40	Blue	Tetragonal	
0.44	Blue	Cubic	3.821
0.56	Violet	Cubic	3.831
0.58	Violet	Cubic	3.832
0.65	Red	Cubic	3.838
0.70	Orange	Cubic	3.842
0.81	Yellow	Cubic	3.851

† E. Banks and A. Wold, *Prep. Inorg. React.*, **4**, 237 (1968).
‡ The lattice parameter of a cubic crystal is the "repeat distance," i.e., the side dimension of the unit cell.

reducing agents. The structure may be considered to be a WO_3 network in which holes are randomly occupied by sodium atoms. Each sodium atom furnishes an electron to a band formed by overlap of the tungsten $5d$ orbitals with the oxygen atomic orbitals. Because this band is only partially filled, the compound exhibits metallic conductivity. Some of the properties of sodium tungsten bronzes are listed in Table 11.5.

Wold[1] has shown that it is possible to prepare materials of the type $WO_{3-x}F_x$ by replacing some of the oxygen atoms of WO_3 by fluorine atoms. These materials are analogous to tungsten bronzes; the extra electrons introduced by the fluorine atoms partially fill the valence band and cause the materials to be good conductors.

PROBLEMS

11.1   The compression of a transition metal by application of pressure inproves the overlap of the valence d orbitals and increases the bonding ability of the d orbitals relative to that of the s and p orbitals. Predict the effect of pressure on the relative stabilities of the type I and II structures of zirconium and the type I and III phases of iron.

11.2   Why would you expect an increase in temperature to stabilize a bcc structure relative to an hcp structure instead of vice versa?

11.3   How would you expect the electrical conductivity of a solid-solution alloy (e.g., an alloy of Cu and Au) to vary with composition?

11.4   What are the structure types (bcc, hcp, ccp, or diamondlike) of the following alloys: CuBe, Ag_5Cd, AgMg, Ag_5In_3, Cu_3Ge, FeAl, and InSb?

[1] A. Wold, unpublished work.

11.5 Give the empirical formulas (or approximate composition ranges) that might be expected for five solid phases in the Ag–Al phase diagram. For each phase, predict the crystal structure type.

11.6 Arrange the following materials in order of hardness: Cs, Na, C(diamond), W, Al, Fe, Mg.

11.7 Can a tantalum crucible be used in the melting of (a) Sr, (b) Re, (c) Pt?

11.8 Why are none of the metals of the second and third transition series ferromagnetic?

11.9 The densities of the lanthanide metals change gradually from 6.19 g cm^{-3} for La to 9.84 for Lu, except for Eu and Yb, which have low densities of 5.24 and 6.98, respectively. Explain.

12
SEMICONDUCTORS[1, 2]

A semiconductor is a crystal with a narrow energy gap between a filled valence band and a conduction band. If the crystal were cooled to absolute zero, the conduction band would be empty and the material would be a perfect non-conductor. However, at ordinary temperatures some electrons are thermally excited from the valence band to the conduction band, enough to give the material an electrical conductivity between that of a metal and that of an insulator. (Typical room-temperature *resistivities* of metals, semiconductors, and insulators are 5, 5×10^7, and 10^{23} $\mu\Omega$ cm, respectively.) The conductivity of a semiconductor is proportional to the number of electrons in the conduction band, which in turn is proportional to the Boltzmann factor $e^{-E_g/RT}$, where E_g is the energy gap. Consequently the conductivity of a semiconductor increases exponentially with increasing temperature.

The semiconducting material of greatest commercial importance is elemental silicon. If one were able to prepare a perfectly pure crystal of silicon, the small number of electrons in the conduction band at any given temperature would equal the number of positive "holes" in the valence band. Both conduction

[1] C. A. Wert and R. M. Thomson, "Physics of Solids," McGraw-Hill, New York, 1964.
[2] L. V. Azaroff, "Introduction to Solids," McGraw-Hill, New York, 1960.

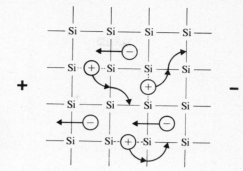

FIGURE 12.1
Schematic drawing of silicon, showing the negative-electron and positive-hole current carriers which give rise to a small intrinsic conductivity in the pure element.

electrons and holes contribute to electrical conductivity. The electrons travel through the interstices of the lattice, and the holes jump from one bond to another, as crudely illustrated in Fig. 12.1. The countercurrent motion of conduction electrons and valence holes in an applied field constitutes the *intrinsic* semiconductivity of the crystal. Generally even highly purified silicon contains enough impurity atoms to increase the semiconductivity far above the hypothetical intrinsic semiconductivity. For example, let us suppose that a minute fraction of the silicon atoms are randomly replaced by atoms containing more than four valence electrons each, such as arsenic atoms. Because the crystal must remain essentially electrically neutral, each arsenic atom contributes an electron which can enter the conduction band. The principal conducting species in the sample of silicon would be the negative electrons, and such a material is called an *n*-type (for *negative*) semiconductor. A schematic illustration of this type of semiconductor and the corresponding density-of-states diagram are shown in Fig. 12.2.

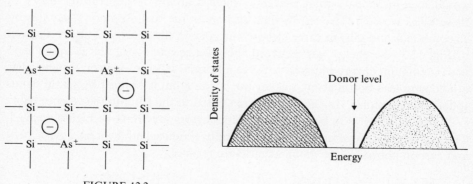

FIGURE 12.2
Schematic drawing of n-type silicon, and the corresponding density-of-states diagram.

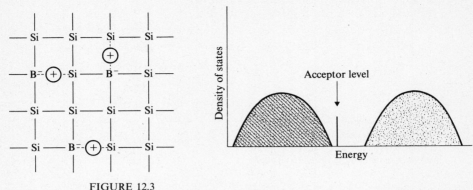

FIGURE 12.3
Schematic drawing of p-type silicon, and the corresponding density-of-states diagram.

At absolute zero, the extra electrons introduced by the arsenic atoms would be localized around the arsenic atoms (which have $+1$ formal charges), but at ordinary temperatures, many of them would be excited to the conduction band. The energy gap between the donor level and the conduction band is the energy required to break an interstitial electron completely away from an arsenic atom.

Now suppose that a very small fraction of the atoms in a crystal of silicon were replaced by atoms containing *fewer* than four valence electrons each, such as boron atoms. The principal conducting species in the material would be positive valence holes, as shown in Fig. 12.3, and such a material is called a *p*-type (for *positive*) semiconductor. At absolute zero, the positive holes introduced by the impurity boron atoms would be localized around the boron atoms (which would have negative formal charges), whereas at finite temperatures many of these holes would be excited so that they were free to hop among the silicon-silicon bonds and thus to carry electric current. In the density-of-states diagram of Fig. 12.3, the energy gap between the valence band and the acceptor level is the minimum energy required to force a normal valence electron to neutralize a hole next to a boron atom, so that the boron atom is bonded by single bonds to four silicon atoms. In effect, holes are excited to the left into the valence band.

Silicon which is to be used in semiconductors must first be highly purified. Crude silicon (~ 98 percent pure) is generally produced by heating quartz rock and carbon above $1800°C$ in an electric-arc furnace.[1]

$$SiO_2 + C \xrightarrow{\Delta} CO_2 + Si$$

[1] *Chem. Eng. News*, Nov. 30, 1970, pp. 46–47.

This silicon has about 20 million times more impurity than can be tolerated. It can be purified by converting it to trichlorosilane with hydrogen chloride, carefully distilling the trichlorosilane, and then decomposing it to give purified silicon. The chemical reactions involved are

$$3HCl + Si \underset{\substack{\text{higher} \\ \text{temp.}}}{\overset{350\ C}{\rightleftharpoons}} SiHCl_3 + H_2$$

If necessary, this silicon can then be further purified by zone refining. In this process a rod of the element is melted near one end, and, by moving the furnace, the short melted zone is moved slowly to the other end of the rod. Because impurities are more soluble in the melt than in the solid, they concentrate in the melt and are carried to one end of the rod. After the process is repeated several times, the impure end is removed.

Sometimes semiconductors can be purified by gas-phase thermal transport processes.[1] For example, germanium can be purified by placing it in one end of an evacuated tube with a small amount of iodine and then heating the tube so that the end containing the impure germanium is at 500°C and the other end is at 350°C. The iodine reacts with the germanium to form vapors of GeI_4 and GeI_2. Soon a steady state is reached in which the following reaction takes place continuously, as written, in the high-temperature zone, and the reverse reaction takes place in the initially empty, low-temperature zone.

$$Ge(s) + GeI_4(g) \underset{350°C}{\overset{500°C}{\rightleftharpoons}} 2GeI_2(g)$$

Thus germanium is transported from one end of the tube to the other, leaving impurities behind.

Purified silicon and germanium crystals can be converted into n-type or p-type semiconductors by high-temperature diffusion of the appropriate impurity elements ("dopants") into the crystals. Arsenic and phosphorus are often used for n-type semiconductors, and boron and aluminum are often used for p-type semiconductors.

BAND GAPS

The band gap of a semiconductor is the energy difference between the top of the valence band and the bottom of the conduction band. Values of the band gap in various substances are listed in Table 12.1.

The binding energies associated with impurities are much smaller than the band gaps. For example, the energy required to excite an electron away from a

---

[1] H. Schäfer, "Chemical Transport Reactions," Academic, New York, 1963.

Table 12.1　BAND GAPS OF SOME SEMICONDUCTORS

Compound	E_g, eV (0 K)	Compound	E_g, eV (0 K)
α-Sn	0.0	CdTe	1.6
InSb	0.2	Se	1.8†
PbTe	0.2	Cu_2O	2.2
Te	0.3	InN	2.4
PbS	0.3	CdS	2.6
InAs	0.4	ZnSe	2.8
ZnSb	0.6	GaP	2.88
Ge	0.7	ZnO	3.4
GaSb	0.8	$SrTlO_3$	3.4
Si	1.1	ZnS	3.9
InP	1.3	AlN	4.6
GaAs	1.5	Diamond	5.4
		BP	6.0

† 300 K.

phosphorus atom impurity in silicon to the conduction band is about 0.044 eV. Similarly, the energy required to excite a hole from an aluminum atom impurity in silicon to the valence band is about 0.057 eV. These excitation energies may be compared with the band gap, 1.1 eV.

DEFECT SEMICONDUCTORS[1]

Many compounds are semiconductors because they are nonstoichiometric. For example, when compounds such as NaCl, KCl, LiH, and δ-TiO are subjected to high-energy radiation or are heated with excess of their constituent metals, the compounds become deficient in the electronegative elements and their compositions may be represented by the general formula MY_{1-x}, where x is a small fraction. The crystal lattice of such a compound has anion vacancies, each of which is usually occupied by an electron. Such electron-occupied holes are called F centers. The electron of an F center can be thermally excited to a conduction band, thus giving rise to n-type semiconduction.

Another class of n-type semiconductor are those compounds which contain excess, interstitial, metal atoms and whose compositions correspond to the formula $M_{1+x}Y$. The compounds ZnO, CdO, Cr_2O_3, and Fe_2O_3 show this type of structural defect. The interstitial metal atoms are readily ionized, allowing their valence electrons to enter a conduction band and leaving interstitial metal ions.

[1] H. J. Eméleus and J. S. Anderson, "Modern Aspects of Inorganic Chemistry," 3d ed., Van Nostrand, New York, 1960.

When a defect oxide of this type is heated in oxygen, its room-temperature conductivity decreases because of the loss of some of the interstitial metal atoms by oxidation.

Compounds which are deficient in metal ions, of general formula $M_{1-x}Y$, are p-type semiconductors. This type of semiconductor can be found in Cu_2O, FeO, NiO, δ-TiO, CuI, and FeS. Electrical neutrality is maintained in these compounds, in spite of cation vacancies, by the presence of metal ions of higher oxidation state. Thus the phase $Fe_{0.95}O$ is more precisely represented by the formula $Fe^{II}_{0.85}Fe^{III}_{0.10}O$. Electrical conductivity is achieved by the hopping of valence electrons from lower-oxidation-state metal ions to higher-oxidation-state metal ions. However, because the migrating unit is essentially a positive hole in a metal ion, the conductivity is of p type. The energy gap for this type of semiconductor corresponds to the energy required to move a positive hole away from the vicinity of a cation vacancy, where it is electrostatically held. When a defect oxide of this type is heated in oxygen, its room-temperature conductivity increases because of oxidation of some of the metal ions and the consequent increase in positive-hole concentration.

CONTROLLED-VALENCE SEMICONDUCTORS[1]

Defect semiconductors are generally difficult to obtain with large deviations from stoichiometry. Hence only limited variations in electrical properties are possible. In addition, compositions are difficult to reproduce exactly. These difficulties can be overcome by the use of "controlled-valence" semiconductors, first prepared by Verwey and his coworkers.

For example, consider the material formed by heating an intimate mixture of NiO and a small amount of Li_2O in air at 1200°C. Oxygen is absorbed, and a single phase of composition $Li_x Ni_{1-x}O$ is formed:

$$\frac{x}{2}Li_2O + (1-x)NiO + \frac{x}{4}O_2 \rightarrow Li_x Ni_{1-x}O$$

Inasmuch as Li^+ and Ni^{2+} ions have similar radii, Li^+ ions can substitute for Ni^{2+} ions in the lattice. One Ni^{3+} ion is formed for every Li^+ ion introduced to preserve electrical neutrality. The resulting compound can be represented more precisely by the expanded formula $Li_x Ni^{III}_x Ni^{II}_{1-2x}O$. The positive holes in the Ni^{2+} ions (that is, the Ni^{3+} ions) can move from one nickel ion to another and thus give rise to p-type semiconduction. By simply controlling the ratio of lithium to nickel, the electrical conductivity can be varied at will. The

[1] W. D. Johnston, *J. Chem. Educ.*, **36**, 605 (1959); P. E. Synder, *Chem. Eng. News*, Mar. 13, 1961, pp. 102ff.

conductivity of pure NiO is about $10^{-10} \ \Omega^{-1} \ cm^{-1}$, whereas the compound in which 10 percent of the nickel atoms have been replaced by lithium atoms has a conductivity of about $1 \ \Omega^{-1} \ cm^{-1}$.

Mixed valence systems can also be obtained with compounds having the perovskite structure such as barium titanate, $BaTiO_3$. (This compound consists of a ccp lattice of Ba^{2+} and O^{2-} ions in which the octahedral holes surrounded exclusively by O^{2-} ions are occupied by Ti^{4+} ions.) If a small number of the barium ions are replaced by $+3$ ions such as lanthanum ions, a corresponding number of Ti^{4+} ions must be reduced to Ti^{3+}, forming the system $La_x Ba_{1-x} Ti^{III}_x Ti^{IV}_{1-x} O_3$. This material exhibits n-type semiconductivity.

Compounds of the type $Li_x Mn_{1-x} O$ are semiconducting, but they cannot be made by heating a mixture of Li_2O and MnO in air, because MnO is easily oxidized in air at high temperatures. In this case, the reaction is carried out in a sealed container, and the oxygen is introduced in the form of lithium peroxide:

$$\frac{x}{2} Li_2O_2 + (1-x)MnO \rightarrow Li_x Mn_{1-x} O$$

APPLICATIONS

Semiconductors have innumerable practical applications which contribute to our high standard of living. We shall describe just three applications: the photovoltaic cell, the rectifier, and the insulated-gate field-effect transistor. Each of these devices involves a semiconductor crystal which is p-type in one part and n-type in another. The boundary region is called a p-n junction and has interesting electrical properties. Such a crystal can be made, for example, by allowing small amounts of boron to diffuse into one side of a slice of n-type silicon at high temperatures. If properly doped, the crystal will then be p-type on one side, n-type on the other, with a region within the slice where the two types of silicon meet: the p-n junction.

The Photovoltaic Cell[1]

Let us suppose that a p-n junction is irradiated with light. If the photon energy exceeds the band gap, electron-hole pairs will be formed in the irradiated region and the n-type region will achieve a negative potential relative to the p-type region. If the n-type and p-type regions are electrically connected to an external circuit, a current will flow. Obviously such a crystal can be used as a kind of battery

---

[1] Wert and Thomson, op. cit.

14

THE STEREOCHEMISTRY OF COORDINATION COMPOUNDS

The term "coordination compound" is usually applied to any compound containing molecules or ions in which metal atoms are surrounded by other atoms or groups. Thus both $CH_3Mn(CO)_5$ and $K_3[Fe(CN)_6]$ are considered coordination compounds. Metal-containing molecules and ions are often called "complexes" or, in the case of ions, "complex ions." In fact, the term "complex ion" is even applied to ions in which the central atom is a nonmetal, such as BCl_4^- and PF_6^-. However, nonmetallic species such as $SiBr_4$ and SF_6 are usually simply referred to as molecules rather than as coordination compounds or complexes.

COORDINATION GEOMETRY[1]

In the following paragraphs we shall discuss the commonly observed geometrical arrangements of ligand donor atoms in complexes. For convenience, we shall classify these structures in terms of the coordination numbers of the metal atoms.

[1] The topic is discussed in J. E. Huheey, "Inorganic Chemistry: Principles of Structure and Reactivity," pp. 371–406, Harper & Row, New York, 1972, and in F. A. Cotton and G. Wilkinson, "Advanced Inorganic Chemistry," 3d ed., pp. 22–30, Interscience, New York, 1972.

Coordination Number 2

Although this coordination number is found in high-temperature, gas-phase species such as $MgCl_2(g)$, it is rather unusual for ordinary, stable complexes. It is found principally in some of the complexes of the ions Cu^+, Ag^+, Au^+, and Hg^{2+}, each of which has a ground-state d^{10} electronic configuration. Typical examples of such complexes are $Cu(NH_3)_2{}^+$, $AgCl_2{}^-$, $Au(CN)_2{}^-$, and $HgCl_2$. All such complexes are linear; that is, the ligand-metal-ligand bond angle is 180°. As a rough approximation, we may describe the bonding as a consequence of the overlap of σ orbitals of the ligands with sp hybrid orbitals of the metal atom. However, a metal d orbital is probably involved in the bonding to some extent.[1] Let us assume that the bonds lie on the z axis of the metal atom and that a small amount of the nonbonding electron density of the metal d_{z^2} orbital is shifted to the metal s orbital. In this case the metal orbitals used in bonding would not be simple sp_z hybrid orbitals but rather hybrid orbitals with 50 percent p_z character, a small amount of d_{z^2} character, and the remainder s character. A pair of electrons would occupy a hybrid nonbonding metal orbital, principally of d_{z^2} character with a little s character. This type of hybridization would shift some nonbonding electron density away from the regions between the metal and the ligands and thus would be energetically favored.

Complexes of this type can often be converted, by treatment with excess of the ligand species, to 4-coordinate complexes. For example, in an aqueous system containing Hg^{2+} and Cl^-, the relative amounts of the species $HgCl^+$, $HgCl_2$, $HgCl_3{}^-$, and $HgCl_4{}^{2-}$ are dependent on the concentration of free chloride ion. The equilibrium constants for the formation of these aqueous complexes are as follows:[2]

$$Hg^{2+} + Cl^- \rightleftharpoons HgCl^+ \qquad K_1 = 5.5 \times 10^6$$

$$HgCl^+ + Cl^- \rightleftharpoons HgCl_2 \qquad K_2 = 3.0 \times 10^6$$

$$HgCl_2 + Cl^- \rightleftharpoons HgCl_3{}^- \qquad K_3 = 8.9$$

$$HgCl_3{}^- + Cl^- \rightleftharpoons HgCl_4{}^{2-} \qquad K_4 = 11.2$$

From these data it can be seen that $HgCl_2$ is very stable with respect to disproportionation to $HgCl^+$ and $HgCl_3{}^-$:

$$\frac{[HgCl^+][HgCl_3{}^-]}{[HgCl_2]^2} = \frac{K_3}{K_2} = 3.0 \times 10^{-6}$$

---

[1] L. E. Orgel, "An Introduction to Transition-metal Chemistry: Ligand-field Theory," 2d ed., pp. 69–71, Wiley, New York, 1966.

[2] The complexation constants are for solutions of ionic strength 0.5 M at 25°C; data from L. G. Sillén, Stability Constants of Metal-ion Complexes, *Chem. Soc.* (*London*), *Spec. Publ.* 17, *sec.* I, 1964.

Coordination Number 3

This coordination number is rare among metal complexes. Some well-established examples are $KCu(CN)_2$ (see page 192), $[(CH_3)_3S^+][HgI_3^-]$, $Cr\{N[Si(CH_3)_3]_2\}_3$, $Fe\{N[Si(CH_3)_3]_2\}_3$, $\{Cu[SC(NH_2)_2]_3\}Cl$, $[Cu(SPPh_3)_3]ClO_4$, and $Pt(PPh_3)_3$. In all these cases the metal atom and the three directly coordinated ligand atoms are coplanar. Most compounds of stoichiometry MX_3 have structures in which the coordination number of M is greater than 3. For example, $CrCl_3$ has an infinite layer lattice in which each Cr atom is coordinated to six Cl atoms (see Fig. 10.12c). In $CsCuCl_3$ each Cu atom is coordinated to four Cl atoms in infinite anionic chains: $-Cl-CuCl_2-Cl-CuCl_2$. And $AuCl_3$ really exists as planar Au_2Cl_6 molecules in which each Au atom is bonded to two bridging Cl atoms and two terminal Cl atoms.

Coordination Number 4

This is a very important coordination number, for which two configurations are commonly observed, *tetrahedral* and *square planar*. Four-coordinate nontransition-element complexes, such as $BeCl_4^{2-}$, BF_4^-, $ZnCl_4^{2-}$, $SnCl_4$, and AlF_4^-, are almost always tetrahedral. However, transition metals from both tetrahedral and square planar complexes. Tetrahedral complexes are generally formed whenever four ligands are coordinated to an ion or atom which does *not* have a d^8 electronic configuration. Thus species such as $Co(CO)_4^-$, $Ni(CO)_4$, $HgCl_4^{2-}$, $FeCl_4^-$, $CoCl_4^{2-}$, MnO_4^-, VO_4^{3-}, and FeO_4^{2-} are tetrahedral. Square planar complexes are characteristic of 4-coordinate metal atoms or ions with d^8 configurations. Examples are $[Rh(CO)_2Cl]_2$, $IrCl(CO)(PEt_3)_2$, $Ni(CN)_4^{2-}$, $PdCl_4^{2-}$, $Pt(NH_2CH_2CH_2NH_2)_2^{2+}$, AgF_4^-, and Au_2Cl_6. Tetrahedral coordination is sometimes found for d^8 metals when the metal atom is small or when the ligand atoms are too bulky to permit square planar configuration. Thus $CoBr(PPh_3)_3$ and $NiBr_4^{2-}$ are tetrahedral.

Coordination Number 5

The structures of 5-coordinate complexes lie between two limiting geometries: trigonal bipyramidal and square pyramidal. These limiting structures are not markedly different, as can be seen by examination of Fig. 14.1. The conversion of one structure into the other requires a relatively slight distortion. In Fig. 14.2, seven different 5-coordinate structures, including two nonmetallic compounds, are illustrated to show the gradual transition from $CdCl_5^{3-}$ (which has almost the ideal trigonal bipyramid structure) to $Ni(CN)_5^{3-}$ (which, in one of its forms,

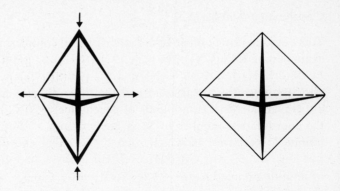

FIGURE 14.1
The distortion of a trigonal bipyramid to a square pyramid. [*Reproduced with permission from E. L. Muetterties and L. J. Guggenberger, J. Amer. Chem. Soc.,* **96,** 1748 (1974). *Copyright © by the American Chemical Society.*]

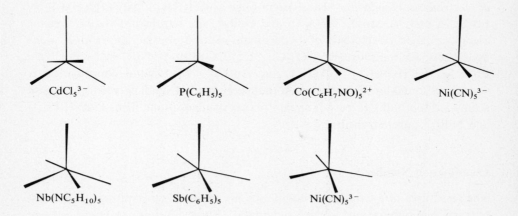

CdCl$_5^{3-}$ P(C$_6$H$_5$)$_5$ Co(C$_6$H$_7$NO)$_5^{2+}$ Ni(CN)$_5^{3-}$

Nb(NC$_5$H$_{10}$)$_5$ Sb(C$_6$H$_5$)$_5$ Ni(CN)$_5^{3-}$

FIGURE 14.2
The 5-coordinate structures found in seven different species. The structures range from almost ideal trigonal bipyramidal to almost ideal square pyramidal. [*Reproduced with permission from E. L. Muetterties and L. J. Guggenberger, J. Amer. Chem. Soc.,* **96,** 1748 (1974). *Copyright © by the American Chemical Society.*]

has almost the ideal square pyramid structure). Several of the intermediate structures are almost equally well described as distorted trigonal bipyramids or as distorted square pyramids. To eliminate such ambiguity, Muetterties and Guggenberger[1] proposed the specification of dihedral angles of the polyhedron as an objective way to describe the shape of a 5-coordinate compound. Dramatic proof of the fact that the various conformations from trigonal bipyramidal to square pyramidal have comparable energies is provided by the compound $[Cr(NH_2CH_2CH_2NH_2)_3][Ni(CN)_5]\cdot 1.5H_2O$, which contains two types of $Ni(CN)_5^{3-}$ ions: One has the geometry of a regular square pyramid, and the other has a geometry practically midway between that of a trigonal bipyramid and a square pyramid.[2]

Caution must be used when interpreting stoichiometric data as evidence for coordination number 5. Thus Cs_3CoCl_5 contains the anions $CoCl_4^{2-}$ and Cl^-, Tl_2AlF_5 contains infinite chain anions of the type $-F-AlF_4-F-AlF_4-$, and $Co(NH_2CH_2CH_2NHCH_2CH_2NH_2)Cl_2$ contains the 6-coordinate ions $Co(NH_2CH_2CH_2NHCH_2CH_2NH_2)_2^{2+}$ and the 4-coordinate ions $CoCl_4^{2-}$.

Coordination Number 6

This is the commonest and most important coordination number for transition-metal complexes. The geometry usually corresponds to six coordinated atoms at the corners of an octahedron or a distorted octahedron. The regular octahedron, shown in the following sketch,

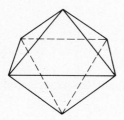

is a highly symmetric polyhedron. The reader can probably recognize the existence of six twofold rotational axes, four threefold rotational axes, three fourfold rotational axes, three mirror planes of one type, six of another type, and a center of symmetry. (See Appendix C for a discussion of symmetry elements.) One common type of distortion of the octahedron, *tetragonal* distortion,

[1] E. L. Muetterties and L. J. Guggenberger, *J. Am. Chem. Soc.,* **96**, 1748 (1974).
[2] K. N. Raymond, P. W. R. Corfield, and J. A. Ibers, *Inorg. Chem.,* **7**, 1362 (1968).

involves either elongation or contraction of the octahedron along one of the fourfold symmetry axes, as shown below.

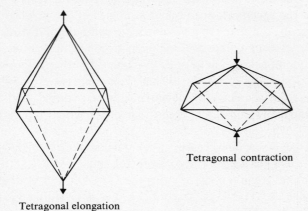

Tetragonal contraction

Tetragonal elongation

Another type of distortion, *trigonal* distortion, involves either elongation or contraction along one of the threefold symmetry axes, as shown below, to form trigonal antiprisms.

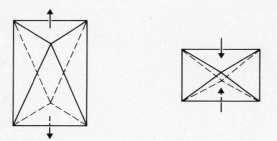

Very rarely one finds a 6-coordinate complex with trigonal prismatic geometry:

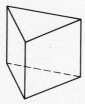

This geometry is probably unusual because repulsions between the coordinated atoms are greater than in the antiprismatic geometry which would be obtained if one rotated one triangular face 60° relative to the other. Trigonal prismatic

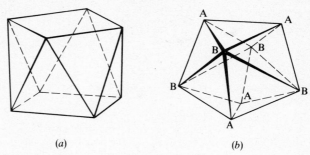

(a)   (b)

FIGURE 14.3

The square antiprism (a) and the dodecahedron (b). Note that whereas all the vertices of the square antiprism are equivalent, the dodecahedron contains two types of vertices (A and B).

coordination is found in a series of complexes of rhenium, molybdenum, tungsten, and vanadium with dithiolate ligands of the type

$$\begin{array}{c} R \diagdown \quad \diagup S^- \\ C \\ \| \\ C \\ R \diagup \quad \diagdown S^- \end{array}$$

Coordination Number 7

This relatively uncommon coordination number has been found with three different geometries: the pentagonal bipyramid (as in $UO_2F_5^{3-}$), the monocapped octahedron (as in $NbOF_6^{3-}$), and the trigonal prism with one of its rectangular faces capped (as in TaF_7^{2-}).

Coordination Number 8

Cubic coordination is almost never found. However, 8-coordinate complexes often have the geometry of a square antiprism, which is obtained by rotating one face of a cube 45° relative to the opposite face. Another commonly found geometry is that of the dodecahedron, which may be looked upon as a combination of an elongated tetrahedron and a flattened tetrahedron. The square antiprism and the dodecahedron are illustrated in Fig. 14.3. These configurations have very similar energies, and, in the case of octacyanocomplexes such as $Mo(CN)_8^{3-\ or\ 4-}$ and $W(CN)_8^{3-\ or\ 4-}$, either geometry can be obtained in solids by appropriate choice of the cation. For example, the $Mo(CN)_8^{3-}$ ion is square antiprismatic in $Na_3Mo(CN)_8 \cdot 4H_2O$ and dodecahedral in $[N(n\text{-}C_4H_9)_4]_3Mo(CN)_8$.

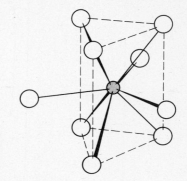

FIGURE 14.4
The geometry of a 9-coordinate complex
such as $ReHg^{2-}$. The ligands are at the
vertices of a trigonal prism which has been
capped on each of its rectangular faces.

Higher Coordination Numbers

Nine-coordinate complexes such as $Nd(H_2O)_9^{3+}$ and ReH_9^{2-} have geometries corresponding to a tricapped trigonal prism, as illustrated in Fig. 14.4. Coordination numbers 10, 11, and 12 are occasionally found in complexes of lanthanide and actinide elements but, in view of their rarity, are not important.

LIGANDS

The ligands of a complex may be classified by the way in which they are coordinated to the central metal atom. A *monodentate* ligand is one which is attached to the metal atom by a bond from only one atom (the donor atom) of the ligand. Ligands such as F^-, Cl^-, O^{2-}, $\overset{*}{P}R_3$, $H_2\overset{*}{O}$, $\overset{*}{C}H_3^-$, $\overset{*}{O}R^-$, and $\overset{*}{C}O$ are monodentate. (The asterisks indicate the donor atoms of the polyatomic ligands.) Frequently such monodentate ligands act as bridges between two or more metal atoms. For example, the complexes $Fe_2(CO)_9$, Au_2Cl_6, and $Cr_2(OH)_2(H_2O)_8^{4+}$ involve the bridging ligands CO, Cl^-, and OH^-, respectively. The structures of these complexes are shown in Fig. 14.5. A *polydentate* ligand (bidentate, tridentate, etc.) is one which can be attached to the metal atom by bonds from two or more donor atoms. Examples of bidentate ligands are ethylenediamine, $NH_2CH_2CH_2NH_2$ (which can coordinate through two nitrogen atoms and is often represented by the symbol en), and *o*-phenylenebis-(dimethylarsine),

As(CH₃)₂

As(CH₃)₂

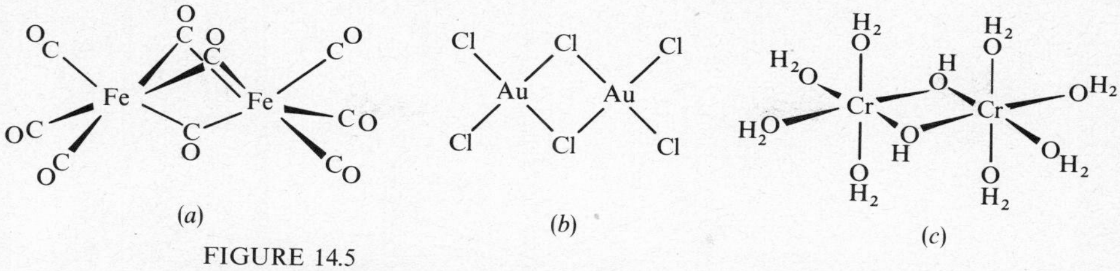

FIGURE 14.5
Some complexes with bridging monodentate ligands. (a) $Fe_2(CO)_9$, (b) Au_2Cl_6, (c) $Cr_2(OH)_2(H_2O)_8^{4+}$.

(which can coordinate through two arsenic atoms and is often represented by the symbol diars). A number of ligands can be either monodentate or bidentate; these include SO_4^{2-}, CO_3^{2-}, and NO_3^-, which can coordinate through one or two oxygen atoms. Examples of tridentate and tetradentate ligands are diethylene-triamine, $NH_2CH_2CH_2NHCH_2CH_2NH_2$ (dien), and nitrilotriacetate, $N(CH_2CO_2)_3^{3-}$ (NTA), respectively. The complex formed by the coordination of a polydentate ligand to a metal atom is called a "chelate complex"; several such complexes are illustrated in Fig. 14.6. Acetylacetone can be readily deprotonated to form a mononegative anion ($acac^-$) which is resonance-stabilized:

$$CH_3-\overset{\overset{\displaystyle O}{\|}}{C}-CH=\overset{\overset{\displaystyle O^-}{|}}{C}-CH_3 \leftrightarrow CH_3-\overset{\overset{\displaystyle O^-}{|}}{C}=CH-\overset{\overset{\displaystyle O}{\|}}{C}-CH_3$$

Very stable chelate complexes can be formed by this ligand:

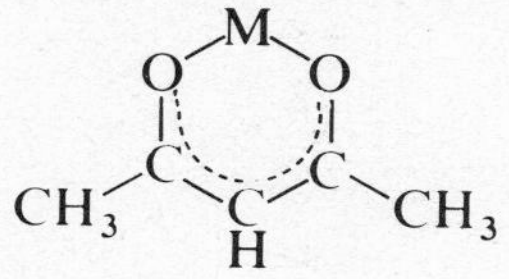

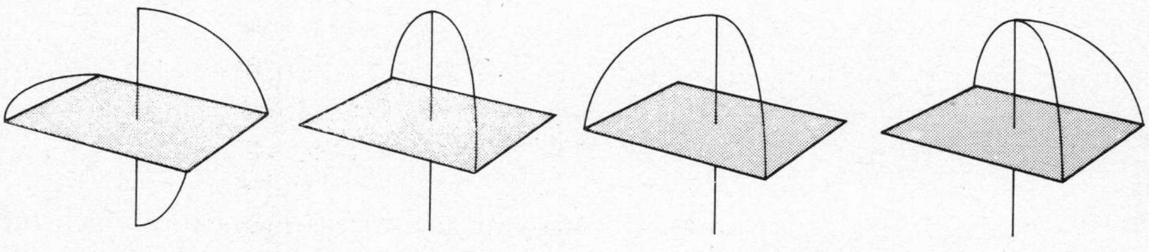

FIGURE 14.6
Schematic representations of chelates. (a) A complex with three bidentate ligands, such as $Co(en)_3^{3+}$; (b) meridional chelation of a tridentate ligand; (c) facial chelation of a tridentate ligand; (d) the chelation of a tetradentate ligand such as NTA. Notice that, in all these cases, adjacent donor atoms of a ligand are coordinated cis to one another.

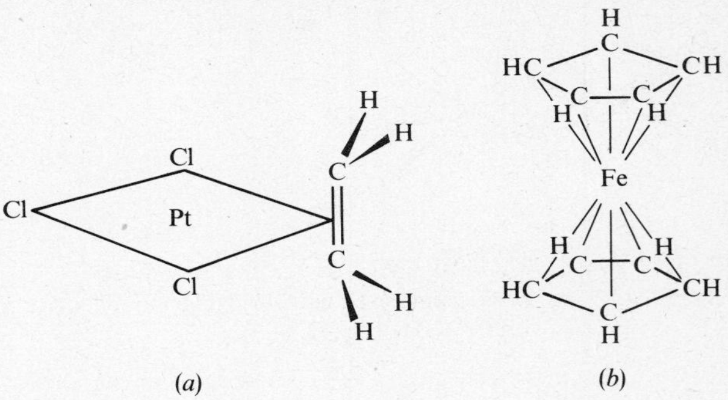

FIGURE 14.7
Structures of two π-bonded organometallic compounds. (a) $PtCl_3(C_2H_4)^-$, (b) $Fe(C_5H_5)_2$, ferrocene.

When the coordination number of the metal ion is twice its charge, it is possible to prepare neutral acetylacetonates, such as $Be(acac)_2$ and $Cr(acac)_3$, which are volatile and soluble in organic solvents. Bidentate ligands frequently act as bridges between metal atoms. For example, in $Cr_2(OAc)_4 \cdot 2H_2O$ the two chromium atoms are linked together by the four bridging acetate ions:

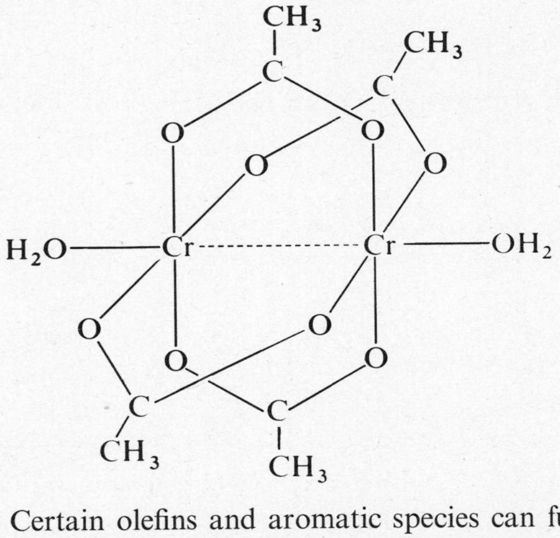

Certain olefins and aromatic species can function as ligands in which the π-bonded carbon atoms coordinate to metal atoms. Ethylene, benzene, the cyclopentadienide ion $(C_5H_5^-)$, and the cycloheptatriene (tropylium) cation $(C_7H_7^+)$ are important examples of such ligands. Structures of coordination compounds of ethylene and the cyclopentadienide ion are shown in Fig. 14.7. The complexes of this type differ from chelate complexes in two important respects: (1) The coordinated atoms of the ligand are bonded together as a chain

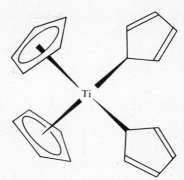

FIGURE 14.8
Structure of $Ti(C_5H_5)_4$. Notice that two
of the rings are σ-bonded, through one
carbon atom (monohapto), and that two
of the rings are π-bonded, through five
carbon atoms (pentahapto).

or ring; (2) the number of σ bonding electron pairs (or *effective* number of coordination sites occupied by the ligand) is less than the number of coordinated atoms. Thus the two carbon atoms of ethylene in effect occupy one of the square planar coordination sites in $C_2H_4PtCl_3{}^-$, and the five carbon atoms of each of the $C_5H_5{}^-$ ions in $Fe(C_5H_5)_2$ in effect occupy three octahedral coordination sites. The details of the bonding in such compounds will be discussed in later chapters.

It is important to point out that it is not necessary that a π-bonded species such as $C_5H_5{}^-$ coordinate so that all the carbon atoms are equivalent and simultaneously bonded to the metal atom. In contrast to the "π-type" bonding in $Fe(C_5H_5)_2$, the $C_5H_5{}^-$ ion can act as a simple σ donor, thus:

$$
\begin{array}{c}
\overset{\displaystyle H}{\underset{\displaystyle }{}}\\[-2pt]
H \diagdown \; C = CH \\
\quad C \qquad\quad | \\
M \diagup \;\diagdown C = CH \\
\; H
\end{array}
$$

Cotton[1] proposed that this type of bonding be referred to as "monohapto" (η-) coordination, and that the π type of bonding in ferrocene be referred to as "pentahapto" (η^5-) coordination. In $Ti(C_5H_5)_4$, both types of bonding are found, as shown in Fig. 14.8. The proton nmr spectrum of this compound at temperatures around $-30°C$ shows two peaks corresponding to the pentahapto C_5H_5 groups and the monohapto C_5H_5 groups. The fact that only one peak is observed for the monohapto groups indicates that an intramolecular rearrangement of each monohapto ring occurs rapidly enough to make the protons equivalent in the time scale of the nmr experiment (roughly, 10^{-3} s). If the sample temperature is raised above $40°C$, the two nmr peaks merge to form a single band. This

[1] F. A. Cotton, *J. Am. Chem. Soc.*, **90**, 6230 (1968).

behavior is ascribed to an increase in the rate of a scrambling process in which the two types of rings interchange their roles, so as to make all the protons equivalent in the nmr time scale.

ISOMERISM

Around the turn of the century, Alfred Werner explained many of the chemical and physical properties of a large number of complexes by means of his revolutionary coordination theory.[1] The type of information at his disposal is typified by the data in Table 14.1. Werner proposed that, in each of these seven compounds, the platinum $+4$ ion is strongly coordinated to six ligands and that, in those cases in which the resulting complex has a net charge, the counter-ions (Cl^- or K^+) are not directly coordinated to the platinum. The formulas given in the fourth column, in which the formulas of the complexes are written inside brackets, readily explain the relative electrical conductivity data of the second column. As expected, the molar conductivity increases with the number of ions in the formula and is zero for the neutral complex $Pt(NH_3)_2Cl_4$.

Cis-Trans Isomerism

The number of isomers, given in column 3 of Table 14.1, was explained by Werner by postulating that the six ligands of each complex are attached to the platinum atom at the vertices of an imaginary octahedron around the platinum

---

[1] A. Werner, *Z. Anorg. Chem.*, **3**, 267 (1893); A. Werner, "Neuere Anschauungen auf der Gebiete der Anorganischen Chemie," Friedrich Vieweg und Sohn, Brunswick, Germany, 1905; G. B. Kauffman, "Classics in Coordination Chemistry, Part I, The Selected Papers of Alfred Werner," Dover, New York, 1968.

Table 14.1 PROPERTIES OF PLATINUM(IV) COMPLEXES OF AMMONIA AND CHLORIDE

Old formula	Relative molar conductivity of aqueous solution	No. of isomers known	Werner's formula
$PtCl_4 \cdot 6NH_3$	523	1	$[Pt(NH_3)_6]Cl_4$
$PtCl_4 \cdot 5NH_3$	404	1	$[Pt(NH_3)_5Cl]Cl_3$
$PtCl_4 \cdot 4NH_3$	228	2	$[Pt(NH_3)_4Cl_2]Cl_2$
$PtCl_4 \cdot 3NH_3$	97	2	$[Pt(NH_3)_3Cl_3]Cl$
$PtCl_4 \cdot 2NH_3$	0	2	$[Pt(NH_3)_2Cl_4]$
$PtCl_4 \cdot KCl \cdot NH_3$	108	1	$K[Pt(NH_3)Cl_5]$
$PtCl_4 \cdot 2KCl$	256	1	$K_2[PtCl_6]$

atom. Thus, in the complex $Pt(NH_3)_4Cl_2{}^{2+}$, the two chlorine atoms can be either at adjacent (cis) positions of the octahedron, as in the following representations,

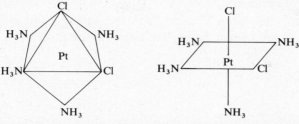

or at opposite (trans) positions, as in the following representations,

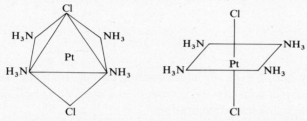

In a completely analogous way the isomerism of $Pt(NH_3)_2Cl_4$ can be explained in terms of cis and trans configurations of the coordinated NH_3 molecules.

One of the isomers of $Pt(NH_3)_3Cl_3{}^+$ has the structure in which the three NH_3 molecules are cis to one another and the three Cl atoms are cis to one another,

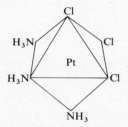

and the other isomer has the structure in which two NH_3 molecules are trans to one another and two Cl atoms are trans to one another,

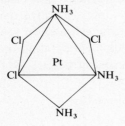

The former, "facially" coordinated, structure has one threefold rotational axis of symmetry, whereas the latter, "meridionally" coordinated, structure has one twofold rotational axis of symmetry.

Cis-trans isomerism is also found in square planar complexes. For example, two types of $Pt(NH_3)_2Cl_2$ are known, having the following structures:

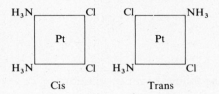

Cis   Trans

Optical Isomerism

A complex is optically active if its structure cannot be superimposed on its mirror image. For example, consider the complex $cis\text{-}Co(en)_2(NO_2)_2{}^+$, which can exist in either of the following configurations:

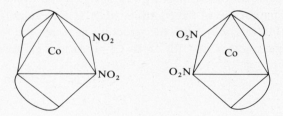

It should be noted that each structure is the mirror image of the other and that the structures are not superimposable. Consequently they are optical isomers. The usual method for the synthesis of $cis\text{-}Co(en)_2(NO_2)_2{}^+$ yields a product containing equal amounts of these isomers, i.e., a racemic mixture which does not rotate polarized light. The isomers cannot be separated by the usual techniques for separating other types of isomers (e.g., cis and trans isomers) because the optical isomers have essentially the same physical[1] and chemical properties, at least in systems which contain no other optically active species. However, optical isomers do interact differently with optically active species, and such specific interactions form the basis of methods for separating optical isomers. For example, if a solution containing the isomers of $Co(en)_2(NO_2)_2{}^+$

---

[1] Of course, optical isomers rotate the plane of polarized light in opposite directions.

is treated with potassium antimonyl-(+)-tartrate, the $[(-)\text{-}cis\text{-}Co(en)_2(NO_2)_2]$ $[(+)\text{-}SbOC_4H_4O_6]$ combination (which is only slightly soluble) precipitates as crystals, leaving the $(+)\text{-}cis\text{-}Co(en)_2(NO_2)_2{}^+$ isomer in solution.

Linkage Isomerism

Some ligands can bond to a metal atom through one type of donor atom in one complex and through another type of donor atom in another complex. This type of isomerism, "linkage isomerism," was first discovered by Jörgensen in 1894.[1] He characterized the nitrito and nitro isomers of the complex having the formula $Co(NH_3)_5NO_2{}^+$. In the red nitritopentaamminecobalt(III) ion, the nitrite ion is coordinated through one of its oxygen atoms, thus: $Co-O-N\diagdown^O$. In the

yellow nitropentaamminecobalt(III) ion, the nitrite ion is coordinated through its

nitrogen atom, thus: $Co-N\diagup^O_{\diagdown O}$. The nitrito isomer is prepared by treatment of

$Co(NH_3)_5H_2O^{3+}$ with a cold buffered $HNO_2\text{-}NO_2{}^-$ solution. The reaction is believed to proceed by the following mechanism:[2]

$$Co(NH_3)_5H_2O^{3+} + OH^- \underset{}{\overset{\text{fast}}{\rightleftharpoons}} Co(NH_3)_5OH^{2+} + H_2O$$

$$2HNO_2 \overset{\text{fast}}{\rightleftharpoons} N_2O_3 + H_2O$$

$$Co(NH_3)_5OH^{2+} + N_2O_3 \longrightarrow \begin{bmatrix} (NH_3)_5Co-O\text{---}H \\ \quad\quad\vdots\quad\quad\vdots \\ \quad O-N\text{---}ONO \\ \quad\quad\downarrow \\ (NH_3)_5Co-ONO^{2+} + HONO \end{bmatrix}$$

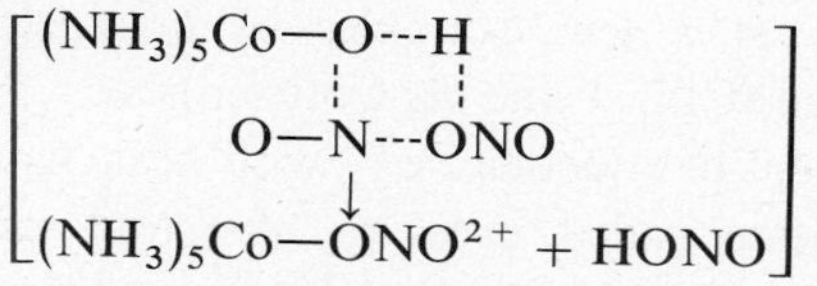

The nitrito complex, when heated, rearranges by an intramolecular process to the more stable nitro complex. The nitrite ion can coordinate in three other ways, as shown below.

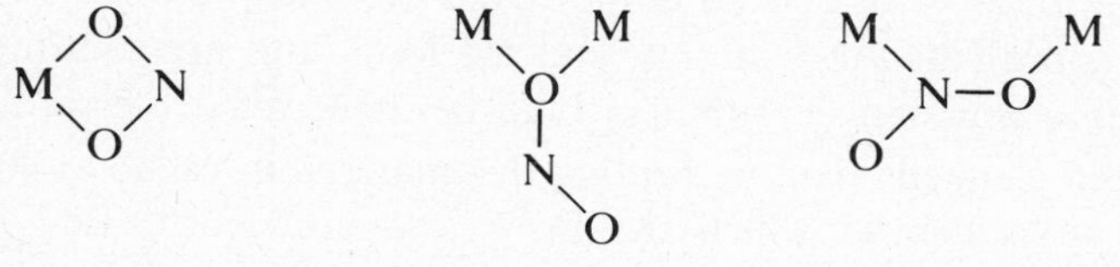

[1] S. M. Jörgensen, *Z. Anorg. Chem.,* **5**, 169 (1894).
[2] J. L. Burmeister and F. Basolo, *Prep. Inorg. React.,* **5**, 1 (1968).

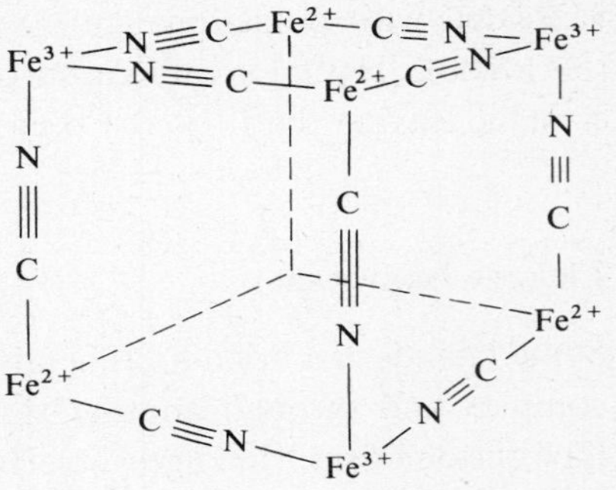

FIGURE 14.9
One-eighth of the unit cell of prussian (Turnbull's) blue. For simplicity, the alkali-metal ions and water molecules are omitted. Each iron atom is octahedrally coordinated.

All three of these types of coordination are found in the complex Ni_3(3-methylpyridine)$_6$(NO$_2$)$_6$, which has the following structure.[1]

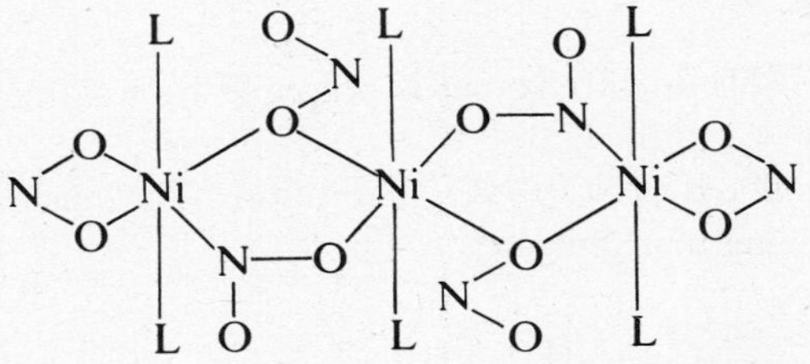

The cyanide ion, CN^-, almost always coordinates through the carbon atom, but a few examples of linkage isomerism are known, such as *cis*-Co(trien)(CN)$_2{}^+$ and *cis*-Co(trien)(NC)$_2{}^+$. Practically every chemist knows that treatment of aqueous Fe^{3+} with hexacyanoferrate(II), $Fe(CN)_6{}^{4-}$, yields a blue precipitate called prussian blue and that treatment of aqueous Fe^{2+} with hexacyanoferrate(III), $Fe(CN)_6{}^{3-}$, yields a blue precipitate called Turnbull's blue. These precipitates are actually the same; the empirical formula is $M^IFe_2(CN)_6 \cdot xH_2O$, where M^I is Na, K, or Rb. The structure (given in Fig. 14.9) consists of a simple cubic array of iron atoms, with the cyanide ions bridging the iron atoms and with M^I ions and water molecules in some of the cubic holes. The Fe^{2+} ions are coordinated to C atoms, and the Fe^{3+} ions are coordinated to N atoms; therefore prussian (Turnbull's) blue is often called "ferric ferrocyanide." It has been claimed[2] that, by heating this material in vacuo at 400°C, it is converted to an isomer in which the Fe^{2+} ions are coordinated to N atoms and the Fe^{3+} ions are coordinated to C atoms, i.e., "ferrous ferricyanide."

---

[1] D. M. L. Goodgame, M. A. Hitchman, and D. F. Marsham, *J. Chem. Soc.*, (*A*), p. 259, 1971.
[2] J. G. Cosgrove, R. L. Collins, and D. S. Murty, *J. Am. Chem. Soc.*, **95**, 1083 (1973).

STRUCTURAL REORGANIZATION

The fact that it is possible to isolate and to characterize structural isomers of the types that we have discussed (cis-trans, optical, and linkage) indicates that the bonds between the metal atoms and ligands in these isomers are not rapidly broken and re-formed under ordinary conditions. If the ligand-to-metal bonds did dissociate rapidly, equilibrium would be rapidly achieved and it would be possible to isolate only the thermodynamically stable isomer (or mixture, in the case of optical isomers). Rapid interconversion of isomers is found in many complexes. For example, aluminum(III) forms a trisacetylacetonato complex, $Al(C_5H_7O_2)_3$, which undoubtedly consists of a mixture of $(+)$ and $(-)$ optical isomers. However, because of the rapid interconversion of these isomers, nobody has thus far succeeded in separating them. Obviously the prediction of the relative kinetic stabilities of coordination compounds is an important problem, one which we shall take up in Chap. 19.

PROBLEMS

14.1 Plot, on one graph, separate curves showing the concentrations of Hg^{2+}, $HgCl^+$, etc., as functions of log (Cl^-) for an aqueous system 0.1 M in mercury(II). What can you conclude about the relative stabilities of the complexes?

14.2 In crystalline AgO, half of the silver atoms are linearly coordinated to two nearest-neighbor oxygen atoms and half are coordinated to four nearest-neighbor oxygen atoms in a square planar configuration. Explain.

14.3 Predict the structures of the following discrete complexes: OsO_4, $VOCl_3$, $Pt(NH_3)_4^{2+}$, $Ag(NH_3)_2^+$, $Pt(PPh_3)_4$, Nb_2Cl_{10}, $Cr_2O_7^{2-}$, Ru_4F_{20}.

14.4 Draw all the possible isomers (including optical) for each of the following complexes. (Be careful not to draw the same structure twice.) (a) $Co(en)_2(H_2O)Cl^{2+}$, (b) $Co(NH_3)_3(H_2O)ClBr^+$, (c) $Pt(NH_3)(NH_2OH)(NO_2)(C_5H_5N)^+$.

14.5 Draw all the possible isomers of $Pt(NH_3)_2Cl_4$, assuming (a) trigonal prismatic geometry and (b) hexagonal planar geometry.

14.6 Which type of octahedral distortion would you expect in $Co(en)_3^{3+}$? In *trans*-$Co(NH_3)_4Cl_2^+$?

14.7 The ^{13}C nmr spectrum of an aqueous solution of $Mo(CN)_8^{4-}$ shows only one peak. What may one conclude about the structure of the aqueous complex ion?

14.8 Although both tetrahedral and square planar complexes of Ni(II) are known, there are no known tetrahedral complexes of Pd(II) and Pt(II). Explain in terms of steric factors.

15

THE BONDING IN COORDINATION COMPOUNDS

ELECTROSTATIC CRYSTAL-FIELD THEORY

Let us consider the bonding in transition-metal complexes from a very simple electrostatic point of view. We may consider an octahedral complex as a metal ion surrounded by either six negative ions or six dipolar groups with their negative ends pointing toward the metal ion. These ligands occupy the vertices of an imaginary octahedron. It is completely arbitrary how we align the cartesian coordinates of the metal ion with respect to the six ligands; for convenience, we suppose that the ligands lie exactly on the x, y, and z axes of the metal atom, as shown in Fig. 15.1. In view of the spatial configurations of the five metal d orbitals (see Fig. 1.8), which are degenerate in the free atom or ion, it is obvious that these d orbitals are no longer degenerate in the octahedral electric field of the ligands. In the complex, the lobes of the d_{xy}, d_{xz}, and d_{yz} orbitals are directed *between* the ligands, and the lobes of the d_{z^2} and $d_{x^2-y^2}$ orbitals are directed *toward* the ligands. Thus the five d orbitals are split by the octahedral field into two groups: a triply degenerate group composed of the d_{xy}, d_{xz}, and d_{yz} orbitals (which are labeled by the group-theoretical symbol t_{2g}) and a doubly degenerate group composed of the d_{z^2} and $d_{x^2-y^2}$ orbitals (labeled by the symbol e_g).

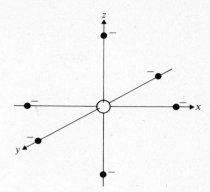

FIGURE 15.1
Octahedral arrangement of negative ligands around a metal ion.

From Fig. 1.8, it should be obvious that, in an octahedral environment, the t_{2g} orbitals are equivalent, but it is not obvious that the e_g orbitals are equivalent; however, this equivalence becomes apparent when one recognizes that the d_{z^2} orbital is a kind of hybrid of hypothetical $d_{z^2-x^2}$ and $d_{z^2-y^2}$ orbitals.

The d electrons of the metal ion would obviously prefer to occupy the t_{2g} set of orbitals than the e_g set of orbitals. That is, the e_g energy level lies above the t_{2g} energy level. Figure 15.2 is a plot of the energies of the d orbitals of a metal atom as a function of the distribution of six negative charges on a spherical shell surrounding the atom. The left side of the diagram corresponds to a uniform distribution of the six charges on the spherical shell; the right side corresponds to placing the negative charges at octahedral positions. Because the total energy of the d-orbital system is independent of the way the negative charges are

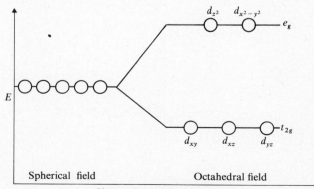

FIGURE 15.2
Splitting of d-orbital energies in an octahedral field.

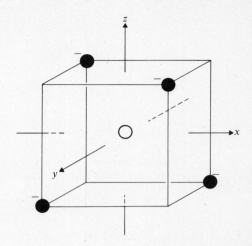

FIGURE 15.3
Tetrahedral arrangement of negative ligands around a metal ion.

distributed on the sphere, the "center of gravity" of the energy levels remains constant. Hence the e_g level is raised 1.5 times as much as the t_{2g} level is lowered. If we call the total octahedral-field splitting Δ_o, then the relative energies of the e_g and t_{2g} levels are $\frac{3}{5}\Delta_o$ and $-\frac{2}{5}\Delta_o$, respectively.

Let us now consider a tetrahedral complex. For convenience, we shall consider the ligands to occupy tetrahedral corners of a cube positioned so that the cartesian coordinates of the metal atom pass through the centers of the cube faces, as shown in Fig. 15.3. The lobes of the d_{xy}, d_{xz}, and d_{yz} orbitals are directed toward the cube edges and come close to the ligands. The lobes of the d_{z^2} and $d_{x^2-y^2}$ orbitals are directed toward the cube faces and bisect the angles between pairs of ligands. Calculations confirm the intuitive inference that electrons in the d_{z^2} and $d_{x^2-y^2}$ orbitals are repelled less by the ligands than are electrons in the

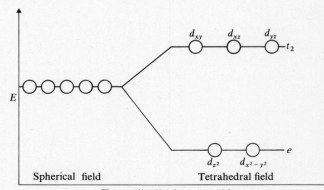

FIGURE 15.4
Splitting of d-orbital energies in a tetrahedral field.

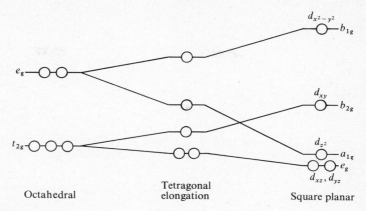

FIGURE 15.5
Splitting d-orbital energies in a tetragonally distorted octahedral field and in a square planar field.

d_{xy}, d_{xz}, and d_{yz} orbitals. As a result, we obtain a splitting of the d-orbital energies into two levels, as shown in Fig. 15.4. The energy of the triply degenerate t_2 set is raised $\frac{2}{5}\Delta_t$, and the energy of the doubly degenerate e set is lowered $\frac{3}{5}\Delta_t$. Calculations show that, if the metal-ligand distances are maintained constant, $\Delta_t = \frac{4}{9}\Delta_o$; this relation is qualitatively predictable from the fact that the total negative charge is less for four ligands than for six ligands.

It is interesting to consider the effect on the d-orbital energy levels of a tetragonal elongation of an octahedral complex. If we pull out the ligands on the z axis and push in the ligands on the x and y axes so as to maintain the total coulomb energy constant, the d_{z^2} and $d_{x^2-y^2}$ orbitals are no longer equivalent, and their energies diverge, as shown in Fig. 15.5. Similarly, the d_{xy} orbital is no longer equivalent to the d_{xz} and d_{yz} orbitals, and its energy is raised relative to that of the latter orbitals. In the limit of a square planar configuration, reached when the axial ligands have been completely withdrawn, the $d_{x^2-y^2}$ orbital definitely has the highest energy, but the relative energies of the other orbitals are somewhat uncertain. Theoretical calculations and experimental data indicate that the energy of the d_{z^2} orbital becomes comparable to that of the d_{xz} and d_{yz} orbitals, as shown in Fig. 15.5.

MOLECULAR ORBITAL THEORY

We have seen that, by application of electrostatic crystal-field theory, we predict that the d-orbital energies of a transition-metal ion are split into two or more levels when the ion is subjected to a nonspherical electrostatic field such as it

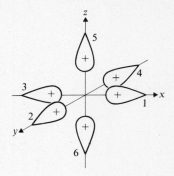

FIGURE 15. 6
Combination of ligand σ orbitals which
can interact with the metal s orbital.

would experience in a coordination compound. Before discussing the various chemical ramifications of such orbital splitting, it will be useful to consider an entirely different approach to the bonding in transition-metal compounds, i.e., molecular orbital theory, to show that the MO approach leads essentially to the same qualitative results. The MO approach is more realistic than the purely electrostatic approach and can explain several phenomena that cannot be explained by the electrostatic approach.

Let us apply simple MO theory to the σ bonding of a regular octahedral complex. The metal atom has, in its valence shell, five d orbitals, an s orbital, and three p orbitals. Each of the six equivalent ligand atoms contributes an atomic orbital, or a hybrid atomic orbital, to the σ bonding system. We are concerned with how these 15 orbitals combine to form molecular orbitals. The combination of ligand orbitals which has the same symmetry as, and can interact with, the metal s orbital to yield a bonding MO and an antibonding MO is represented by the following group orbital wave function,

$$\phi(a_{1g}) = \frac{1}{\sqrt{6}} (\sigma_1 + \sigma_2 + \sigma_3 + \sigma_4 + \sigma_5 + \sigma_6)$$

where σ_1, σ_2, etc., stand for the ligand atomic orbital wave functions. This ligand combination is pictured in Fig. 15.6. The ligand-orbital combination which can interact with the metal d_{z^2} orbital is represented by the following function,

$$\phi(e_g) = \frac{1}{2\sqrt{3}} (2\sigma_5 + 2\sigma_6 - \sigma_1 - \sigma_2 - \sigma_3 - \sigma_4)$$

and is pictured in Fig. 15.7. The ligand-orbital combination which can interact with the metal $d_{x^2-y^2}$ orbital is represented as follows,

$$\phi(e_g) = \tfrac{1}{2}(\sigma_1 - \sigma_2 + \sigma_3 - \sigma_4)$$

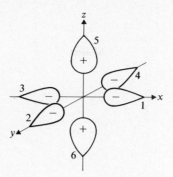

FIGURE 15.7
Combination of ligand σ orbitals which
can interact with the metal d_{z^2} orbital.

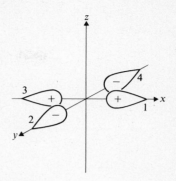

FIGURE 15.8
Combination of ligand σ orbitals which
can interact with the metal $d_{x^2-y^2}$ orbital.

and is pictured in Fig. 15.8. The combinations suitable for interacting with the
metal p orbitals are given in the following functions:

$$\phi(t_{1u}) = \frac{1}{\sqrt{2}}(\sigma_1 - \sigma_3) \qquad \text{for } p_x$$

$$\phi(t_{1u}) = \frac{1}{\sqrt{2}}(\sigma_2 - \sigma_4) \qquad \text{for } p_y$$

$$\phi(t_{1u}) = \frac{1}{\sqrt{2}}(\sigma_5 - \sigma_6) \qquad \text{for } p_z$$

These combinations are shown in Fig. 15.9. The metal d_{xz}, d_{xy}, and d_{yz} orbitals
have no net overlap with the ligand σ orbitals and therefore these d orbitals,

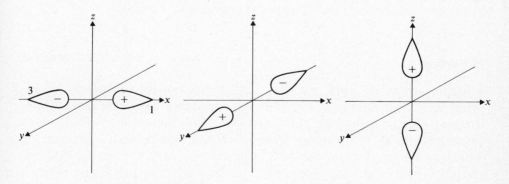

FIGURE 15.9
Combinations of ligand σ orbitals which can interact with the metal p orbitals.

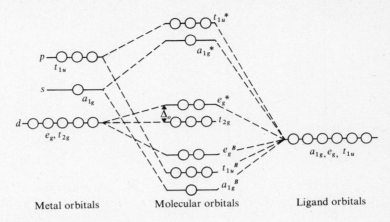

FIGURE 15.10
Molecular orbital energy level diagram for an octahedral complex with only σ bonding.

which have t_{2g} symmetry, are nonbonding. The energy level diagram for the MOs which arise from the interactions which we have just described is shown in Fig. 15.10. Notice that the separation between the t_{2g} and $e_g{}^*$ levels is identified with Δ_0. Each ligand contributes two electrons to the MOs; these 12 electrons can be used to fill the bonding a_{1g}, t_{1u}, and e_g MOs. The metal ion contributes as many valence electrons as it possesses. Thus an octahedral complex of a d^1 metal ion, such as $Ti(H_2O)_6{}^{3+}$, has one electron in one of the nonbonding t_{2g} orbitals. An octahedral complex of a d^8 metal ion, such as $Ni(H_2O)_6{}^{2+}$, has six electrons in the t_{2g} orbitals and two unpaired electrons in the $e_g{}^*$ orbitals. Completely analogous descriptions of the d electrons are possible, using the energy level diagram of Fig. 15.2, based on crystal-field theory.

Pi Bonding

A ligand such as H^-, NH_3, or $CH_3{}^-$ can engage only in σ bonding, and an energy level diagram such as Fig. 15.10 is adequate for describing the bonding. However, a ligand such as Cl^-, O^{2-}, or $CO_3{}^{2-}$ has electrons in essentially nonbonding p orbitals of the donor atom which can interact with appropriate d orbitals of the metal atom. Similarly, a ligand such as CN^-, CO, or PR_3 has empty π orbitals which can interact with metal d orbitals. In the case of CN^- and CO, the empty π orbitals are the antibonding π MOs of these diatomic species; in the case of PR_3, the empty π orbitals are the valence-shell d orbitals

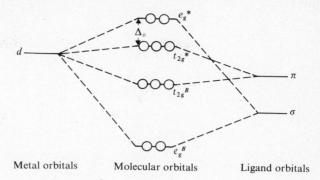

Metal orbitals   Molecular orbitals   Ligand orbitals

FIGURE 15.11
Energy level diagram showing the interaction of the t_{2g} d orbitals with filled π orbitals of the ligands. The nonbonding ligand π orbitals are not shown. Notice that the interaction tends to decrease Δ_o.

of the phosphorus atom. All these ligand-metal interactions are of the π type, as indicated in the following sketch.

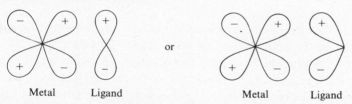

In a complex in which each ligand donor atom has two π orbitals that can interact with the metal orbitals, the ligand π orbitals may be combined into four group orbitals of symmetry t_{1g}, t_{2g}, t_{1u}, and t_{2u}. The t_{1g} and t_{2u} combinations are nonbonding because there are no metal orbitals having those symmetries, and the t_{1u} combinations are essentially nonbonding because the metal p orbitals, which have this symmetry, are engaged in strong σ bonding with the ligands. However, the t_{2g} combinations can interact significantly with the metal t_{2g} orbitals. Two types of interaction are possible, differing as to whether the ligand π orbitals are filled and of lower energy than the metal t_{2g} orbitals, or empty and of higher energy than the metal t_{2g} orbitals. The first type of interaction, such as found in $MnCl_6{}^{4-}$, is illustrated in the energy level diagram of Fig. 15.11. Notice that, in this case, the $p\pi$-$d\pi$ interaction causes a decrease in Δ_o. The second type of interaction, such as found in $Fe(CN)_6{}^{4-}$, is illustrated in the energy level diagram of Fig. 15.12. In this case, the $d\pi \rightarrow p\pi$ "back bonding" causes an increase in Δ_o. Such back bonding serves to transfer electrons from the metal atom (which would be excessively negative in the absence of

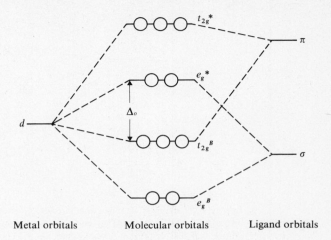

Metal orbitals   Molecular orbitals   Ligand orbitals

FIGURE 15.12
Energy level diagram showing the interaction of the t_{2g} d orbitals with empty π orbitals of the ligands. The ligand π orbital combinations of other symmetries are not shown. Notice that the interaction tends to increase Δ_o.

back bonding) to the ligand. The combination of σ donor bonding and π back bonding involved in the coordination of species such as CN^-, CO, N_2, etc., is often referred to as synergic bonding. The σ basicity of N_2 is much too low to account for the strength of the many transition-metal–N_2 bonds that are known. However, the combination of σ and π bonding is remarkably strong, perhaps because it does not require a large interatomic transfer of charge.

The bonding of a transition-metal atom to an olefin can be similarly described in terms of a combination of a σ olefin→metal bond and a π metal→ olefin bond, as indicated in the following sketches:

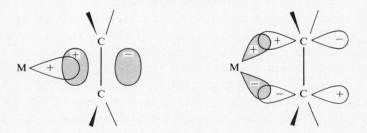

It is also acceptable to assume that the two metal d orbitals involved in the bonding form two equivalent hybrid orbitals and to describe the bonding as analogous to that in cyclopropane:

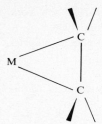

WEAK- AND STRONG-FIELD COMPLEXES

An octahedral complex having 4, 5, 6, or 7 d electrons can have either of two ground-state electronic configurations. In a weak field, the maximum number of electrons are unpaired, as indicated by the following energy level diagrams:

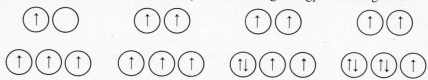

These configurations are referred to as "high-spin." In a strong field, the energy gap Δ_o is greater than the energy required to pair electrons in the same orbital and the t_{2g} level is filled as far as possible, as shown by the following diagrams:

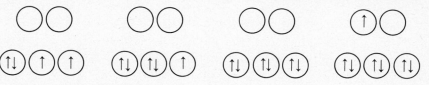

These configurations are referred to as "low-spin." In tetrahedral complexes, Δ_t is never great enough to convert a high-spin configuration into a low-spin configuration; in other words, all tetrahedral complexes are "weak-field."

GENERAL REFERENCES

COTTON, F. A., and G. WILKINSON: "Advanced Inorganic Chemistry," 3d ed., chap. 20, Interscience, New York, 1972.

HUHEEY, J. E.: "Inorganic Chemistry," chap. 8, Harper & Row, New York, 1972.

EMELÉUS, H. J., and A. G. SHARPE: "Modern Aspects of Inorganic Chemistry," chap. 15, Wiley, New York, 1973.

ORGEL, L. E.: "An Introduction to Transition-metal Chemistry. Ligand-field Theory," Wiley, New York, 1966.

BALLHAUSEN, C. J.: "Introduction to Ligand Field Theory," McGraw-Hill, New York, 1962.

FIGGIS, B. N.: "Introduction to Ligand Fields," Interscience, New York, 1966.

JØRGENSEN, C. K.: "Modern Aspects of Ligand Field Theory," American Elsevier, New York, 1971.

COTTON, F. A.: Ligand Field Theory, *J. Chem. Educ.*, **41,** 466 (1964).

GRAY, H. B.: Molecular Orbital Theory for Transition Metal Complexes, *J. Chem. Educ.*, **41,** 2 (1964).

PROBLEMS

15.1   Sketch the *d*-orbital energy level diagrams for complexes of the following type: (*a*) 2-coordinate, linear; (*b*) 5-coordinate, trigonal bipyramidal; (*c*) 5-coordinate, square pyramidal, with the metal atom in the basal plane.

15.2   Rationalize the fact that Δ_o increases in the order $CrCl_6^{3-}$, $Cr(NH_3)_6^{3+}$, $Cr(CN)_6^{3-}$.

15.3   Rationalize the fact that Δ_o increases in the order $Co(H_2O)_6^{2+}$, $Co(H_2O)_6^{3+}$, $Rh(H_2O)_6^{3+}$.

15.4   Assuming that $d\pi \rightarrow p\pi$ back bonding occurs to the fullest possible extent in $Fe(CN)_6^{4-}$, use valence bond theory to estimate the average bond orders of the Fe—C and C—N bonds. What are the formal charges of the atoms?

PHYSICAL CHEMICAL PROPERTIES OF COORDINATION COMPOUNDS

SPECTRA[1-6]

d-d Transitions

First we shall discuss electronic transitions between d orbitals for octahedral and tetrahedral complexes. These transitions are often referred to as d-d or ligand-field transitions. When such transitions occur in complexes with centers of symmetry, such as regular octahedral complexes, the intensities of the absorption bands are low. However, relatively strong absorption bands are obtained for d-d transitions in complexes which lack centers of symmetry, such as cis complexes of the type MA_4B_2 and tetrahedral complexes. The change in intensity of d-d transitions on going from octahedral to tetrahedral complexes can be demonstrated

[1] F. A. Cotton and G. Wilkinson, "Advanced Inorganic Chemistry," 3d ed., Interscience, New York, 1972.
[2] J. E. Huheey, "Inorganic Chemistry: Principles of Structure and Reactivity," Harper & Row, New York, 1972.
[3] C. J. Ballhausen, "Introduction to Ligand Field Theory," McGraw-Hill, New York, 1962.
[4] B. N. Figgis, "Introduction to Ligand Fields," Interscience, New York, 1966.
[5] H. J. Emeléus and A. G. Sharpe, "Modern Aspects of Inorganic Chemistry," Wiley, New York, 1973.
[6] C. J. Jørgensen, "Modern Aspects of Ligand Field Theory," pp. 343–353, American Elsevier, New York, 1971.

by adding excess concentrated hydrochloric acid to an aqueous solution of a Co(II) salt. The color changes from the pale red of $Co(H_2O)_6{}^{2+}$ to the intense blue of $CoCl_4{}^{2-}$.

Consider a d^1 octahedral complex, for which one observes a single absorption band corresponding to the process $t_{2g}{}^1 e_g{}^{*0} \rightarrow t_{2g}{}^0 e_g{}^{*1}$. This transition is represented in spectroscopic notation by the symbols $^2T_{2g} \rightarrow {}^2E_g$, where the left superscripts indicate the spin multiplicities of the states. (The spin multiplicity is $2S + 1$, where S is the total spin of the state.) In the case of a d^1 tetrahedral complex, which has an energy level diagram that is essentially the inverse of that for a d^1 octahedral complex, the notation for the transition is $^2E \rightarrow {}^2T_2$. The ligand-field transition of a d^9 complex may be treated in essentially the same way as that of a d^1 complex if one recognizes that a d^9 configuration corresponds to a positive hole in a filled, d^{10} configuration. The electrostatic behavior of a hole is the opposite of that of an electron, and so for a d^9 octahedral complex one simply inverts the d^1 octahedral energy level diagram. The electronic transition is $^2E_g \rightarrow {}^2T_{2g}$ (the same as that for a d^1 tetrahedral complex, except for the g subscripts). For a d^9 tetrahedral complex, the transition is $^2T_2 \rightarrow {}^2E$. In general, a d^n octahedral complex has electronic states analogous to those of a d^{10-n} tetrahedral complex, and vice versa.

If a d^5 metal ion is subjected to an octahedral or tetrahedral field which is weak enough not to cause electron pairing, so that each d orbital has one electron, the metal ion will be essentially spherically symmetric, like a d^0 or d^{10} ion. Removal of a d electron from such a complex, to form a "weak-field" d^4 complex, is analogous to removal of an electron from a d^{10} complex, and addition of a d electron, to form a "weak-field" d^6 complex, is analogous to addition of an electron to a d^0 complex. Thus a d^6 weak-field octahedral complex undergoes a $^5T_{2g} \rightarrow {}^5E_g$ transition analogous to the $^2T_{2g} \rightarrow {}^2E_g$ transition of a d^1 octahedral complex. In general, weak-field d^{n+5} complexes have transitions analogous to those of the corresponding d^n complexes.

Thus far, we have discussed one-electron and pseudo-one-electron systems. In such cases there is only one conceivable ligand-field transition, and only one ligand-field band is observed. The situation is much more complicated when more than one electron must be considered. That is, the energy level diagrams shown in Figs. 15.2, 15.4, and 15.10 are oversimplified. In general, there are several possible electronic states corresponding to a particular distribution of two or more electrons in d-orbital levels. These states arise because of the interaction of the electrons with one another. The states have different energies, causing relatively complicated spectra in some cases. If the ligand field of the complex were so strong as to make interelectronic interactions negligible by comparison, then only one d-d transition would be observed for a complex, for example,

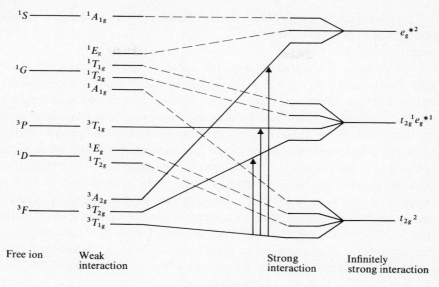

FIGURE 16.1
Correlation diagram for an octahedral d^2 complex.

$t_{2g}^2 e_g^{*0} \to t_{2g}^1 e_g^{*1}$. Actually, of course, the electrons interact appreciably, and the energy level for each MO configuration is split into various terms. If we were able to decrease gradually the ligand-field strength to zero, we would observe that each term would gradually approach an energy corresponding to one of the states of the free ion. Thus the states of the complex can be correlated with those of the free metal ion. In Fig. 16.1 the correlation diagram for a d^2 octahedral complex is shown. The detailed significance and methods of construction of the diagram[1] are beyond the scope of this book, but for the purposes of this discussion certain features of the diagram are noteworthy. Notice that no matter how strong or weak the field, the ground state is always $^3T_{1g}$. Now, one of the rules of spectroscopy is that a transition involving a change in multiplicity is forbidden. Therefore, inasmuch as the ground state of the complex is a triplet state, the only allowed transitions are those to other triplet states. Notice that in Fig. 16.1 only the energy levels for triplet states are drawn with solid lines. The allowed transitions, marked with vertical arrows, are $^3T_{1g}(F) \to {}^3T_{2g}$, $^3T_{1g}(F) \to {}^3A_{2g}$, and $^3T_{1g}(F) \to {}^3T_{1g}(P)$. The two different $^3T_{1g}$ levels are distinguished by indicating, in parentheses, the free-ion states from which they are derived. Inasmuch as the

[1] F. A. Cotton, "Chemical Applications of Group Theory," 2d ed., pp. 254–260, Wiley-Interscience, New York, 1971.

$^3A_{2g}$ level crosses the $^3T_{1g}(P)$ level, the highest-energy transition of a d^2 octahedral complex can be either $^3T_{1g}(F) \rightarrow {}^3T_{1g}(P)$ or $^3T_{1g}(F) \rightarrow {}^3A_{2g}$, depending on the magnitude of Δ_o.

A correlation diagram of the type shown in Fig. 16.1 can be drawn for each electronic configuration in the series d^1 to d^9 for both octahedral and tetrahedral ions. However, when only qualitative information regarding allowed spectral transitions is desired, the necessary data can be summarized as shown in Table 16.1. This table was constructed by taking advantage of the hole formalism, the spherical symmetry of d^5 ions in weak fields, and the fact that the splitting of a given free-ion state is inverted on going from an octahedral to a tetrahedral field. The table lists the symbols of the ground state and all higher states having the same multiplicity. The states are given in the order of increasing energy. States having the same term symbol are distinguished by parenthetical indication of their free-ion states. The energies of weak-field states derived from free-ion ground states are given in brackets, in units of Δ relative to the energies of the free-ion ground states. To use the table, it is only necessary to know the number of d electrons, whether the complex is octahedral (O_h symmetry) or

Table 16.1 SYMBOLS OF GROUND STATES AND HIGHER STATES OF THE SAME MULTIPLICITY FOR OCTAHEDRAL AND TETRA-HEDRAL COMPLEXES

No. of d electrons	Symmetry	Field strength	State symbols
1, 6 4, 9	O_h T_d	Weak	$T_2[-\frac{2}{5}\Delta]$, $E[\frac{3}{5}\Delta]$
2, 7 3, 8	O_h T_d	Weak	$T_1(F)[-\frac{3}{5}\Delta]$, $T_2[\frac{1}{5}\Delta]$, $A_2[\frac{6}{5}\Delta]\dagger$, $T_1(P)$
3, 8 2, 7	O_h T_d	Weak	$A_2[-\frac{6}{5}\Delta]$, $T_2[-\frac{1}{5}\Delta]$, $T_1(F)[\frac{3}{5}\Delta]$, $T_1(P)$
4, 9 1, 6	O_h T_d	Weak	$E[-\frac{3}{5}\Delta]$, $T_2[\frac{2}{5}\Delta]$
5	O_h, T_d	Weak	$^6A_1$
4	O_h	Strong	$^3T_{1g}$, ^3E_g, $^3T_{2g}$
5	O_h	Strong	$^2T_{2g}$, $^2T_{1g}$, $^2A_{2g}$, ^2E_g, $^2A_{1g}$
6	O_h	Strong	$^1A_{1g}$, $^1T_{1g}$, $^1T_{2g}$, ^1E_g, $^1A_{2g}$
7	O_h	Strong	^2E_g, $^2T_{1g}$, $^2T_{2g}$, $^2A_{1g}$

$\dagger$ If Δ is high enough, this state can be the highest-energy state.

tetrahedral (T_d symmetry), and, in the case of d^4, d^5, d^6, and d^7 octahedral complexes, whether the complex is weak-field (high-spin) or strong-field (low-spin). The subscript g (for *gerade*) should be added to the state symbols for octahedral complexes, in cases where these subscripts have been omitted to permit the use of the same symbols for both octahedral and tetrahedral complexes. The multiplicities of the states should be indicated by left superscripts. As a simple example of the use of Table 16.1, let us determine the spectroscopic notation for the lowest-energy ligand-field transition of the octahedral $Ni(H_2O)_6{}^{2+}$ ion. Because the complex is a d^8 system, we obtain the ground-state and first excited-state terms from the third row of the table. The complex has two unpaired electrons, corresponding to a multiplicity of 3, and so the transition is $^3A_{2g} \rightarrow {}^3T_{2g}$. From the bracketed terms, we see that the ground state lies $\frac{6}{5}\Delta$ below the free-ion energy and that the first excited state lies $\frac{1}{5}\Delta$ below the free-ion energy. Hence the transition energy is the difference Δ.

The Ligand-Field Splitting, Δ

Before we apply the data of Table 16.1 to some actual spectroscopic data, it is necessary to have some understanding of the factors which determine the magnitudes of Δ_o and Δ_t. The following generalizations have been drawn from a large body of spectroscopic data.

1 For complexes of the first transition series, Δ_o ranges from 7500 to 12,500 cm^{-1} for $+2$ ions and from 14,000 to 25,000 cm^{-1} for $+3$ ions.
2 For metal ions of the same group and with the same charge, Δ_o increases by 30 to 50 percent from the first transition series to the second, and from the second to the third.
3 We have already pointed out that, other things being equal, $\Delta_t \approx \frac{4}{9}\Delta_o$.
4 Ligands may be arranged in a series in the order of their abilities to split d-orbital levels, i.e., in the order of the Δ values for the complexes of a given metal ion. This series is called the "spectrochemical series." For some common ligands, it is $I^- < Br^- < Cl^- < F^- < OH^- < C_2O_4{}^{2-} \sim H_2O < -NCS^- < $ pyridine $\sim NH_3 < $ en $< $ bipyridyl $< o$-phenanthroline $< NO_2{}^- < CN^-$. Slightly different orders are found for different metal ions, and this series should be used with caution.

The spectrochemical series constitutes experimental evidence for the superiority of the molecular orbital treatment of bonding in complexes over the electrostatic crystal-field treatment. Although the order of the halide ions, $I^- < Br^- < Cl^- < F^-$, is quite logical, the electrostatic theory alone cannot explain why F^- does not give the strongest field of all ligands, as would be expected because of its small size. Undoubtedly $p\pi$-$d\pi$ repulsions are responsible for the relatively low Δ values of halide complexes. Ammonia and the amines have no $p\pi$ orbitals and therefore give relatively high Δ values. The water molecule has only one $p\pi$ orbital in contrast to two for the OH^- ion; consequently water gives greater Δ values than OH^-. The very high Δ values of CN^- and NO_2^- are due, of course, to $d\pi \rightarrow p\pi^*$ back bonding. The valence bond representations of the bonding of these ligands to metal ions containing d electrons are resonance hybrids of the following type:

$$M^- - C \equiv N \;\longleftrightarrow\; M = C = N^-$$

$$M^- - \overset{+}{N}\!\!\Big\langle{\overset{\displaystyle O^{-\frac{1}{2}}}{O^{-\frac{1}{2}}}} \;\longleftrightarrow\; M = \overset{+}{N}\!\!\Big\langle{\overset{\displaystyle O^-}{O^-}}$$

Jørgensen has shown that Δ_o can be roughly estimated from the empirical relation $\Delta_o = fg$, where f is a parameter characteristic of the ligand and g is a parameter characteristic of the metal.[1] The relation can sometimes be helpful when experimental spectral data are not available. Some f and g values are given in Table 16.2.

[1] Jørgensen, op. cit.

Table 16.2 VALUES OF f AND g FOR ESTIMATING Δ_o†

Ligand	f	Metal ion	$g \times 10^{-3}$, cm^{-1}
Br^-	0.72	Mn^{2+}	8.0
Cl^-	0.78	Ni^{2+}	8.7
$POCl_3$	0.82	Co^{2+}	9
N_3^-	0.83	V^{2+}	12.0
F^-	0.9	Fe^{3+}	14.0
$(CH_3)_2SO$	0.91	Cu^{3+}	15.7
$(CH_3)_2CO$	0.92	Cr^{3+}	17.4
C_2H_5OH	0.97	Co^{3+}	18.2
$C_2O_4^{2-}$	0.99	Ru^{2+}	20
H_2O	1.00	Ag^{3+}	20.4
$-NCS^-$	1.02	Ni^{4+}	22
CH_3CN	1.22	Mn^{4+}	24
C_5H_5N	1.23	Mo^{3+}	24.6
NH_3	1.25	Rh^{3+}	27.0
$-SO_3^{2-}$	1.3	Pd^{4+}	29
$-NO_2^-$	~ 1.4	Tc^{4+}	31
$-CN^-$	~ 1.7	Ir^{3+}	32
		Pt^{4+}	36

† C. K. Jørgensen, "Modern Aspects of Ligand Field Theory," pp. 343–353, American Elsevier, New York, 1971.

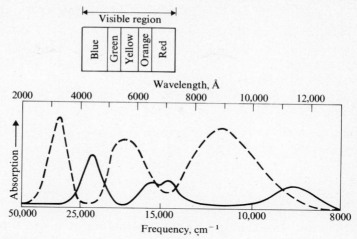

FIGURE 16.2
Absorption spectra of $Ni(H_2O)_6^{2+}$ (solid line) and $Ni(en)_3^{2+}$ (dashed line). The colors of the visible region of the spectrum are indicated above. [*Reproduced with permission from F. A. Cotton, J. Chem. Educ.*, **41**, 466 (1964).]

Spectral Calculations

Figure 16.2 shows the absorption spectra of the $Ni(H_2O)_6^{2+}$ and $Ni(en)_3^{2+}$ complex ions. The spectrum of $Ni(H_2O)_6^{2+}$ has two main bands in the visible region, with a "window" in the green region (~ 5500 Å); hence this complex is green. The spectrum of $Ni(en)_3^{2+}$ has one band in the visible region, with windows in the blue and red regions; hence this complex is purple. Each spectrum has three bands, as expected for d^8 octahedral complexes. The $Ni(en)_3^{2+}$ bands are at higher frequencies than the corresponding $Ni(H_2O)_6^{2+}$ bands, in accord with the spectrochemical series. On the basis of Table 16.1, one would assign the lowest-frequency band in each spectrum to a $^3A_{2g} \rightarrow \,^3T_{2g}$ transition, the intermediate-frequency band to a $^3A_{2g} \rightarrow \,^3T_{1g}(F)$ transition, and the highest-frequency band to a $^3A_{2g} \rightarrow \,^3T_{1g}(P)$ transition. The plausibility of these assignments can be verified by simple calculations. In Table 16.1, we see that the energy of the $^3A_{2g} \rightarrow \,^3T_{2g}$ transition is Δ_o and that the energy of the $^3A_{2g} \rightarrow \,^3T_{1g}(F)$ transition should be $\frac{9}{5}\Delta_o$. Thus by multiplying the frequency of the first band by $\frac{9}{5}$, we can obtain a "calculated" value for the frequency of the second band. In Table 16.1 we also see that the energy of the third transition should be $\frac{6}{5}\Delta_o$ plus the energy of the $^3T_{1g}$ state derived from the free-ion ^3P state. Now a P state is not split by an octahedral field, and as a rough approximation we may assume that the $^3T_{1g}(P)$ energy is the same as that of the ^3P state in the free

Table 16.3 ENERGY DIFFERENCE BE-
TWEEN THE GROUND STATE
AND THE P STATE OF THE
SAME MULTIPLICITY FOR
SOME GASEOUS FREE IONS,
IN WAVE NUMBERS

Metal	+2	+3
V	11,300	12,900
Cr	12,200	13,800
Mn	12,900	14,500
Fe	13,800	15,200
Co	14,600	16,000
Ni	15,500	16,700

Ni^{2+} ion. In Table 16.3 we see that this energy is 15,500 cm^{-1}, and therefore we calculate, for the energy of the $^3A_{2g} \rightarrow {}^3T_{1g}(P)$ transition, $\frac{6}{5}\Delta_o + 15{,}500$ cm^{-1}. The experimental frequencies of the three bands, and the calculated values for the second and third bands, are given in Table 16.4 for both $Ni(H_2O)_6^{2+}$ and $Ni(en)_3^{2+}$. The agreement between the experimental and calculated values, although only approximate, is adequate to confirm the assignments.

The $Co(H_2O)_6^{2+}$ ion has absorption bands at 8350 and 19,000 cm^{-1} which have been assigned to the transitions $^4T_{1g}(F) \rightarrow {}^4T_{2g}$ and $^4T_{1g}(F) \rightarrow {}^4T_{1g}(P)$, respectively.[1] On the basis of this assignment, the higher energy band should have a frequency of approximately $\frac{3}{4}(8350) + 14{,}600 = 20{,}900$ cm^{-1}, which is in fair agreement with the actual value. The frequency of the $^4T_{1g}(F) \rightarrow {}^4A_{2g}$ band is estimated to be $\frac{9}{4}(8350) = 18{,}800$ cm^{-1}. It is believed that the latter band is weak and that it is obscured by the relatively strong $^4T_{1g}(F) \rightarrow {}^4T_{1g}(P)$ band at essentially the same frequency.

Let us now consider the tetrahedral $CoCl_4^{2-}$ complex. Although from Table 16.1 we predict three transitions, only two have been observed,[1] at 6300

[1] Ballhausen, op. cit.

Table 16.4 SPECTRAL DATA FOR $Ni(H_2O)_6^{2+}$ AND $Ni(en)_3^{2+}$ TRANSITION ENERGIES, IN WAVE NUMBERS

Transition	$Ni(H_2O)_6^{2+}$		$Ni(en)_3^{2+}$	
	Exptl.	Calcd.	Exptl.	Calcd.
$^3A_{2g} \rightarrow {}^3T_{2g}$	9000	......	11,000	......
$^3A_{2g} \rightarrow {}^3T_{1g}(F)$	14,000	16,200	18,500	19,800
$^3A_{2g} \rightarrow {}^3T_{1g}(P)$	25,000	26,300	30,000	28,700

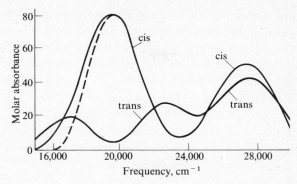

FIGURE 16.3

Absorption spectra of *cis*- and *trans*-Co(en)$_2$F$_2^+$. The splitting of the $^1A_{1g} \rightarrow {}^1T_{1g}$ band is very apparent in the spectrum of the trans complex and is evidenced only by a band asymmetry in the spectrum of the cis complex. The dashed curve shows the probable shape of the high-frequency component of the split band for the cis isomer. [*Reproduced with permission from F. Basolo et al., Acta Chem. Scand.,* **9**, 810 (1955).]

and 15,000 cm^{-1}. We shall attempt to assign these transitions by comparing the frequencies with those estimated for the three allowed transitions. From the data for Co(H$_2$O)$_6^{2+}$, we estimate $\Delta_o = \frac{5}{4}(8350) = 10{,}400$ cm^{-1}. Because chloride gives about 0.78 as strong a ligand field as water (see Table 16.2) and because $\Delta_t \approx \frac{4}{9}\Delta_o$, we predict Δ_t for CoCl$_4^{2-}$ to be $0.78(\frac{4}{9})10{,}400 = 3600$ cm^{-1}. Using data from Tables 16.1 and 16.3, we estimate frequencies of 3600, 6500, and 18,900 cm^{-1} for the transitions $^4A_2 \rightarrow {}^4T_2$, $^4A_2 \rightarrow {}^4T_1(F)$, and $^4A_2 \rightarrow {}^4T_1(P)$, respectively. It seems reasonable to assign the observed bands to the latter two transitions and to assume that the band predicted to be at 3600 cm^{-1} (in the infrared region) has not been observed because of the inconvenient spectral region.

Almost all cobalt(III) complexes are strong-field octahedral complexes. Two transitions, from the $^1A_{1g}$ ground state to the $^1T_{1g}$ and $^1T_{2g}$ states, are generally observed. (Transitions to the ^1E_g and $^1A_{2g}$ states are of extremely high energy and are generally not observable.) In a cobalt(III) complex of type CoA$_4$B$_2$, the symmetry is lowered and the $^1A_{1g} \rightarrow {}^1T_{1g}$ band is split into two components. The splitting of the band of a trans complex is about twice that of a cis complex. Because a cis complex lacks a center of symmetry, its bands are more intense than those of the corresponding trans complex. The spectra of *cis*- and *trans*-Co(en)$_2$F$_2^+$, shown in Fig. 16.3, illustrate these effects.[1]

The modest success achieved in the sample calculations described in the preceding paragraphs is partly a consequence of the cancellation of errors. One

[1] F. Basolo, C. J. Ballhausen, and J. Bjerrum, *Acta Chem. Scand.,* **9**, 810 (1955).

poorly justified approximation was the assumption that the free-ion values of the energy differences between the ground-state metal ion and the excited P state of the same multiplicity can be used in calculations for complexes. It has been shown that interelectronic repulsions are smaller in complexes than in free ions. The decreased repulsions are due to an expansion of the orbitals through metal-ligand interactions and a consequent increased distance between electrons. The energy differences between different states of an ion can be reduced by as much as 40 percent in a complex.[1] Another poor approximation was the neglect of repulsive interactions between states of the same symmetry. For example, in d^8 octahedral complexes, the $^3T_{1g}(F)$ and $^3T_{1g}(P)$ states interact in such a way as to increase the energy separation between the states. Methods of properly accounting for these and other complications are discussed in various texts.[2, 3]

Charge-Transfer Spectra

An electronic transition between orbitals that are centered on different atoms is called a charge-transfer transition, and the absorption band is usually very strong. Such bands are often prominent in the spectra of complexes in which there are electrons in π orbitals of the ligands. An energy level diagram for a complex of this type is shown in Fig. 15.11. The spectra of $RuCl_6^{2-}$ and $IrBr_6^{2-}$ (d^4 and

[1] Jorgensen, op. cit.
[2] Ballhausen, op. cit.
[3] Figgis, op. cit.

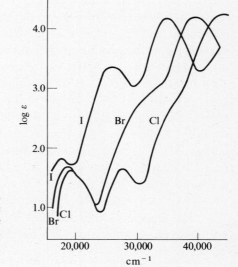

FIGURE 16.4
The spectra of the $Co(NH_3)_5X^{2+}$ ions, where X is a halogen. *Reproduced with permission from W. L. Jolly, "The Synthesis and Characterization of Inorganic Compounds," Prentice-Hall, Englewood Cliffs, N.J., 1970.*

d^5 complexes, respectively) show two sets of bands that have been assigned to transitions from the weakly bonding π orbitals on the ligands to the antibonding $t_{2g}{}^*$ and $e_g{}^*$ orbitals of the metal atom. In $IrBr_6{}^{3-}$ (a d^6 complex), the $t_{2g}{}^*$ orbitals are filled, and only the transitions to the $e_g{}^*$ orbitals can be observed.

In the halogenopentaammine complexes of the type $Co(NH_3)_5X^{2+}$, strong charge-transfer bands are observed in the ultraviolet region, as shown in Fig. 16.4. These bands appear at progressively lower frequencies on going from the chloro to the bromo to the iodo complex, as one might expect from the trend in reduction potentials for these halogens. Indeed, in $Co(NH_3)_5I^{2+}$, the charge-transfer bands largely obscure the weaker d-d transitions.

PARAMAGNETISM

Magnetic Moment

When a substance is subjected to a magnetic field H, a magnetization I is induced. The ratio I/H is called the "volume susceptibility" κ and can be measured by a variety of techniques, including the Gouy balance method, the Faraday method, and an nmr method.[1] The volume susceptibility is simply related to the "gram susceptibility" χ and the "molar susceptibility" χ_M,

$$\chi = \frac{\kappa}{d} \qquad \chi_M = \frac{\kappa M}{d}$$

where d and M are the density and molecular weight of the substance, respectively. For a paramagnetic substance, κ, χ, and χ_M are positive quantities. The *effective magnetic moment* μ_{eff} is calculated from the relation

$$\mu_{\text{eff}} = 2.83\sqrt{\chi_M T}$$

where T is the absolute temperature and the constant 2.83 is a combination of various fundamental constants. For most paramagnetic compounds, μ_{eff} is practically the same as the true magnetic moment μ, but small differences between these quantities are found in some cases. The importance of μ_{eff} or μ in transition-metal chemistry lies in the fact that, for most compounds, the magnetic moment is theoretically calculable from a knowledge of the structure and type of bonding. Often the magnitude of the calculated moment is markedly changed by a change in the assumed structure or a change in the assumed type of bonding. In such cases comparison of experimental μ_{eff} values with calculated μ values can be of considerable value in the characterization of the compounds.

[1] W. L. Jolly, "The Synthesis and Characterization of Inorganic Compounds," pp. 369–384, Prentice-Hall, Englewood Cliffs, N.J., 1970.

The magnetic moment of a transition-metal complex is a combination of spin and orbital moments. In many complexes, the ligand field almost completely quenches the orbital contribution, and the magnetic moment can be calculated by the following "spin-only" formula,

$$\mu = 2\sqrt{S(S + 1)}$$

where S is the total spin of the complex. Because in the ground state S is one-half the number of unpaired electrons, n, we may write

$$\mu = \sqrt{n(n + 2)}$$

In Table 16.5, the experimental μ_{eff} values for various octahedral complexes of the first transition series are listed with the μ values calculated by the spin-only formula. The data permit us to distinguish readily between strong- and weak-field complexes. For example, $[Cr(H_2O)_6]SO_4$ is a weak-field d^4 complex with $\mu_{eff} = 4.8$, whereas $[Cr(dipy)_3]Br_2 \cdot 4H_2O$ is a strong-field d^4 complex with $\mu_{eff} = 3.3$. Similar pairs of weak- and strong-field complexes, respectively, are the d^5 complexes $K_2[Mn(H_2O)_6](SO_4)_2$ ($\mu_{eff} = 5.9$) and $K_4[Mn(CN)_6] \cdot 3H_2O$ ($\mu_{eff} = 2.2$) and the d^7 complexes $(NH_4)_2[Co(H_2O)_6](SO_4)_2$ ($\mu = 5.1$) and $K_2Pb[Co(NO_2)_6]$ ($\mu_{eff} = 1.8$). The d^6 weak-field complex $(NH_4)_2[Fe(H_2O)_6](SO_4)_2$ ($\mu_{eff} = 5.5$) may be compared with the strong-field $K_4[Fe(CN)_6]$, which is not listed in the table because it is diamagnetic.

Table 16.5   EXPERIMENTAL AND CALCULATED VALUES OF THE MAGNETIC MOMENT†

No. of unpaired electrons, n	Compound	Ground state	Experimental μ_{eff}	$[n(n + 2)]^{1/2}$
1	$CsTi(SO_4)_2 \cdot 12H_2O$	$^2T_{2g}$	1.8	1.73
	$K_4Mn(CN)_6 \cdot 3H_2O\ddagger$	$^2T_{2g}$	2.2	1.73
	$K_2PbCo(NO_2)_6\ddagger$	^2E_g	1.8	1.73
	$(NH_4)_2Cu(SO_4)_2 \cdot 6H_2O$	^2E_g	1.9	1.73
2	$(NH_4)V(SO_4)_2 \cdot 12H_2O$	$^3T_{1g}$	2.7	2.83
	$Cr(dipy)_3Br_2 \cdot 4H_2O\ddagger$	$^3T_{1g}$	3.3	2.83
	$(NH_4)_2Ni(SO_4)_2 \cdot 6H_2O$	$^3A_{2g}$	3.2	2.83
3	$KCr(SO_4)_2 \cdot 12H_2O$	$^4A_{2g}$	3.8	3.87
	$(NH_4)_2Co(SO_4)_2 \cdot 6H_2O$	$^4T_{1g}$	5.1	3.87
4	$CrSO_4 \cdot 6H_2O$	^5E_g	4.8	4.90
	$(NH_4)_2Fe(SO_4)_2 \cdot 6H_2O$	$^5T_{2g}$	5.5	4.90
5	$K_2Mn(SO_4)_2 \cdot 6H_2O$	$^6A_{1g}$	5.9	5.92

† W. L. Jolly, "The Synthesis and Characterization of Inorganic Compounds," pp. 369–384, Prentice-Hall, Englewood Cliffs, N.J., 1970.
‡ Strong-field (low-spin) complexes.

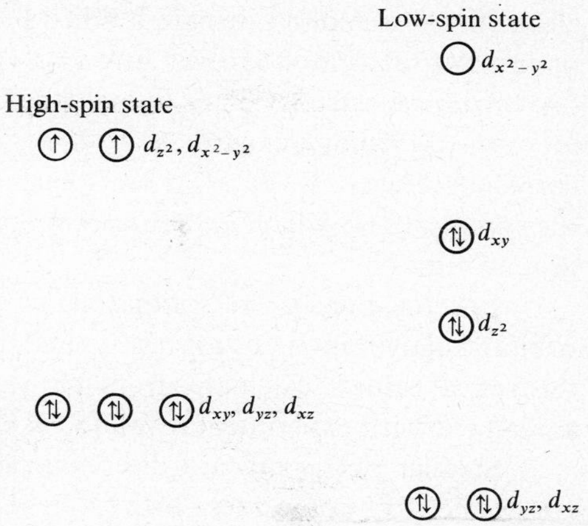

Low-spin state

$\bigcirc d_{x^2-y^2}$

High-spin state

$\uparrow$ $\uparrow$ $d_{z^2}, d_{x^2-y^2}$

$\uparrow\downarrow d_{xy}$

$\uparrow\downarrow d_{z^2}$

FIGURE 16.5
Energy level diagrams for regular and strongly tetragonally distorted octahedral d^8 complexes, showing the electron occupancies.

$\uparrow\downarrow$ $\uparrow\downarrow$ $\uparrow\downarrow d_{xy}, d_{yz}, d_{xz}$

$\uparrow\downarrow$ $\uparrow\downarrow d_{yz}, d_{xz}$

Consider the effect on a d^8 octahedral complex of a tetragonal elongation. Two possible electron configurations are shown in Fig. 16.5. In the case of a regular octahedral complex, there are two unpaired electrons. However, in the case of strong distortion or a square planar complex, a zero-spin state results. Indeed, all square planar d^8 complexes are diamagnetic unless the ligands have unpaired electrons. In tetrahedral complexes, high-spin and low-spin states are, in principle, possible for the configurations d^3, d^4, d^5, and d^6, but because of the relatively small magnitude of Δ_t, there are no known low-spin tetrahedral complexes.

Orbital Angular Momentum[1]

About half of the experimental μ_{eff} values listed in Table 16.5 are close to the theoretical μ values calculated from the spin-only formula. However, the remaining μ_{eff} values differ significantly from the simple theoretical values. The deviations are generally due to incomplete quenching of the orbital magnetic moment by the ligand field.

An electron must be able to circulate about an axis if it is to have orbital angular momentum. Therefore an orbital must be available, in addition to the orbital containing the electron, which has the following properties. It must have the same shape and energy as the orbital containing the electron. It must be superimposable with the orbital containing the electron by rotation about an axis. And, finally, it cannot contain an electron having the same spin as the first

---

[1] B. N. Figgis and J. Lewis, *Prog. Inorg. Chem.*, **6**, 37 (1964).

electron. These conditions are fulfilled in the case of d electrons in an octahedral or tetrahedral field whenever any two t_2 or t_{2g} orbitals (d_{xy}, d_{xz}, d_{yz}) contain one or three electrons. Thus unquenched orbital angular momentum will remain for a metal complex with 1, 2, 4, or 5 electrons in the t_2 or t_{2g} orbitals, corresponding to T_1, T_2, T_{1g}, or T_{2g} ground states. In Table 16.5 practically all the complexes for which μ_{eff} deviates markedly from the spin-only μ value have T ground states.

For the purpose of estimating μ values for complexes with unquenched orbital angular momentum, the values in Table 16.6 can be used. These are theoretical values,[1] calculated from free-ion parameters, and one should not expect agreement with experimental values for complexes to better than ± 0.2.

Smaller but significant discrepancies between μ_{eff} and spin-only moments are found for complexes with A_2, A_{2g}, E, and E_g ground states. These discrepancies are caused by angular momentum introduced to the ground state by a mixing in of higher T states. Theory shows that the magnetic moment of such a complex can be calculated from the relation

$$\mu = \left(1 - \frac{\lambda'}{\Delta}\right)(\text{spin-only } \mu) \tag{16.1}$$

Appropriate values of λ' for metal ions in octahedral and tetrahedral environments are given in Table 16.7.

Magnetic data may be used to determine the type of bonding in cobalt(II) complexes. Tetrahedral Co(II) complexes have room-temperature magnetic moments that range from 4.40 to 4.88. For example, $\mu_{\text{eff}} = 4.59$ for $CoCl_4^{2-}$.

---

[1] Figgis, op. cit.

Table 16.6 CALCULATED VALUES OF μ AT 300 K FOR OCTAHEDRAL AND TETRAHEDRAL STEREOCHEMISTRIES, ASSUMING FREE-ION VALUES OF SPIN-ORBIT COUPLING†

$^2T_{2(g)}$ States		$^3T_{1(g)}$ States		$^4T_{1(g)}$ States		$^5T_{2(g)}$ States	
Ti^{3+}	1.9	V^{3+}	2.7	Cr^{3+}	3.4	Cr^{2+}	4.7
Mo^{5+}	1.0	Cr^{2+}	3.5	Co^{2+}	5.1	Mn^{3+}	4.5
Mn^{2+}	2.55	Mo^{4+}	1.9			Fe^{2+}	5.65
Fe^{3+}	2.45	W^{4+}	1.5			Co^{3+}	5.75
Ru^{3+}	2.1	Mn^{3+}	3.65				
Os^{3+}	1.85	Re^{3+}	2.4				
Ir^{4+}	1.8	Fe^{4+}	3.55				
Cu^{2+}	2.2	Ru^{4+}	3.2				
		Os^{4+}	1.9				
		Ni^{2+}	4.0				

† B. N. Figgis, "Introduction to Ligand Fields," Interscience, New York, 1966.

Table 16.7 VALUES OF χ FOR SEVERAL METAL IONS†

Metal ion	Symmetry	Ground state	χ, cm^{-1}
V^{4+}	T_d	^2E	500
V^{2+}	O_h	$^4A_{2g}$	228
Cr^{3+}	O_h	$^4A_{2g}$	368
Cr^{2+}	O_h	^5E_g	116
Mn^{4+}	O_h	$^4A_{2g}$	552
Mn^{3+}	O_h	^5E_g	178
Fe^{2+}	T_d	^5E	−200
Co^{2+}	T_d	$^4A_2$	−688
Co^{2+}	O_h	^2E_g	−1030
Ni^{2+}	O_h	$^3A_{2g}$	−1260
Cu^{2+}	O_h	^2E_g	−1660

† B. N. Figgis, "Introduction to Ligand Fields," Interscience, New York, 1966.

We can show that this is a reasonable value by use of Eq. 16.1:

$$\mu = \left(1 + \frac{688}{3600}\right)(3.88) = 4.61$$

Octahedral weak-field Co(II) complexes such as $(NH_4)_2[Co(H_2O)_6](SO_4)_2$ have room-temperature moments of 5.0 to 5.1, in remarkably close agreement with the value of 5.1 from Table 16.6. Octahedral strong-field Co(II) complexes appear to be very rare. However, $K_2Pb[Co(NO_2)_6]$ has been reported to have $\mu_{eff} = 1.8$, in fair agreement with the value estimated by using Eq. 16.1:

$$\mu = \left(1 + \frac{1030}{12,600}\right)(1.73) = 1.87$$

JAHN-TELLER DISTORTIONS

A regular octahedral environment is the most stable one for a spherically symmetric metal ion surrounded by six ligand atoms. For metal ions with certain d electron configurations which are not spherically symmetric, the regular octahedral configuration is not the most stable. The situation can be expressed in a general and rigorous way by the Jahn-Teller theorem:[1] Any nonlinear molecule in a degenerate electronic state will undergo distortion to remove the degeneracy and to lower the energy. For example, consider a high-spin d^4 complex. In a regular octahedral environment the e_g* electron is doubly

[1] H. A. Jahn and E. Teller, *Proc. Roy. Soc., A,* **161,** 220 (1937).

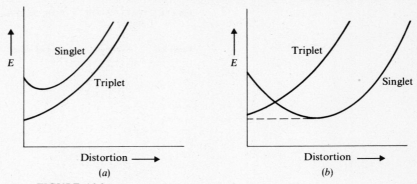

FIGURE 16.6

Potential-energy curves illustrating (a) stability of the high-spin regular octahedral d^8 complex, (b) stability of the highly distorted (essentially square planar) low-spin d^8 complex. (*Adapted with permission from L. E. Orgel, "An Introduction to Transition-Metal Chemistry. Ligand Field Theory," 2d ed., Chapman and Hall, London, 1966.*)

degenerate; it can occupy either the d_{z^2} or the $d_{x^2-y^2}$ orbital. However, if the complex undergoes a tetragonal distortion, the $e_g{}^*$ levels are split and the electron can then occupy the lower of the two orbitals (the d_{z^2} orbital in the case of tetragonal elongation; the $d_{x^2-y^2}$ orbital in the case of tetragonal contraction). The same sort of distortion is expected in the case of d^7 and d^9 complexes. Examples of Jahn-Teller distortions of octahedral coordination are found in CrF_2 and MnF_3 (d^4 configurations), $NaNiO_2$ (low-spin d^7 configuration), and in many copper(II) compounds, such as $CuCl_2$ and CuF_2 (d^9 configurations). In each of these cases the metal ion is surrounded by six anions at the vertices of an elongated octahedron. Complexes with d^8 configurations are a special case. In general, the complexes are either regular octahedral (with two unpaired electrons) or square planar and diamagnetic. The reason for the lack of d^8 octahedral complexes with small or intermediate tetragonal elongation can be seen by reference to Fig. 16.6. The potential-energy diagram will be either like that in Fig. 16.6a or like that in Fig. 16.6b. In either case, the triplet state is much more stable than the singlet state in the undistorted octahedral complex. In Fig. 16.6a no degree of distortion will make the complex more stable than the undistorted triplet-state complex. In Fig. 16.6b, a large tetragonal distortion is required to stabilize the singlet state relative to the undistorted triplet state.

Degeneracies arising from partially filled t_{2g} levels should also produce Jahn-Teller distortions, but the effects are relatively small because the t_{2g} orbitals are relatively unaffected by the ligands. Evidence for Jahn-Teller distortions and the consequent splitting of energy levels is found in the spectra of many complexes. For example, consider the absorption spectrum of $Ti(H_2O)_6{}^{3+}$,

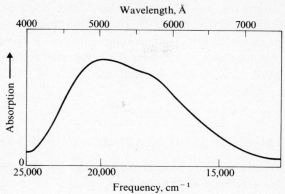

FIGURE 16.7
The visible absorption spectrum of $Ti(H_2O)_6^{3+}$. [*Reproduced with permission from F. A. Cotton, J. Chem. Educ.,* **41**, 466 (1964).]

shown in Fig. 16.7. This band is obviously not a simple symmetric band such as one might expect for a $^2T_{2g} \rightarrow {}^2E_g$ transition. The breadth and asymmetry of the band are evidence of transitions to two excited states caused by distortion of the complex.

Relatively weak distortions are expected for tetrahedral complexes with d^3, d^4, d^8, and d^9 configurations.[1] In the d^3 and d^8 complexes, one of the t_2 orbitals has one more electron than the other two. This situation should cause an elongation of the tetrahedron, as shown in Fig. 16.8a. In the d^4 and d^9

[1] L. E. Orgel, "An Introduction to Transition-metal Chemistry. Ligand Field Theory," 2d ed., Wiley, New York, 1966.

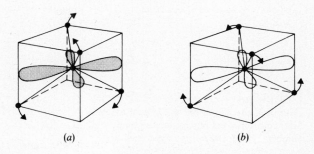

(a)   (b)

FIGURE 16.8
Distortions of tetrahedral coordination. (a) Elongation caused by $t_2{}^1$ and $t_2{}^4$ configurations; (b) flattening caused by $t_2{}^2$ and $t_2{}^5$ configurations. (*Adapted with permission from L. E. Orgel, "An Introduction to Transition-Metal Chemistry. Ligand Field Theory,"* 2d ed., Chapman and Hall, London, 1966.)

complexes, one of the t_2 orbitals has one fewer electron than the other two (i.e., an electron hole). This situation should cause a flattening of the tetrahedron, as shown in Fig. 16.8b. In agreement with these predictions, it has been found that the $CuCl_4{}^{2-}$ complex has a flattened tetrahedral configuration and that the oxide environment around the Ni^{2+} in $NiCr_2O_4$ is that of an elongated tetrahedron.

LIGAND-FIELD STABILIZATION ENERGY

When a transition-metal ion is placed in an octahedral field, three of the d orbitals are stabilized by $\frac{2}{5}\Delta_o$ and two of the d orbitals are destabilized by $\frac{3}{5}\Delta_o$. In the case of an octahedral complex with the configuration $t_{2g}{}^p e_g{}^{*q}$, the net stabilization energy is $(\frac{2}{5}p - \frac{3}{5}q)\Delta_o$. For example, a d^3 complex such as $Cr(H_2O)_6{}^{3+}$ is stabilized by $\frac{6}{5}\Delta_o$, that is (using $\Delta_o = 17{,}400 \text{ cm}^{-1}$) by $20{,}900 \text{ cm}^{-1}$ or $59.7 \text{ kcal mol}^{-1}$, relative to a hypothetical $Cr(H_2O)_6{}^{3+}$ ion in which the three d electrons have a spherically symmetric disposition. Obviously these stabilization energies are of great importance in determining the relative thermodynamic properties of transition-metal complexes.

Consider the hydration energies of the $+2$ ions of the first transition series

$$M^{2+}(g) \rightarrow M^{2+}(aq) \qquad \Delta H^\circ = -\Delta H_{\text{hyd}}$$

These energies, obtained from thermochemical data, are plotted as a function of atomic number in Fig. 16.9. If one draws a smooth curve through the three points corresponding to the spherically symmetric metal ions (Ca^{2+}, Mn^{2+}, and Zn^{2+}), for which one would expect no net ligand-field stabilization, it is obvious that

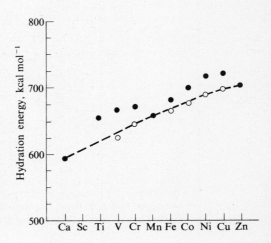

FIGURE 16.9
Hydration energies of the $+2$ ions of the first transition series plotted versus atomic number. Solid circles are experimental points; open circles correspond to values from which spectrally evaluated ligand-field stabilization energies have been subtracted. [*Reproduced with permission from F. A. Cotton, J. Chem. Educ.*, **41**, 466 (1964).]

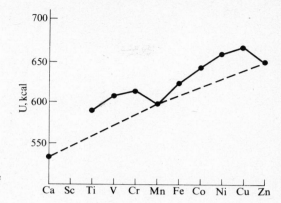

FIGURE 16.10
Lattice energies of the dichlorides of the
metals of the first transition series.

all the other points lie above the curve. It is reasonable to assume that the deviations from the curve are due to the extra ligand-field stabilization energies of the $M(H_2O)_6^{2+}$ ions. If we subtract the appropriate values of $(\frac{2}{5}p - \frac{3}{5}q)\Delta_o$ (calculated from spectral Δ_o data) from the hydration energies, we obtain the points indicated by open circles, which fall on the smooth curve. This result gives strong support to the concept of the splitting of d orbitals in octahedral fields.

In Fig. 16.10 the lattice energies of the dichlorides of the first-transition-series metals are plotted as a function of atomic number. The plot shows that the lattice energies of the salts other than $CaCl_2$, $MnCl_2$, and $ZnCl_2$ are exceptionally large, undoubtedly because of ligand-field stabilization energy. Similar plots are obtained for the lattice energies of other MX_2 salts and for MX_3 salts.

In the absence of ligand-field stabilization effects the ionic radii of the transition-metal ions would be expected to decrease steadily with increasing atomic number. However, from the festoons which appear in the plot of the radii of the $+2$ ions versus atomic number, shown in Fig. 16.11, it is clear that the stabilization energies correlate with the distances between atoms in ionic solids. The greater the ligand-field stabilization energy, the smaller is the ionic radius, relative to the hypothetical spherically symmetric ion.

Ligand-field stabilization effects are clearly seen in the structures of various spinels. The spinel structure, named after the mineral $MgAl_2O_4$, is adopted by a large number of compounds of the type $M^{II}M^{III}_2O_4$. The structure consists of a close-packed lattice of oxide ions with one-third of the metal ions in tetrahedral holes and two-thirds of the metal ions in octahedral holes. In a "normal" spinel, the M^{II} ions occupy tetrahedral holes and the M^{III} ions occupy octahedral holes, thus: $[M^{II}]_{tet}[M^{III}_2]_{oct}O_4$. In an "inverse" spinel, half of the

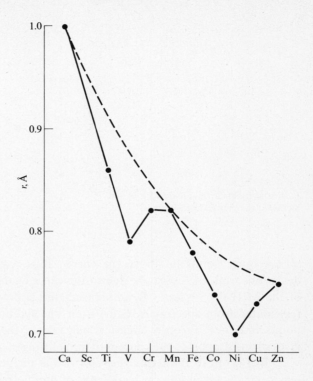

FIGURE 16.11
Ionic radii of the +2 ions of the first transition series, as in high-spin compounds.

M^{III} ions exchange positions with the M^{II} ions, thus: $[M^{III}]_{tet}[M^{II}M^{III}]_{oct}O_4$. The oxides Mn_3O_4 and Co_3O_4 are normal spinels, and Fe_3O_4 and $NiAl_2O_4$ are inverse spinels. These differences can be explained in terms of ligand-field stabilization energies. Because $\Delta_o > \Delta_t$, octahedral site stabilization energies are usually much greater than tetrahedral site stabilization energies. In Mn_3O_4, the Mn^{3+} ions, but not the Mn^{2+} ions, are ligand-field-stabilized, and therefore the Mn^{3+} ions occupy the sites which provide the maximum stabilization. In Co_3O_4, the Co^{3+} ions occupy octahedral sites as low-spin d^6 ions; if the structure were to invert, half of the Co^{3+} ions would be transferred to tetrahedral sites where they would probably become high-spin and lose a tremendous amount of ligand-field stabilization energy. In Fe_3O_4, the Fe^{2+} ions, but not the Fe^{3+} ions, are ligand-field-stabilized; hence the Fe^{2+} ions occupy octahedral sites. Similarly, in $NiAl_2O_4$, only the Ni^{2+} ions are ligand-field-stabilized, and therefore they occupy octahedral sites.

Spinels containing ions with magnetic moments have interesting magnetic properties. In the spinel structure, these ions are close enough to interact with one another, and one often observes ferromagnetism and antiferromagnetism. For example, in the black mineral magnetite, Fe_3O_4, all the ions in the octahedral holes have their magnetic spins aligned, and the Fe^{3+} ions in the tetrahedral holes have their spins aligned in the opposite direction, thus:

$$\underset{[Fe^{III}]_{tet}}{\downarrow}\underset{[Fe^{III}Fe^{II}]_{oct}}{\uparrow\ \uparrow}O_4$$

Hence the Fe^{3+} ions are antiferromagnetically coupled; their mutual interaction yields no magnetic moment. The net ferromagnetism of the compound is entirely attributable to the Fe^{2+} ions. If the Fe^{2+} ions are replaced with diamagnetic ions, as in $ZnFe_2O_4$, the resulting spinel is normal and completely antiferromagnetic, with no net magnetic moment.

PROBLEMS

16.1 The permanganate ion is a d^0 complex, and yet it is intensely colored. Comment.

16.2 The $Mn(H_2O)_6{}^{2+}$ ion has an extremely pale pink color, attributable to transitions which are formally forbidden. Explain why there are no completely allowed d-d transitions.

16.3 Two d-d bands can be seen, at 17,000 and 26,000 cm^{-1}, in the absorption spectrum of $V(H_2O)_6{}^{3+}$. Assign the bands.

16.4 Predict the colors of the following complexes, which have absorption bands at the indicated frequencies: $CrF_6{}^{3-}$ (14,900, 22,700 cm^{-1}), $Cu(NH_3)_4{}^{2+}$ (14,000 cm^{-1}), $FeO_4{}^{2-}$ (12,700, 19,600 cm^{-1}), $Co(en)_3{}^{2+}$ (21,400, 29,600 cm^{-1}).

16.5 Predict the d-d absorption spectra (frequencies and spectroscopic assignments for all bands) of $CoF_6{}^{3-}$ (high-spin), $NiCl_4{}^{2-}$, and $Fe(H_2O)_6{}^{2+}$ (high-spin). Estimate parameters when necessary.

16.6 The following compounds have effective magnetic moments of 5.85 to 5.95: $Fe(NO_3)_3 \cdot 9H_2O$, $FeCl_3 \cdot 6H_2O$, $K_3Fe(C_2O_4)_3 \cdot 3H_2O$, $Fe(urea)_6Cl_3 \cdot 6H_2O$, Na_3-FeF_6, $[Fe(dipy)_2Cl_2][FeCl_4]$, $[Fe(phen)_2Cl_2][FeCl_4]$, $Fe(acac)_3$. The following have μ_{eff} values of 2.3 to 2.4: $[Fe(phen)_2(CN)_2]NO_3 \cdot 4H_2O$, $[Fe(dipy)_2(CN)_2]NO_3$, $H_3O[Fe(phen)(CN)_4] \cdot H_2O$, $K_3Fe(CN)_6$, $Fe(dipy)_3(ClO_4)_3 \cdot 3H_2O$, $Fe(phen)_3$-$(ClO_4)_3 \cdot 3H_2O$. Estimate the energy required to go from a $t_{2g}{}^3 e_g{}^{*2}$ configuration to a $t_{2g}{}^5$ configuration in Fe^{3+}.

16.7 Which would you expect to have the greater magnetic moment, $CoCl_4{}^{2-}$ or $CoI_4{}^{2-}$? Why?

16.8 The complex $Cr_2(OAc)_4 \cdot 2H_2O$ (see structure, page 250) is diamagnetic. What can you say regarding the nature of the bonding in this compound?

16.9 The following compounds have the indicated effective magnetic moments. Describe the structure and bonding of these compounds on the basis of the μ_{eff} values. K_2NiF_6, 0.0; $Ni(NH_3)_2Cl_2$, 3.3; $Ni(PEt_3)_2Cl_2$, 0.0; $Ni(Ph_3AsO)_2Cl_2$, 3.95.

16.10 $Ni(C_5H_5)_2$ has the same type of structure as that of ferrocene. Would you expect $Ni(C_5H_5)_2$ to be paramagnetic or diamagnetic? Why?

16.11 Although a d^8 tetrahedral complex is somewhat stabilized by flattening of the tetrahedron (show this by consideration of the energy level diagram), more stabilization is obtained by elongation of the tetrahedron. Similar, but opposite, statements apply to d^9 complexes. Explain.

16.12 Equilibrium constants for the stepwise formation of mono-, bis-, and tris-(ethylenediamine) complexes of Co^{2+}, Ni^{2+}, and Cu^{2+} in aqueous solution are as follows:

	log K_1	log K_2	log K_3
Co^{2+}	5.89	4.83	3.10
Ni^{2+}	7.52	6.28	4.26
Cu^{2+}	10.55	9.05	−1.0

$$M(H_2O)_6{}^{2+} + en \rightleftharpoons M(H_2O)_4(en)^{2+} + 2H_2O \qquad K_1$$

$$M(H_2O)_4(en)^{2+} + en \rightleftharpoons M(H_2O)_2(en)_2{}^{2+} + 2H_2O \qquad K_2$$

$$M(H_2O)_2(en)_2{}^{2+} + en \rightleftharpoons M(en)_3{}^{2+} + 2H_2O \qquad K_3$$

Explain the anomalously low value of K_3 for Cu^{2+}.

16.13 Prepare a table in which you indicate, for each of the following complexes, the number of unpaired electrons, the estimated magnetic moment, whether or not one would expect an appreciable Jahn-Teller distortion, the number of d-d transitions, and whether or not one would expect crystal-field stabilization: $Co(CO)_4{}^-$, $Cr(CN)_6{}^{4-}$, $Fe(H_2O)_6{}^{3+}$, VCl_4, $Co(NO_2)_6{}^{4-}$, $Co(NH_3)_6{}^{3+}$, $Ir(CO)Cl(PPh_3)_2$, $CuCl_2{}^-$, $MnO_4{}^{2-}$, $Cu(H_2O)_6{}^{2+}$.

16.14 Suggest a reason for the fact that a number of tetrahedral Co(II) complexes are stable, whereas the corresponding Ni(II) complexes are not.

16.15 Indicate whether the following spinels are normal or inverse: $CuFe_2O_4$, $CuCr_2O_4$, $NiFe_2O_4$, $MnCr_2O_4$, $MgFe_2O_4$.

16.16 Suggest an explanation for the fact that MnO (NaCl structure) has no net magnetic moment.

16.17 Calculate the difference in ligand-field stabilization energy (in units of Δ_o) between octahedral and tetrahedral coordination for high-spin configurations from d^1 to d^{10}. Assume that $\Delta_t = \frac{4}{9}\Delta_o$.

16.18 Can you explain why it might be possible to observe the effects of Jahn-Teller distortion in the electronic spectrum of a crystal even though an X-ray diffraction study shows no evidence of distortion?

16.19 Explain the fact that the high-spin d^6 complex $CoF_6{}^{3-}$ shows two absorption bands in the visible spectral region.

THE 18-ELECTRON RULE

In Chap. 2 we discussed the basis and application of the Lewis octet theory. The theory ascribes special stability to an electronic configuration in which each nontransition-element atom achieves a number of valence electrons equal to that of a rare gas atom, i.e., eight valence electrons. Because transition-metal atoms have five d orbitals in addition to an s and three p orbitals in their valence shells, extension of the Lewis theory to transition-metal compounds corresponds to ascribing special stability to an electronic configuration in which each transition-metal atom achieves 18 valence electrons. Thus we have the 18-electron rule (alias the rare-gas, nine-orbital, or effective atomic number rule). This rule is not rigorously followed by some transition-metal compounds; in fact, Mitchell and Parish[1] pointed out that transition-metal compounds fall into three groups:

1 Compounds in which the electronic configurations are *completely unrelated* to the 18-electron rule
2 Compounds in which transition-metal atoms *never have more than* 18 valence electrons
3 Compounds which generally conform to the rule

[1] P. R. Mitchell and R. V. Parish, *J. Chem. Educ.*, **46**, 811 (1969).

The first group, the completely nonconforming compounds, consists of compounds with low Δ values. For example, in the case of octahedral complexes with low Δ values, there is essentially no restriction (except the Pauli restriction) on the number of electrons which can occupy the nonbonding t_{2g} and weakly antibonding $e_g{}^*$ orbitals. In principle, the number of valence electrons can range from 12 to 22. Some examples of such complexes are given in Table 17.1. It can be seen that all these examples involve first-row transition elements in low or medium oxidation states, with no strongly back-bonding ligands such as CO, olefins, or arenes. Notice that, although $Mn(CN)_6{}^{3-}$ is low-spin, it does not conform to the 18-electron rule.

The second group of compounds, in which the number of transition-metal valence electrons never *exceeds* 18, consists of compounds with relatively high Δ values but with ligands which do not engage in strong back bonding. In octahedral complexes of this type, there is essentially no restriction on the number of nonbonding t_{2g} electrons, but because of the high energy of the $e_g{}^*$ orbitals, electrons are forbidden to occupy these orbitals. Therefore the number of valence electrons can range only from 12 to 18. Some examples are given in Table 17.2. It can be seen that all the examples are second- and third-row transition-metal complexes, which have higher Δ values than the corresponding first-row metal complexes. There are no well-established complexes of these heavier metals with more than 18 valence electrons.

The third group of compounds, those which fairly rigorously conform to the 18-electron rule, consists of compounds with high Δ values and ligands which strongly back-bond. In octahedral complexes of this type, the t_{2g} orbitals are *bonding*, and therefore it is energetically favorable for these orbitals to be completely filled. Obviously the 18-electron rule is most useful as a predictive

Table 17.1 OCTAHEDRAL COMPLEXES WITH RELATIVELY LOW Δ VALUES

Complex	No. of d electrons	No. of valence electrons
$TiF_6{}^{2-}$	0	12
$VCl_6{}^{2-}$	1	13
$V(C_2O_4)_3{}^{3-}$	2	14
$Cr(NCS)_6{}^{3-}$	3	15
$Mn(CN)_6{}^{3-}$	4	16
$Fe(C_2O_4)_3{}^{3-}$	5	17
$Fe(H_2O)_6{}^{2+}$	6	18
$Co(H_2O)_6{}^{2+}$	7	19
$Ni(en)_3{}^{2+}$	8	20
$Cu(NH_3)_6{}^{2+}$	9	21
$Zn(en)_3{}^{2+}$	10	22

Table 17.2 COMPLEXES WITH HIGH Δ VALUES BUT WITH NO STRONG BACK BONDING†

Complex	No. of d electrons	No. of valence electrons
$ZrF_6{}^{2-}$	0	12
$ZrF_7{}^{3-}$	0	14
$Zr(C_2O_4)_4{}^{4-}$	0	16
$WCl_6{}^-$	1	13
$TcF_6{}^{2-}$	3	15
$OsCl_6{}^{2-}$	4	16
$W(CN)_8{}^{3-}$	1	17
$W(CN)_8{}^{4-}$	2	18
PtF_6	4	16
$PtF_6{}^-$	5	17
$PtF_6{}^{2-}$	6	18
$PtCl_4{}^{2-}$	8	16

† P. R. Mitchell and R. V. Parish, *J. Chem. Educ.*, **46**, 811 (1969).

guide for compounds of this category, particularly for π-bonded organometallic and carbonyl compounds. Some simple examples are given in Table 17.3.

Complexes with d^8 metal atoms sometimes have 18 valence electrons and sometimes only 16 valence electrons. The 18-electron complexes are found when the ligands are strongly back-bonding and can remove much of the electron density contributed to the metal atom by σ bonding. Thus one finds complexes such as $Fe(CO)_5$, $Fe(CNR)_5$, and $Pt(SnCl_3)_5{}^{3-}$. The 16-electron complexes are found with ligands which do not back-bond as strongly and cannot remove much electron density from the metal atom. Examples are $AuCl_4{}^-$, $PdCl_4{}^{2-}$, and $Ni(C_4H_7N_2O_2)_2$ [bis(dimethylglyoximato)nickel(II)]. The cyanide complexes of nickel(II) are a borderline case; both $Ni(CN)_4{}^{2-}$ and $Ni(CN)_5{}^{3-}$ are known.

Table 17.3 COMPLEXES WITH BACK BONDING WHICH CONFORM TO THE 18-ELECTRON RULE

Complex	No. of d electrons	No. of valence electrons
$V(CO)_6{}^-$	6	18
$Mo(CO)_3(PF_3)_3$	6	18
$HMn(CO)_5$	7	18
$Ni(CN)_5{}^{3-}$	8	18
$Fe(CO)_5$	8	18
$CH_3Co(CO)_4$	9	18
$Co(CO)_4{}^-$	10	18
$Ni(CNR)_4$	10	18

The 18-electron rule is so valuable that a detailed discussion of its application is warranted. In the following paragraphs we shall show how one can rationalize the structure and bonding of a series of compounds of various degrees of complexity. First let us consider a simple example, $Cr(CO)_6$. The chromium atom furnishes 6 valence electrons, and the six carbon monoxide ligands furnish 12 electrons.

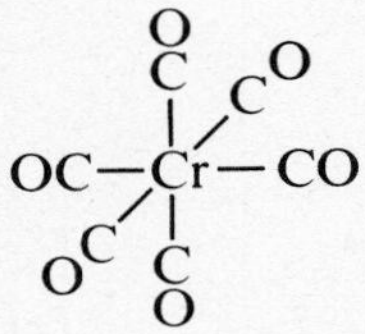

$$
\begin{array}{ll}
\text{Cr:} & 6 \\
\text{6 CO:} & \underline{12} \\
& 18
\end{array}
$$

In $Mn_2(CO)_{10}$, two $Mn(CO)_5$ groups are joined by an Mn—Mn bond. The formation of this bond effectively adds one valence electron to each manganese atom:

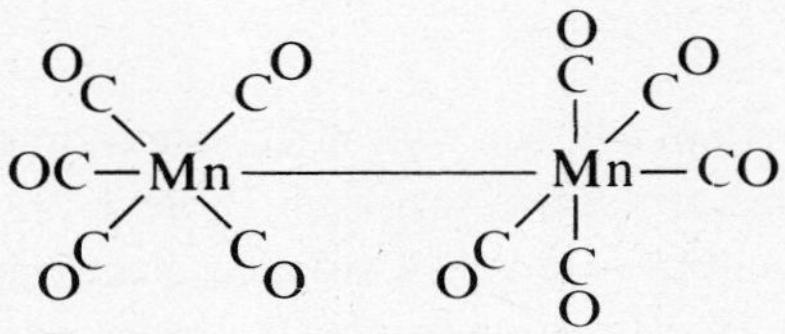

$$
\begin{array}{ll}
\text{Mn:} & 7 \\
\text{5 CO:} & 10 \\
\text{Mn—Mn bond:} & \underline{1} \\
& 18
\end{array}
$$

The Mn—Mn bond of $Mn_2(CO)_{10}$ can be easily cleaved by treatment with an alkali metal in an ether to give the $Mn(CO)_5^-$ ion. We may look upon this as a reduction of the Mn(0) to Mn($-$I).

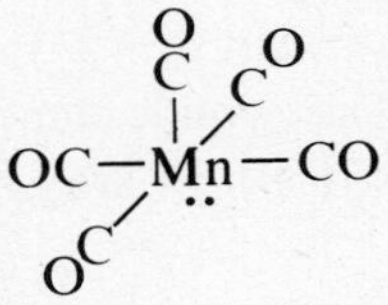

$$
\begin{array}{ll}
\text{Mn}^-: & 8 \\
\text{5 CO:} & \underline{10} \\
& 18
\end{array}
$$

The $Mn(CO)_5^-$ ion reacts with alkyl halides to give organic derivatives such as $CH_3Mn(CO)_5$. The CH_3 group may be considered as a one-electron donor.

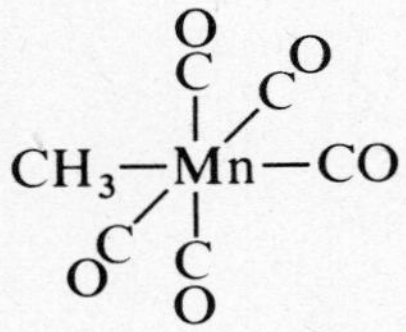

$$
\begin{array}{ll}
\text{Mn:} & 7 \\
\text{5 CO:} & 10 \\
\text{CH}_3: & \underline{1} \\
& 18
\end{array}
$$

Alternatively, we may consider the compound as a complex of Mn(I) and a CH_3^- ion:

$$
\begin{array}{lr}
\text{Mn}^+: & 6 \\
\text{5 CO}: & 10 \\
\text{CH}_3{}^-: & \underline{2} \\
& 18
\end{array}
$$

In $Fe_2(CO)_9$, we have bridging carbon monoxide ligands (which in effect contribute one electron to each metal atom) and an Fe—Fe single bond.

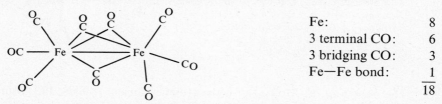

$$
\begin{array}{lr}
\text{Fe}: & 8 \\
\text{3 terminal CO}: & 6 \\
\text{3 bridging CO}: & 3 \\
\text{Fe—Fe bond}: & \underline{1} \\
& 18
\end{array}
$$

Ferrocene, $Fe(C_5H_5)_2$, may be looked upon as a complex of Fe(II) with two $C_5H_5{}^-$ ions (each with six π electrons) or as a complex of Fe(0) with two C_5H_5 radicals (each with five π electrons).

$$
\begin{array}{lr}
\text{Fe}^{2+}: & 6 \\
\text{2 C}_5\text{H}_5{}^-: & \underline{12} \\
& 18 \\
\\
\text{Fe}: & 8 \\
\text{2 C}_5\text{H}_5: & \underline{10} \\
& 18
\end{array}
$$

It is convenient to consider bisbenzenechromium as a complex of Cr(0) and benzene ligands:

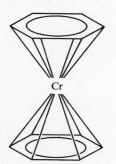

$$
\begin{array}{lr}
\text{Cr}: & 6 \\
\text{2 C}_6\text{H}_6: & \underline{12} \\
& 18
\end{array}
$$

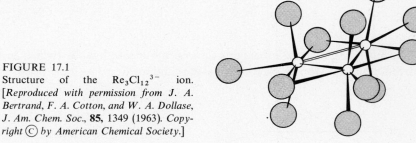

FIGURE 17.1
Structure of the $Re_3Cl_{12}^{3-}$ ion.
[*Reproduced with permission from J. A.
Bertrand, F. A. Cotton, and W. A. Dollase,
J. Am. Chem. Soc.,* **85,** 1349 (1963). *Copy-
right* © *by American Chemical Society.*]

The reaction of $Mo(CO)_6$ with cycloheptatriene yields the complex $C_7H_8Mo(CO)_3$, in which three essentially localized double bonds are coordinated to the molybdenum atom.

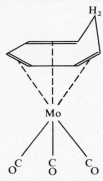

Mo:	6
C_7H_8:	6
3 CO:	6
	18

When this compound is treated with the triphenylcarbonium ion, $C(C_6H_5)_3^+$, a hydride ion is abstracted, yielding a cationic complex, $C_7H_7Mo(CO)_3^+$, containing the planar, aromatic $C_7H_7^+$ ring:

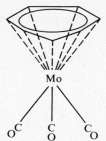

Mo:	6
$C_7H_7^+$:	6
3 CO:	6
	18

Consider the $Re_3Cl_{12}^{3-}$ ion, illustrated in Fig. 17.1. A molecular orbital treatment of the bonding[1] shows that the Re—Re bonds are double bonds.

---

[1] F. A. Cotton and G. Wilkinson, "Advanced Inorganic Chemistry," 3d ed., pp. 551–552, Interscience, New York, 1972.

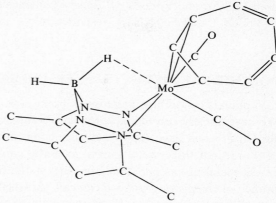

FIGURE 17.2

Skeletal representation of $[H_3B(3,5\text{-dimethylpyrazolyl})]Mo(CO)_2C_7H_7$. (*Reproduced with permission from F. A. Cotton et al., J. Chem. Soc. Chem. Commun., p. 777, 1972.*)

Indeed, such double bonding must also be assumed in order to obtain agreement with the 18-electron rule.

Re^{3+}:	4
$5\ Cl^-$:	10
2 Re=Re bonds:	4
	18

Figure 17.2 shows a skeletal representation of a molybdenum complex containing the cycloheptatrienyl group, C_7H_7, and the 3,5-dimethylpyrazolyl-borato anion, $H_2B(C_5H_7N_2)_2^-$. To rationalize this structure in terms of the 18-electron rule, it is necessary to assume that the C_7H_7 ring is bound in *trihapto* fashion and that the H atom near the molybdenum atom is engaged in a three-center B—H—Mo bond.

Mo^+:	5
2 N donor atoms:	4
B—H—Mo:	2
2 CO:	4
trihapto C_7H_7:	3
	18

Exceptions to the 18-electron rule are known, even for compounds with strongly back-bonding ligands. However, most of the exceptional compounds are readily converted to compounds which follow the rule, and, in a sense, this reactivity can be looked upon as conformity to the rule. For example, both $V(CO)_6$ and $Fe(C_5H_5)_2{}^+$ have 17 valence electrons and are easily reduced to $V(CO)_6{}^-$ and $Fe(C_5H_5)_2$, respectively.

POLYHEDRAL METAL CLUSTER COMPOUNDS

Compounds containing metal-metal bonds are called metal cluster compounds. We have already considered three such complexes [$Mn_2(CO)_{10}$, $Fe_2(CO)_9$, and $Re_3Cl_{12}{}^{3-}$] and have shown that they conform to the 18-electron rule. Clusters of more than three metal atoms generally consist of polyhedral arrangements of the metal atoms and thus are part of the fascinating and rapidly growing branch of chemistry which we can call "polyhedral cluster chemistry."

When applying the 18-electron rule to polyhedral clusters, it is usually convenient to treat the entire cluster as a unit. Assuming that each transition-metal atom conforms to the 18-electron rule, we write, for a given metal atom,

$$18 = v' + l' + 6 - c \tag{17.1}$$

Here v' is the number of valence electrons of the free metal atom, l' is the number of electrons donated by the ligands attached to the metal, and c is the number of electrons contributed by the metal atom to the cluster bonds. It is assumed that three of the metal orbitals are engaged in bonding to other cluster atoms; therefore $6 - c$ is the net number of electrons donated to the metal atom by the other cluster atoms. By rearrangement of Eq. 17.1 we obtain

$$c = v' + l' - 12$$

The total number of electrons involved in cluster bonding is the sum of the values of c for each transition-metal atom and the number of electrons contributed by any nontransition-element atoms in the cluster. It can be readily shown that the number of cluster electrons, C, is given by the following expression,

$$C = v + l + m - 12t - q \tag{17.2}$$

where v = total number of valence electrons of transition elements in cluster
l = total number of electrons donated by ligands to transition-element atoms
m = total number of electrons contributed to cluster by nontransition-element cluster atoms
t = number of transition-element atoms in cluster
q = charge of cluster

In Chap. 9, we pointed out that the polyhedral boranes, carboranes, and

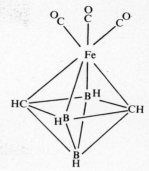

FIGURE 17.3
Proposed structure of $C_2B_3H_5Fe(CO)_3$.
(*R. N. Grimes, Am. N.Y. Acad. Sci., from Symposium on New Horizons in Organometallic Chemistry, August, 1973.*)

metallocarboranes generally contain $2n + 2$ framework electrons, where n is the number of framework atoms. The same rule has been shown[1-4] to apply to many metal cluster compounds which are based on triangulated polyhedra with polar axes—that is, based on polyhedra such as the trigonal bipyramid, the octahedron, the pentagonal bipyramid, and the icosahedron.

For example, the rhodium atoms in $Rh_6(CO)_{16}$ have the configuration of an octahedron. Each metal atom is bonded to two terminal CO groups, and four of the octahedral faces are capped by CO groups which bridge three metal atoms. In accord with the $2n + 2$ rule, the cluster has 14 electrons.

$$
\begin{aligned}
v &= 6 \times 9 = 54 \\
l &= 16 \times 2 = \underline{32} \\
  & 86 \\
-12t &= -12 \times 6 = \underline{-72} \\
  & 14
\end{aligned}
$$

It has been proposed[5] that $C_2B_3H_5Fe(CO)_3$ has an octahedral cluster, as shown by the structure in Fig. 17.3. Again, the cluster has 14 electrons:

$$
\begin{aligned}
v &= 8 \\
l = 3 \times 2 &= 6 \\
m\begin{cases} C: & 2 \times 3 = 6 \\ B: & 3 \times 2 = 6 \end{cases}& \\
&\underline{26} \\
-12t &= \underline{-12} \\
&14
\end{aligned}
$$

[1] K. Wade, *Inorg. Nucl. Chem. Letters,* **8,** 559, 563 (1972).
[2] K. Wade, *J. Chem. Soc. Chem. Commun.,* p. 792, 1971.
[3] D. M. P. Mingos, *Nat. Phys. Sci.,* **236,** 99 (1972).
[4] C. J. Jones, W. J. Evans, and M. F. Hawthorne, *J. Chem. Soc. Chem. Commun.,* p. 543, 1973.
[5] R. N. Grimes, *Ann. N. Y. Acad. Sci.,* from Symposium on New Horizons in Organometallic Chemistry, August, 1973.

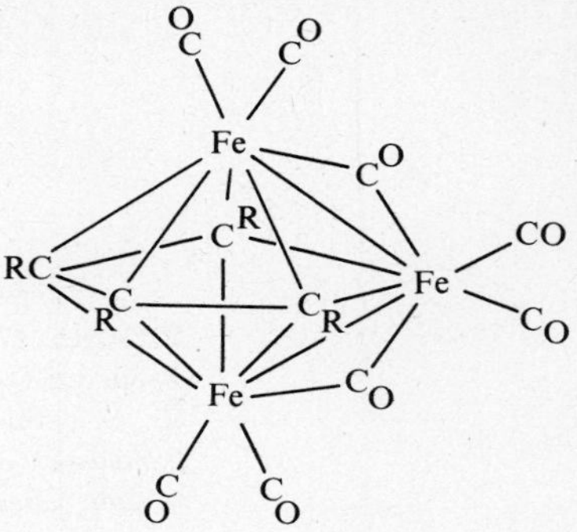

FIGURE 17.4
Structure of $(C_2R_2)_2Fe_3(CO)_8$. $R = C_6H_5$.

The triangulated polyhedron characteristic of seven-atom clusters is the pentagonal bipyramid. The bis(acetylene) complex $(C_2Ph_2)_2Fe_3(CO)_8$, whose structure[1] is shown in Fig. 17.4, is an example of a compound containing such a cluster. In agreement with the rule, the cluster contains 16 electrons:

$$
\begin{aligned}
v &= & 3 \times 8 &= & 24 \\
l &= & 8 \times 2 &= & 16 \\
m &= & 4 \times 3 &= & 12 \\
& & & & \overline{52} \\
-12t &= & -12 \times 3 &= & -36 \\
& & & & \overline{16}
\end{aligned}
$$

In the anion $(C_2B_9H_{11})Fe(\pi\text{-}C_5H_5)^-$, two carbon atoms, nine boron atoms, and an iron atom form the vertices of an icosahedron. This cluster contains 26 electrons, as expected:

$$
\begin{aligned}
v &= & 8 \\
l &= & 5 \\
m \begin{cases} C: 2 \times 3 &= & 6 \\ B: 9 \times 2 &= & 18 \end{cases} \\
& & \overline{37} \\
-12t &= & -12 \\
& & \overline{25} \\
-q &= & 1 \\
& & \overline{26}
\end{aligned}
$$

The $2n + 2$ rule may be applied to a polyhedron lacking a polar axis if the polyhedron is formally derivable from an n-vertex polyhedron with a polar

[1] R. P. Dodge and V. Schomaker, *J. Organomet. Chem.*, **3**, 274 (1965).

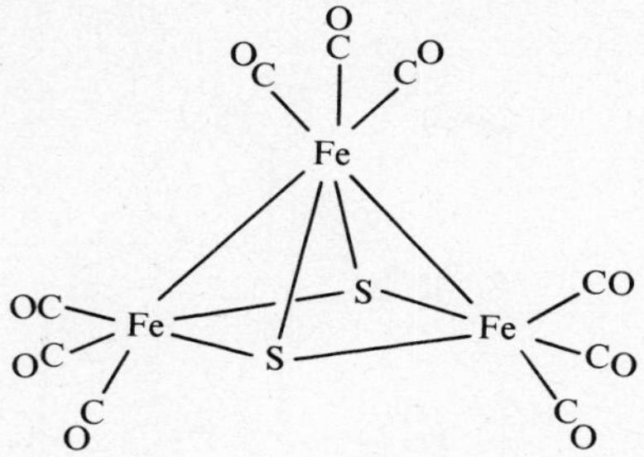

FIGURE 17.5
Structure of $Fe_3S_2(CO)_9$.

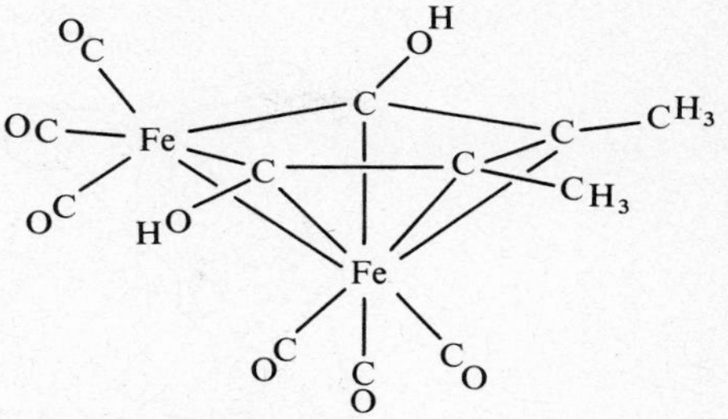

FIGURE 17.6
Structure of $Fe_2(CO)_6[C_6H_6(OH)_2]$.

axis by the removal of a vertex (to form a "nido" structure) or by the capping of a face. In such cases the number of cluster electrons is $2n + 2$, where n is the number of vertices of the polyhedron with a polar axis from which the cluster is formally derived. Thus a tetrahedral cluster is regarded as a derivative of a trigonal bipyramidal cluster, and 12 cluster electrons are expected. As a first example, let us consider $Ir_4(CO)_{12}$.

$$
\begin{aligned}
v = \quad & 4 \times 9 = \quad 36 \\
l = \quad & 12 \times 2 = \quad 24 \\
\hline
& \qquad\qquad 60 \\
-12t = & -12 \times 4 = -48 \\
\hline
& \qquad\qquad 12
\end{aligned}
$$

A second example is $(\pi\text{-}C_5H_5)RhFe_3(CO)_{11}$.[1]

$$
\begin{aligned}
v = \quad & 3 \times 8 + 9 = \quad 33 \\
l = \quad & 5 + 11 \times 2 = \quad 27 \\
\hline
& \qquad\qquad 60 \\
-12t = & -12 \times 4 \quad = -48 \\
\hline
& \qquad\qquad 12
\end{aligned}
$$

The cluster in $Os_6(CO)_{18}$ ($C = 12$) is a capped trigonal bipyramid, and the cluster in $Os_7(CO)_{21}$ ($C = 14$) is a capped octahedron.[2] The complex $Fe_3S_2(CO)_9$ ($C = 14$) has a structure based on a square pyramid,[3] as shown in Fig. 17.5, and $Fe_2(CO)_6[C_6H_6(OH)_2]$ ($C = 16$) has a structure based on a pentagonal pyramid,[4] as shown in Fig. 17.6.

[1] M. R. Churchill and M. V. Veidis, *J. Chem. Soc. Chem. Commun.*, p. 1470, 1970.
[2] R. Mason, K. M. Thomas, and D. M. P. Mingos, *J. Am. Chem. Soc.*, **95**, 3802 (1973).
[3] C. H. Wei and L. F. Dahl, *Inorg. Chem.*, **4**, 493 (1965).
[4] A. A. Hock and O. S. Mills, *Acta Cryst.*, **14**, 139 (1961).

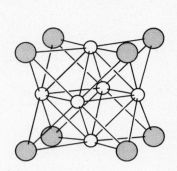

FIGURE 17.7
Structure of the $Mo_6Cl_8^{4+}$ complex.

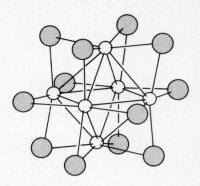

FIGURE 17.8
Structure of the $Nb_6Cl_{12}^{2+}$ complex.

Applications

The largely empirical relations between cluster geometry and the number of cluster electrons have permitted the systematization and prediction of the structures of many unusual compounds. But perhaps even more important, these concepts have been successfully used to design syntheses of new types of compounds. Thus, recognition of the analogy between $(\pi\text{-}C_5H_5)Co$ and BH (each contributes two electrons and three orbitals to polyhedral bonding) and the analogy between $(\pi\text{-}C_5H_5)Ni$ and CH (each contributes three electrons and three orbitals) has led to the preparation of novel metallocarboranes such as $(\pi\text{-}C_5H_5Ni)_2B_{10}H_{10}$ and $(\pi\text{-}C_5H_5)_2NiCoCB_7H_8$.[1,2] Wade[3] has suggested many novel synthetic reactions based on similar analogies.

Exceptions

It is important to point out that the $2n + 2$ rule, like the 18-electron rule, does not always work. We shall now consider a few important exceptions.

The complex $Mo_6Cl_8^{4+}$, pictured in Fig. 17.7, consists of an octahedron of molybdenum atoms with a chlorine atom over each face. This complex cannot be rationalized by the usual method of applying the $2n + 2$ rule. However, it is perhaps significant that if all the chloride ions are stripped from the complex, leaving the Mo_6^{12+} cluster, the number of valence electrons is 24, or just enough to permit the formation of an Mo—Mo bond along each edge of the octahedron.

[1] K. Wade, *J. Chem. Soc. Chem. Commun.*, p. 792, 1971.

[2] C. G. Salentine and M. F. Hawthorne, unpublished work.

[3] K. Wade, *Inorg. Nucl. Chem. Letters*, **8**, 559, 563 (1972); *J. Chem. Soc.* Chem. Commun., p. 792, 1971.

The complex $Nb_6Cl_{12}{}^{2+}$, pictured in Fig. 17.8, consists of an octahedron of niobium atoms with a chlorine atom over each edge. Again, the $2n + 2$ rule does not seem to be applicable. However, the $Nb_6{}^{14+}$ cluster has 16 valence electrons, enough to form eight three-center Nb—Nb—Nb bonds on the faces of the octahedron.

The clusters $Bi_5{}^{3+}$, $Bi_8{}^{2+}$, and $Bi_9{}^{5+}$ have been prepared by heating mixtures of elemental bismuth and $BiCl_3$ in molten salts.[1] Although $Bi_5{}^{3+}$ conforms to the $2n + 2$ rule, the other species have more electrons than correspond to $2n + 2$. Perhaps these discrepancies are due to the fact that excessively high positive charges would be attained if the Bi_8 and Bi_9 species were to conform to the rule.

Obviously the $2n + 2$ rule for polyhedral clusters must be used with discretion. Our present treatment of the bonding in these systems is much like the use which chemists made of rudimentary periodic tables in the nineteenth century. We can hope that in the next decade our understanding of these systems will improve and that the empirical rules will be refined or replaced with more reliable rules.

OXIDATION-STATE AMBIGUITY

Nitrosyls

The following reactions illustrate four methods for the preparation of transition-metal nitrosyls (i.e., complexes containing the NO group as a ligand).

$$Fe_2(CO)_9 + 4NO \rightarrow 2Fe(NO)_2(CO)_2 + 5CO$$

$$Ir(PPh_3)_2(CO)Cl + NO^+ \rightarrow [Ir(PPh_3)_2(CO)(NO)Cl]^+$$

$$CoI_2 + Co + 4NO \rightarrow [Co(NO)_2I]_2$$

$$Ni(CN)_4{}^{2-} + NO^- \rightarrow Ni(CN)_3NO^{2-} + CN^-$$

In many nitrosyl complexes, the metal—N—O bond angle is very close to $180°$, and it is convenient to assume that the ligand is the $:N{\equiv}O:^+$ ion, which is isoelectronic with CO. In other complexes, the metal—N—O bond angle is around $130°$, and it is reasonable to assume that the ligand is $\overset{..}{N}{=}\overset{..}{O}:^-$, isoelectronic with O_2. However, these assumptions are somewhat arbitrary. For example, one may look upon $trans$-$Co(diars)_2(NCS)NO^+$ as a complex of Co(III) and NO^-,

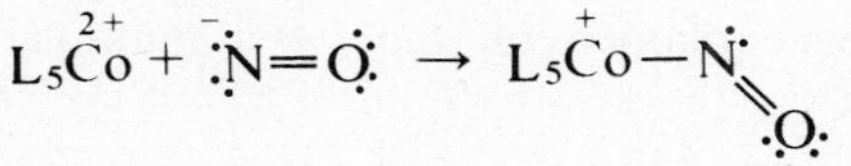

$$L_5\overset{2+}{Co} + \overset{-}{:}N{=}\overset{..}{O}: \rightarrow L_5\overset{+}{Co}{-}N\diagdown\overset{}{\underset{:O:}{}}$$

[1] J. D. Corbett, *Inorg. Chem.*, **7**, 198 (1968).

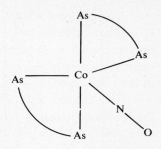

FIGURE 17.9
Structure of the $Co(diars)_2NO^{2+}$ ion. The C and H atoms of the diarsine ligands have been omitted for clarity. [*From J. H. Enemark and R. D. Feltham, Proc. Natl. Acad. Sci. U.S.*, **69**, 3534 (1972).]

or as a complex of Co(I) and NO^+,

$$L_5Co: + :N\equiv O:^+ \rightarrow L_5\overset{+}{Co}-\overset{..}{N}\underset{:O:}{\diagdown}$$

In the former case, NO^- acts as a σ donor; in the latter case, NO^+ acts as a σ acceptor. In either case, it is easy to predict the bent configuration for the Co—NO linkage; a linear configuration would force an extra pair of electrons onto the cobalt atom, in violation of the 18-electron rule. The closely related complex, $Co(diars)_2NO^{2+}$, may be looked upon as a complex of Co(I) and NO^+,

$$L_4\overset{..}{Co}^+ + :N\equiv O:^+ \rightarrow L_4\overset{..}{Co}-\overset{+}{N}\equiv O:^+$$

or as a complex of Co(III) and NO^-,

$$L_4Co^{3+} + {}^-:\overset{..}{N}=\overset{..}{O}: \rightarrow L_4\overset{..}{Co}-\overset{+}{N}\equiv O:^+$$

The linear geometry is readily predicted if one assumes that the cobalt atom must have 18 valence electrons. A bent configuration would remove a pair of electrons from the valence shell of the cobalt atom, leaving only 16 electrons. Of course, considerable back bonding occurs,

$$L_4\overset{..}{Co}-\overset{+}{N}\equiv O:^+ \leftrightarrow L_4\overset{+}{Co}=\overset{+}{N}=\overset{..}{O}:$$

but this interaction does not remove electrons from the valence shell of the cobalt atom.

The $Co(diars)_2NO^{2+}$ ion has a trigonal bipyramidal arrangement of donor atoms[1] (shown in Fig. 17.9), with the linear nitrosyl group occupying an equatorial position. The fact that the nitrosyl group occupies an equatorial position, rather than an axial position, can be explained on the basis that $d\pi$ back

---

[1] J. H. Enemark and R. D. Feltham, *Proc. Natl. Acad. Sci. U.S.*, **69**, 3534 (1972).

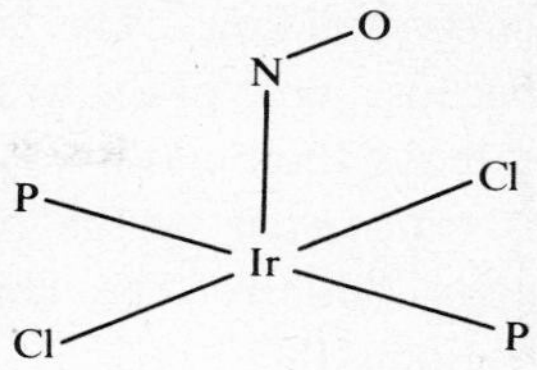

FIGURE 17.10
Structure of $IrCl_2(NO)(PPh_3)_2$. The C
and H atoms have been omitted for clarity.
[*From D. M. P. Mingos and J. A. Ibers,*
Inorg. Chem., **10**, 1035 (1971).]

bonding is favored at an equatorial position relative to an axial position. The $d_{x^2-y^2}$ and d_{xy} orbitals are available only to the three equatorial ligands, corresponding to $\frac{2}{3}$ orbital per equatorial ligand. The d_{xz} and d_{yz} orbitals are available to all five ligands. If we apportion these orbitals to the axial and equatorial ligands in proportion to the number of symmetry-allowed π bonds that can be formed,[1] each axial ligand gets only $\frac{4}{7}$ orbital whereas each equatorial ligand gets a total of $(\frac{2}{3} + \frac{2}{7})$ or $\frac{20}{21}$ orbital. It is reasonable to assume that a linear nitrosyl group, which, in the absence of back bonding, has two atoms with $+1$ formal charges, would engage in stronger back bonding than an organoarsine ligand. Therefore one predicts that the stable structure would be that with an equatorial nitrosyl group.

When $Co(diars)_2NO^{2+}$ is treated with thiocyanate ion, *trans*-$Co(diars)_2$-$(NCS)NO^+$ is formed. The addition of the electron pair of the NCS^- ion to the cobalt atom would cause the 18-electron rule to be violated if no electronic rearrangement took place. Actually, the addition causes an electron pair to transfer from the cobalt to a nonbonding orbital of the nitrogen atom, thus enforcing the bent nitrosyl configuration:

$$SCN^- + [L_4Co{=}N{=}O]^{2+} \rightarrow \left[(SCN)L_4Co{-}\overset{\displaystyle\cdot\cdot}{N}\underset{\displaystyle O}{\diagdown\diagdown}\right]^+$$

Enemark and Feltham[2] have called this a "stereochemical control of valence."

The 5-coordinate compound $IrCl_2(NO)(PPh_3)_2$, pictured in Fig. 17.10, is interesting because, although it is isoelectronic with $Co(diars)_2NO^{2+}$, it has an entirely different structure.[3] We have already pointed out that the $Co(diars)_2$ NO^{2+} structure corresponds to a complex of d^8Co^+ and NO^+, with 18 valence electrons. Although most d^8 complexes have only 16 valence electrons, 18 valence electrons are allowed in this case because the ligands engage in strong

---

[1] For example, the d_{xz} orbital is suited for π-bonding to both of the two axial ligands, but only to $1\frac{1}{2}$ of the three equatorial ligands. Thus we assign $\frac{1}{3\frac{1}{2}}$ or $\frac{2}{7}$ of the orbital to a given axial ligand. Similar comments apply to the d_{yz} orbital.

[2] Enemark and Feltham, op. cit.

[3] D. M. P. Mingos and J. A. Ibers, *Inorg. Chem.*, **10**, 1035 (1971).

back bonding. The analogous structure for $IrCl_2(NO)(PPh_3)_2$ is unstable because most of the ligands do not engage in strong back bonding. To compensate for the buildup of negative charge on the iridium atom, a pair of electrons is transferred to the NO^+ ion, converting it into an NO^- ion with a bent coordination. The compound may be looked upon as a d^6 complex of iridium(III).

The strength of the N—O bond and the charge on the nitrogen atom in a metal nitrosyl complex would be expected to depend on whether the NO coordination is bent or linear, and, in the case of linear coordination, on the degree of $d\pi \rightarrow \pi^*$ back bonding. On the basis of valence bond structures for linear co-

ordination $\left(\bar{M}—\overset{+}{N}\equiv O\!:^+ \longleftrightarrow M=\overset{+}{N}=\overset{..}{O}\!: \right)$ and bent coordination $\left(M—\overset{..}{N}\diagdown\overset{}{\underset{..}{O}} \right)$

one would expect the N—O bond to be weaker in the bent complexes and in those complexes with strong back bonding. Strong back bonding should be favored in complexes of metal atoms with relatively high electron density. The nitrogen atom charge would be expected to be lower (more negative) in the bent complexes. The N—O bond strength can be measured by the N—O stretching frequency and the N—O bond distance, and the nitrogen atom charge can be measured by the nitrogen $1s$ electron binding energy. In Table 17.4 these quantities are tabulated for various transition-metal nitrosyl complexes.[1] As expected, the low stretching frequencies (corresponding to weak bonds) are found

[1] P. Finn and W. L. Jolly, *Inorg. Chem.*, **11**, 893 (1972).

Table 17.4   PROPERTIES OF SOME TRANSITION-METAL NITROSYL COMPLEXES

Complexes with linearly coordinated NO groups	Oxidation state of metal, assuming NO^+	NO stretching frequency, cm^{-1}	$N1s\ E_B$, eV	r_{N-O}, Å
$Na_2[Fe(NO)(CN)_5]$	Fe(II)	1939	403.3	1.13
trans-$[FeCl(NO)(diars)_2](ClO_4)_2$	Fe(II)	1865	402.9	
$[Ir(NO)_2(PPh_3)_2]PF_6$	Ir(−I)	1740	400.2	1.21
$K_3[Mn(NO)(CN)_5]$	Mn(I)	1725	399.7	1.21
$Rh(NO)(PPh_3)_3$	Rh(−I)	1650	400.8	
$K_3[Cr(NO)(CN)_5]$	Cr(I)	1645	400.7	1.21

Complexes with bent NO groups	Oxidation state of metal, assuming NO^-	NO stretching frequency, cm^{-1}	$N1s\ E_B$, eV	r_{N-O}, Å
$[Co(NO)(NH_3)_5]Cl_2$	Co(III)	1620	400.7	1.15
$RhI_2(NO)(PPh_2CH_3)_2$	Rh(III)	1628	400.3	~1.23
trans-$[CoCl(NO)(diars)_2]Cl$	Co(III)	1550	400.5	

for the bent nitrosyls and for the linear nitrosyls of metals in very low oxidation states. The tabulated stretching frequencies, which range from 1550 to 1939 cm^{-1}, may be compared with the frequencies for free NO^+ (2200 cm^{-1}), free NO (1840 cm^{-1}), and organic nitroso compounds, RNO (~ 1550 cm^{-1}). The low nitrogen $1s$ binding energies (corresponding to relatively negative nitrogen atom charges) are found for the bent nitrosyls (as expected) and for the linear nitrosyls of metals in very low oxidation states. The correlation with oxidation state is reasonable inasmuch as one would expect the charge on the nitrogen atom to be affected by the charge on the adjacent metal atom. The N—O bond-distance data for the bent nitrosyls are inconclusive, but the data for the linear nitrosyls are consistent with weaker bonds for the nitrosyls of metals in low oxidation states.

Bipyridyl Complexes[1]

Bipyridyl (bipy) acts as a bidentate ligand toward many metal atoms.

The free molecule reacts in ether solutions with alkali metals to form the anionic species $bipy^-$ (paramagnetic) and $bipy^{2-}$ (diamagnetic), in which the extra electrons presumably occupy the π^* orbitals of the molecule. Transition-metal complexes of bipyridyl can similarly be reduced. For example, $Cr(bipy)_3^{2+}$ can be reduced to $Cr(bipy)_3^+$, $Cr(bipy)_3$, $Cr(bipy)_3^-$, $Cr(bipy)_3^{2-}$, $Cr(bipy)_3^{3-}$, and even $Cr(bipy)_3^{6-}$, as indicated by the formulas in Table 17.5 In all these species,

[1] W. R. McWhinnie and J. D. Miller, *Adv. Inorg. Chem. Radiochem.*, **12**, 135 (1969).

Table 17.5   MAGNETIC   MOMENTS   OF CHROMIUM BIPYRIDYL COMPLEXES†

Compound	μ_{eff}
$Cr(bipy)_3I_2$	2.9
$Cr(bipy)_3I$	2.07
$Cr(bipy)_3$	0.0
$Li[Cr(bipy)_3]\cdot 4THF$	1.83
$Na_2[Cr(bipy)_3]\cdot 7THF$	~ 2.85
$Na_3[Cr(bipy)_3]\cdot 7THF$	~ 3.85
$Ca_3[Cr(bipy)_3]\cdot 7NH_3$	~ 2.46

† W. R. McWhinnie and J. D. Miller, *Adv. Inorg. Chem. Radiochem.*, **12**, 135 (1969).

there is ambiguity as to the oxidation state of the chromium. Thus $Cr(bipy)_3$ could be formulated as a $Cr(III)$ complex of three bipy$^-$ anions, as a $Cr(0)$ complex of three bipy ligands, or as several other possible combinations. The magnetic data of Table 17.5 help to rule out some of the possibilities. Inasmuch as $Cr(bipy)_3$ is diamagnetic, it is reasonable to describe it as a d^6 chromium(0) complex in which much of the electron density of the metal has been delocalized onto the ligands by $d\pi \rightarrow \pi^*$ back bonding. In the anionic complexes, the unpaired electron density is probably principally in the π^* orbitals of the bipyridyl groups.

Metal Dithienes[1]

The reaction of metallic nickel with sulfur and diphenylacetylene yields the planar 4-coordinate complex indicated in the following reaction:

$$Ni + 4S + 2C_2Ph_2 \rightarrow$$

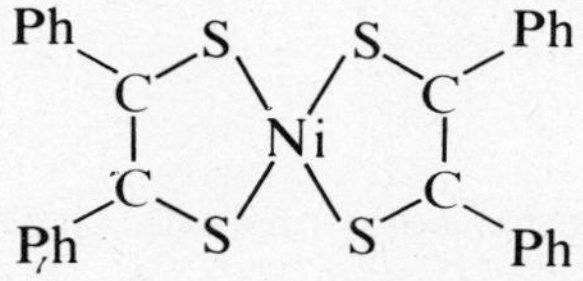

This complex can be reduced to a monoanion, $Ni(S_2C_2Ph_2)_2{}^-$, and a dianion, $Ni(S_2C_2Ph_2)_2{}^{2-}$, which have essentially the same structure. There is ambiguity in the assignment of oxidation states to the nickel atoms in these complexes. For example, the neutral complex can be formulated (1) as a nickel(0) complex of two dithioketone ligands,

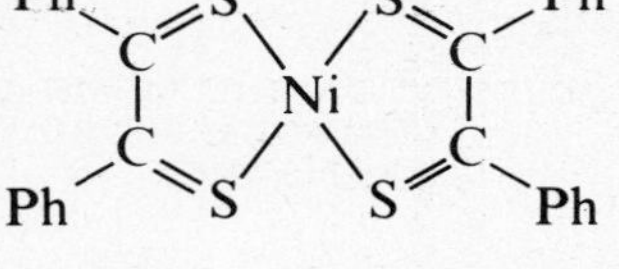

(2) as a nickel(II) complex of a dithioketone and a dithiolate anion,

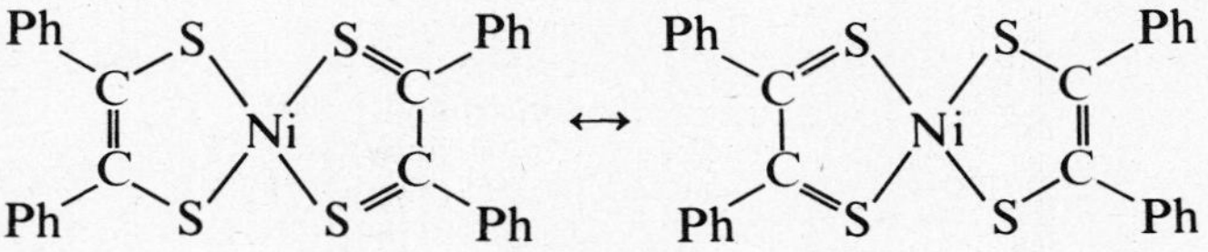

---

[1] G. N. Schrauzer, *Acc. Chem. Res.*, **2**, 72 (1969).

or (3) as nickel(IV) complex of two dithiolate anions,

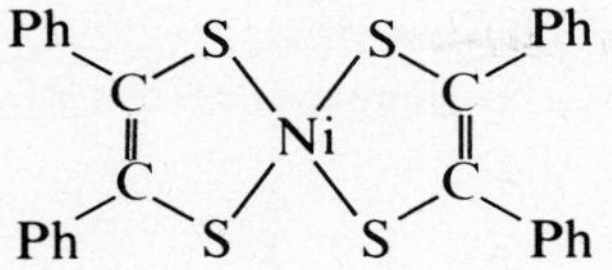

As a kind of noncommital compromise, these compounds have been labeled "dithiene" complexes.

Some experimental evidence favors considering all three species as Ni(II) species, on which basis the reduction is formulated as follows:

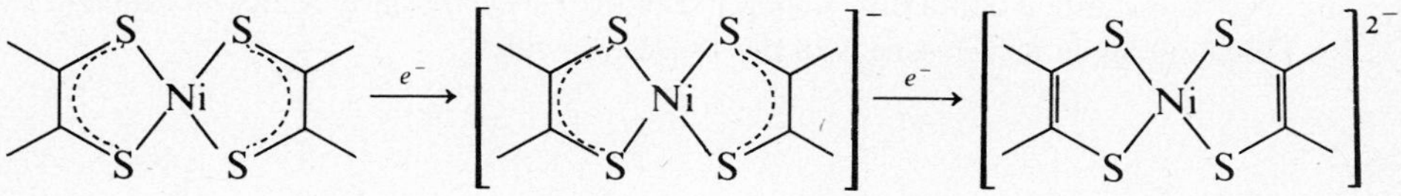

On successive reduction of the neutral complex, the stretching frequency of the C=C bond increases and that of the C=S bond decreases. The infrared and visible absorption spectra of the -2 complex are similar to those of organo-substituted ethylenedithiolates. The nickel $2p_{3/2}$ binding energies for all three species are practically the same (852.9, 852.5, and 852.8 eV, respectively), corresponding to similar charges on the nickel atom in all three species.[1]

PROBLEMS

17.1 Propose a formula and structure for an iron carbonyl complex of cyclobutadiene. Show that the 18-electron rule and the polyhedral cluster rules are obeyed.

17.2 Propose a structure for the acetylene complex $(C_6H_5)_2C_2Fe_3(CO)_9$ which satisfies the 18-electron and polyhedral cluster rules.

17.3 Rationalize the bonding in the molecule $H_6Cu_6(PPh_3)_6$, which contains an octahedral Cu_6 cluster.

17.4 Propose a formula and structure for $Os_5(CO)_x$. Show that the 18-electron and polyhedral cluster rules are satisfied.

17.5 Suggest a synthetic method to cap the rings of ferrocene to produce two pentagonal bipyramidal clusters joined by a common iron atom. Show that the 18-electron and polyhedral cluster rules would be satisfied in the product.

17.6 Which of the following complexes are likely to be unstable: $ZrCl_4$, diamagnetic $NiCl_4^{2-}$, $Ti(CO)_4^{4+}$, $Cd(CO)_3$, $Fe_3(CO)_{12}$, AuF_5^{2-}? Why?

17.7 Which is a stronger oxidizing agent, $Fe(CN)_6^{3-}$ or $Fe(H_2O)_6^{3+}$? Why?

[1] S. O. Grim, L. J. Matienzo, and W. E. Swartz, *J. Am. Chem. Soc.*, **94**, 5116 (1972).

17.8   Predict the structures of $Fe(CO)_4CN^-$, $Mn(CO)_4NO$, $Co(NO)_3$, and $[Fe(NO)_2Br]_2$.

17.9   For the "brown ring" species, $Fe(H_2O)_5NO^{2+}$, the N—O stretching frequency is 1745 cm^{-1} and $\mu_{eff} = 3.9$. Describe the bonding in this complex.

17.10   Show how oxidation-state ambiguities could arise in complexes of the following dianion:

17.11   Nickelocene, $Ni(C_5H_5)_2$, has a structure analogous to that of ferrocene, but the Ni—C distance is about 0.16 Å longer than the Fe—C distance. Can you rationalize the bonding in nickelocene with the 18-electron rule?

CATALYSTS AND SOME BIOLOGICAL SYSTEMS

CATALYSTS[1–5]

General Reaction Types

A reaction is said to be catalyzed when its rate is increased by the presence of a substance which does not appear in the equation for the net, overall reaction. In general, a catalyzed reaction is the sum of a sequence of different reactions. In the case of reactions catalyzed by transition-metal complexes, these constituent reactions can usually be categorized into three broad types. We shall discuss these three types of reactions before describing the detailed mechanisms of particular catalyzed reactions.

1 The transition-metal atom involved in the catalysis often has a variable coordination number. For example, many d^8 metal complexes can, by gain

[1] F. A. Cotton and G. Wilkinson, "Advanced Inorganic Chemistry," 3d ed., pp. 770–800, Interscience, New York, 1972.

[2] H. J. Emeléus and A. G. Sharpe, "Modern Aspects of Inorganic Chemistry," pp. 621–631, Wiley, New York, 1973.

[3] J. Halpern, Oxidative Addition Reactions, *Acc. Chem. Res.*, **3**, 386 (1970).

[4] J. P. Collman, Reactions Related to Homogeneous Catalysis, *Acc. Chem. Res.*, **1**, 136 (1968).

[5] R. Cramer, Rhodium-promoted Reactions of Olefins, *Acc. Chem. Res.*, **1**, 186 (1968).

or loss of ligands, readily convert between 4-coordination and 5-coordination. Thus the following type of reversible reaction is commonly involved in catalysis:

$$\underset{d^8}{L_4M} + X \rightleftharpoons \underset{d^8}{L_4MX}$$

Here X is a species such as CO or an olefin, which can donate an electron pair to the metal atom. We shall refer to the forward reaction as a "simple addition" and to the reverse reaction as a "dissociation."

2 Sometimes when a molecule adds to a complex, the molecule is cleaved into two fragments which are bonded separately to the metal atom. The oxidation state of the metal effectively increases by two units, and, unless the metal atom loses a ligand in the process, the coordination number also increases by two units. This type of reaction, called "oxidative addition," often occurs with d^8 complexes:

$$\underset{d^8}{L_4M} + YZ \rightleftharpoons \underset{d^6}{L_4M}\!\!\begin{array}{c} {\diagup Y} \\ {\diagdown Z} \end{array}$$

The added molecule is typically a small molecule such as H_2, RX, HX, X_2, etc.

3 A third important type of reaction involved in catalysis is *intramolecular rearrangement*. Often this reaction amounts to the *insertion* of one ligand between the metal atom and another ligand. Two examples follow:

In each of these examples, the coordination number of the metal atom decreases by 1.

Olefin Isomerization

Many transition-metal compounds catalyze the migration of double bonds in alkenes. The mechanism involves the reversible transfer of a hydrogen atom from the transition metal to the coordinated olefin to give a σ-bonded alkyl group. For example, the $HCo(CO)_3$-catalyzed isomerization of allyl alcohol to propionaldehyde is believed to proceed by the following mechanism:

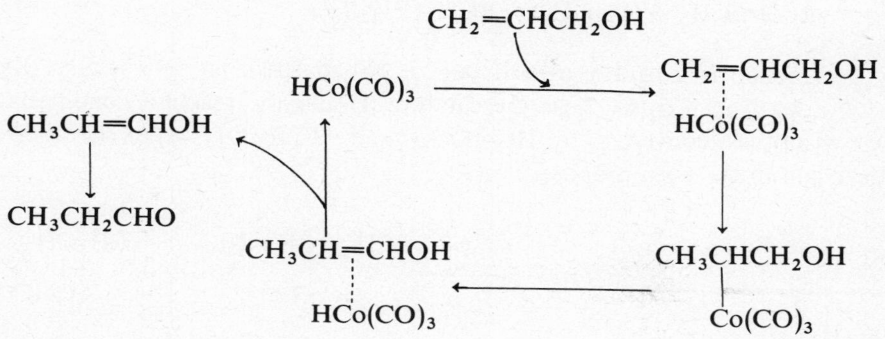

$$CH_2{=}CHCH_2OH$$

$$HCo(CO)_3 \longrightarrow \quad \begin{array}{c} CH_2{=}CHCH_2OH \\ \vdots \\ HCo(CO)_3 \end{array}$$

$$CH_3CH{=}CHOH$$
$$\downarrow$$
$$CH_3CH_2CHO$$

$$\begin{array}{c} CH_3CH{=}CHOH \\ \vdots \\ HCo(CO)_3 \end{array} \longleftarrow \begin{array}{c} CH_3CHCH_2OH \\ | \\ Co(CO)_3 \end{array}$$

Olefin Polymerization

The Ziegler-Natta process for the polymerization of olefins uses a catalyst formed from $TiCl_4$ and $AlEt_3$. The $AlEt_3$ reduces the $TiCl_4$ to a fibrous form of $TiCl_3$ and replaces some of the surface chlorine atoms with ethyl groups. The essential step in the catalysis is a four-center reaction between an alkyl group and a coordinated olefin to give a longer alkyl group, as shown in the following scheme:

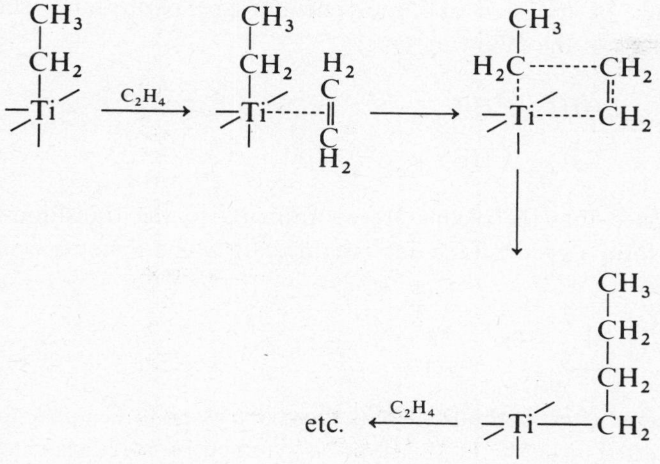

$$\begin{array}{c} CH_3 \\ | \\ CH_2 \\ | \\ {\rightharpoondown}Ti \end{array} \xrightarrow{C_2H_4} \begin{array}{c} CH_3 \\ | \\ CH_2 \\ | \\ {\rightharpoondown}Ti\cdots \begin{array}{c} H_2 \\ C \\ \| \\ C \\ H_2 \end{array} \end{array} \longrightarrow \begin{array}{c} CH_3 \\ | \\ H_2C\cdots\cdots CH_2 \\ {\rightharpoondown}Ti\cdots\cdots CH_2 \\ | \end{array}$$

$$\downarrow$$

$$\begin{array}{c} CH_3 \\ | \\ CH_2 \\ | \\ CH_2 \\ | \end{array}$$

$$\text{etc.} \xleftarrow{C_2H_4} \begin{array}{c} {\rightharpoondown}Ti{-}CH_2 \\ | \end{array}$$

Hydroformylation

The hydroformylation reaction is the addition of H_2 and CO to an olefin to form an aldehyde:

$$RCH{=}CH_2 + H_2 + CO \rightarrow RCH_2CH_2CHO$$

The reaction is generally carried out at 100 to 200°C at pressures of 100 to 400 atm, using $Co_2(CO)_8$ as the catalyst. Under the reaction conditions, the $Co_2(CO)_8$ is converted to $HCo(CO)_4$ and $HCo(CO)_3$, and the following mechanism has been proposed:

$$
\begin{array}{ccccc}
 & & C_2H_4 & CH_2{=}CH_2 & CH_3CH_2 \\
 & HCo(CO)_3 \longrightarrow & & | & | \\
\overset{O}{\underset{\parallel}{C_2H_5CH}} & \longleftarrow & & H\overset{.}{C}o(CO)_3 \longrightarrow & Co(CO)_3 \\
 & \uparrow & & & \downarrow CO \\
 & \overset{O}{\underset{\parallel}{C_2H_5CCoH_2(CO)_3}} & \overset{H_2}{\longleftarrow} & \overset{O}{\underset{\parallel}{C_2H_5CCo(CO)_3}} \longleftarrow & \overset{CH_3CH_2}{\underset{\underset{Co(CO)_4}{|}}{}}
\end{array}
$$

Side reactions of this process are the hydrogenation of the olefin to the saturated hydrocarbon and the reduction of the aldehyde to the alcohol.

Olefin Disproportionation

Certain compounds of tungsten and molybdenum are particularly effective catalysts for reactions of the following type:

$$
\begin{array}{ccc}
CH_2{=}CHR & & CH_2 \quad CHR \\
+ & \rightleftharpoons & \parallel \ + \parallel \\
CH_2{=}CHR & & CH_2 \quad CHR
\end{array}
$$

It has been suggested that the olefins are coordinated to the transition-metal atom, where they form a cyclobutane-like intermediate which then dissociates:

$$
\left\| \cdot\cdot M \cdot\cdot \right\| \rightarrow \boxed{M} \rightarrow \overline{\overline{\underset{\cdot\cdot}{\overset{\cdot\cdot}{M}}}}
$$

By use of orbital symmetry methods such as those discussed in Chap. 4, we can show that a concerted reaction of this type is symmetry-forbidden. The two olefin molecules must be considered as a reacting unit, or "molecule." Hence we must consider the two different combinations of the carbon $p\pi$ orbitals which constitute the two bonding π MOs:

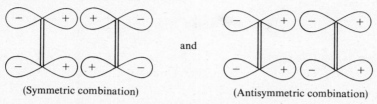

and

(Symmetric combination)   (Antisymmetric combination)

The symmetric combination has the same symmetry as (and a finite overlap with) the symmetric bonding combination of the horizontal $p\sigma$ atomic orbitals of the cyclobutane molecule:

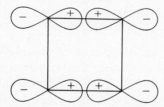

However, the antisymmetric combination does not have the symmetry[1] of (and has zero net overlap with) the antisymmetric bonding combination of the horizontal $p\sigma$ atomic orbitals of the cyclobutane molecule:

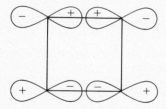

In order for the reaction to proceed, a pathway must be provided for transferring the pair of electrons from the antisymmetric π MO to the antisymmetric σ MO. It has been proposed[2] that a d^2 tungsten(IV) complex can provide this pathway in the form of its d orbitals. If we assume that the two olefin molecules are parallel to the y axis of the tungsten atom, that the common plane of the carbon atoms is perpendicular to the z axis of the tungsten atom, and that a pair of electrons occupies the d_{yz} orbital, the "forbidden" electron transfer can

---

[1] Notice that the symmetry designation for the π MOs are given with respect to the axis which is perpendicular to the p orbitals and that those for the σ MOs are given with respect to the axis which is parallel to the p orbitals.

[2] F. D. Mango and J. H. Schachtschneider, *J. Am. Chem. Soc.*, **93**, 1123 (1971).

take place by the simultaneous shift of electrons from the antisymmetric π MO to the d_{xz} orbital and from the d_{yz} orbital to the antisymmetric σ MO:

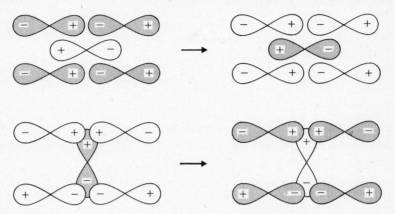

This ingenious postulated mechanism may be moot, however, if olefin reactions of this type do not actually proceed by concerted processes. Cassar, Eaton, and Halpern[1] have obtained evidence which suggests that the rhodium-catalyzed isomerization of cubane to *sym*- tricyclooctadiene proceeds by a nonconcerted process involving oxidative addition of the cubane to the rhodium complex:

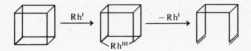

This type of reaction is symmetry-allowed, and other, similar, rearrangements may involve analogous mechanisms.[2]

Template Syntheses[3]

The term "template synthesis" has been suggested for reactions in which a central metal ion serves as a "template" on which to coordinate ligands which then react to form chelate rings. In some cases this process can surround the metal ion with a fused cyclic ring system, i.e., a "macrocyclic" ligand. A common procedure involves the cyclic condensation of an α-diketone with a 1,2- or 1,3-diamine. Thus the following planar nickel complex has been prepared

[1] L. Cassar, P. E. Eaton, and J. Halpern, *J. Am. Chem. Soc.*, **92**, 3515 (1971).
[2] For another type of mechanism for olefin disproportionation see T. J. Katz and J. McGinnis, *J. Am. Chem. Soc.*, **97**, 1592 (1975) and E. L. Muettereits, *Inorg. Chem.*, **14**, 951 (1975).
[3] L. F. Lindoy and D. H. Busch, *Prep. Inorg. React.*, **6**, 1 (1971).

by treating 2 mol of 1,3-diaminopropane monohydrochloride with 2 mol of biacetyl, followed by 1 mol of nickel acetate.

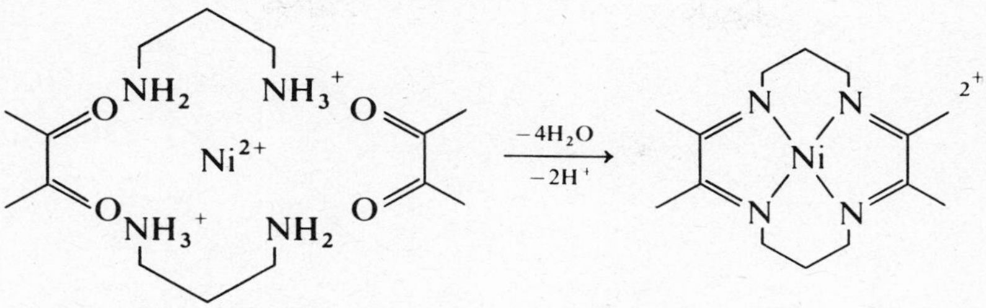

Mercaptoethylamine reacts with biacetyl in the presence of nickel ion to form a Schiff-base complex which in turn can be condensed with α,α'-dibromo-o-xylene to form a macrocyclic ligand:

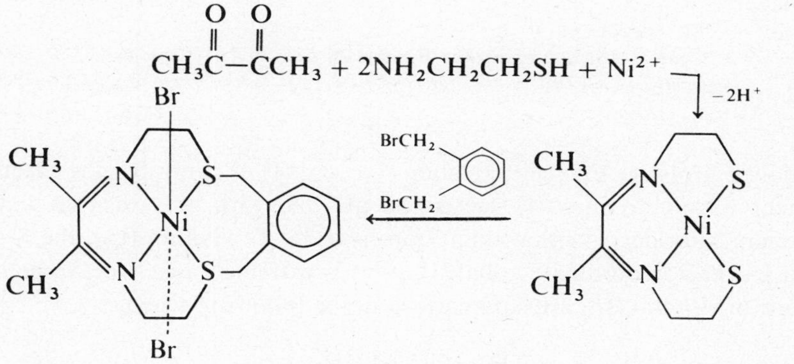

SOME BIOLOGICAL SYSTEMS

Vitamin B$_{12}$[1]

This compound and its derivatives are cobalt complexes with the general structure shown in Fig. 18.1. In the isolated vitamin, R is the 5-deoxyadenosyl(5,6-dimethylbenzimidazole) group (pictured in Fig. 18.2), Y is a nitrogen atom of the imidazole ring, and X is a cyanide ion. However, in vivo, the cyanide ion is not present, and X is probably a loosely bound water molecule. In both the cyano and aquo forms of the vitamin, the cobalt is present in the $+3$ oxidation state. The cobalt(III) can be reduced to cobalt(II) (vitamin B$_{12r}$) by catalytic hydrogenation, by treatment with chromium(II) acetate at pH 5, or by controlled

---

[1] H. A. O. Hill, J. M. Pratt, and R. J. P. Williams, *Chem. Br.*, **5**, 156 (1969); F. Wagner, *Ann. Rev. Biochem.*, **35**, 405 (1966).

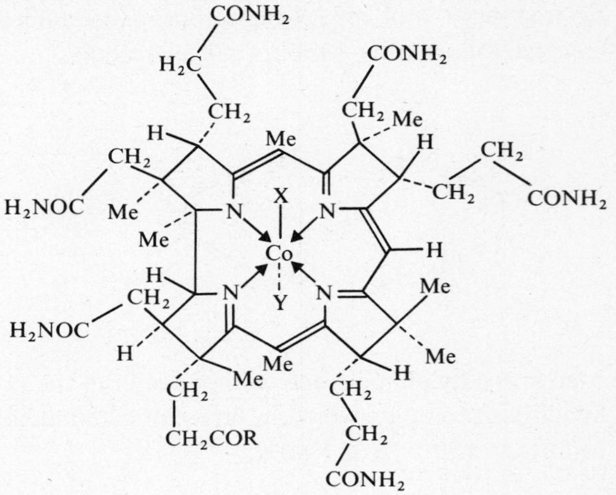

FIGURE 18.1
General structure of vitamin B_{12} and its derivatives. [*Reproduced with permission from H. A. O. Hill, J. M. Pratt, and R. J. P. Williams, Chem. Br., 5, 156 (1969).*]

potential electrolysis. Further reduction to cobalt(I) (vitamin B_{12s}) is effected by treatment with chromium(II) acetate at pH 9.5, with zinc dust in aqueous ammonium chloride, or with sodium borohydride. In vitamin B_{12s}, the X site is vacant, and the 5-coordinate cobalt(I) atom is extremely reactive. Some of the reactions of vitamin B_{12s} are summarized in the following scheme:

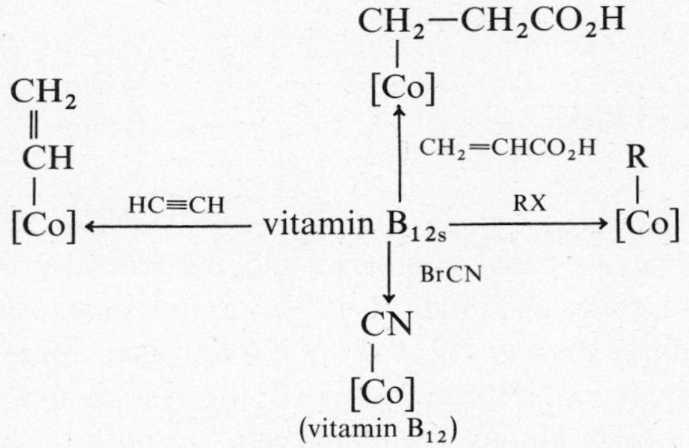

Vitamin B_{12s} is probably intimately involved in biological functions, where it can act as a methyl transfer agent (e.g., in the biosynthesis of methionine) or as a reducing agent (e.g., in the reduction of ribose).

The reaction of adenosine triphosphate with vitamin B_{12s} forms a cobalt-carbon bond, as shown in Fig. 18.3. The resulting molecule is known as vitamin

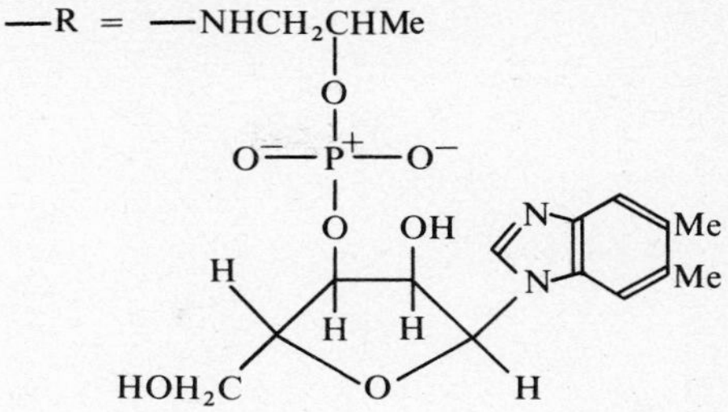

$$-R = -NHCH_2CHMe$$

FIGURE 18.2
The 5-deoxyadenosyl(5,6-dimethylbenzi-
midazole) group. The nitrogen atom of the
imidazole ring can coordinate at the Y
site.

B_{12} coenzyme and was the first organometallic compound discovered in living
systems. This coenzyme is required in various biological processes, but the
detailed mechanisms of these processes are unknown. Many of these processes
are rearrangements in which an R group and an H atom exchange positions.
For example, in conjunction with glutamate mutase, the following reaction is
catalyzed:

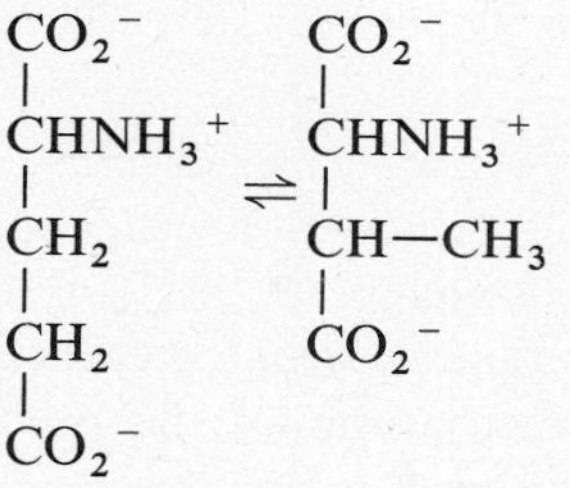

$$\begin{array}{ccc}
CO_2^- & & CO_2^- \\
| & & | \\
CHNH_3^+ & \rightleftharpoons & CHNH_3^+ \\
| & & | \\
CH_2 & & CH-CH_3 \\
| & & | \\
CH_2 & & CO_2^- \\
| & & \\
CO_2^- & &
\end{array}$$

Certain relatively simple cobalt complexes, such as bis(dimethylglyoximato)-
cobalt complexes, mimic many of the reactions of vitamin B_{12}.[1]   It is hoped that
the study of such simple, model complexes will, by the use of analogy, clarify
many of the reactions of vitamin B_{12}.

[1] G. N. Schrauzer, *Acc. Chem. Res.*, **1**, 97, (1968); *Inorg. Synth.*, **11**, 61 (1968).

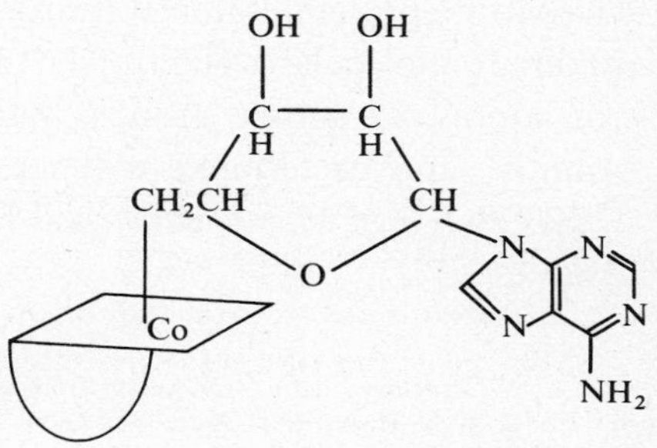

FIGURE 18.3
Schematic structure of vitamin B_{12} coenzyme, showing the group coordinated at
the X site.

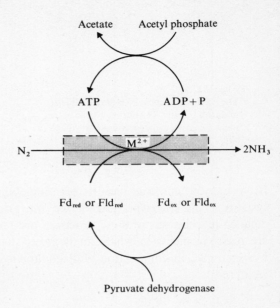

FIGURE 18.4
Schematic representation of the nitroge-
nase-catalyzed reduction of nitrogen. Fd
stands for ferredoxin, Fld for flavodoxin.

Iron-Sulfur Proteins and Nitrogen Fixation[1, 2]

The enzyme "nitrogenase," present in various bacteria, catalyzes the reduction of
molecular nitrogen to ammonia. Although the mechanism of the process is
unknown, enough is known about this system to permit reasonable conjecture
about the mechanism. The nitrogenase reaction is schematically represented by
the diagram in Fig. 18.4. The electrons required in the reduction of the N_2 are
transferred to the reaction system by the reduced forms of electron-transfer agents
such as ferredoxins or flavodoxins. The source of these electrons is the oxidation
of pyruvate. Energy for the process is provided by the hydrolysis of ATP to
ADP in the presence of a divalent metal ion such as Mg^{2+}. Nitrogenase is
composed of two proteins, a molybdenum-iron protein and an iron protein.
The Mo-Fe protein obtained from the bacterium *Clostridium pasteurianum* has a
minimum molecular weight of 110,200, contains 1 molybdenum atom, about 12
iron atoms, 12 sulfide groups, and 15 titratable SH groups. The Fe protein
from the same bacterium is a dimer of a subunit which has a molecular weight of
27,500, two iron atoms, two "acid-labile" sulfide groups, and six "free" cysteine
residues.

[1] W. Lovenberg (ed.), "Iron-Sulfur Proteins," vols. 1, 2, Academic, New York, 1973.
[2] R. W. F. Hardy, R. C. Burns, and G. W. Parshall, *Adv. Chem. Ser.*, **100,** 219 (1971).

One proposed mechanism for the reduction of N_2 is illustrated by the following scheme:[1]

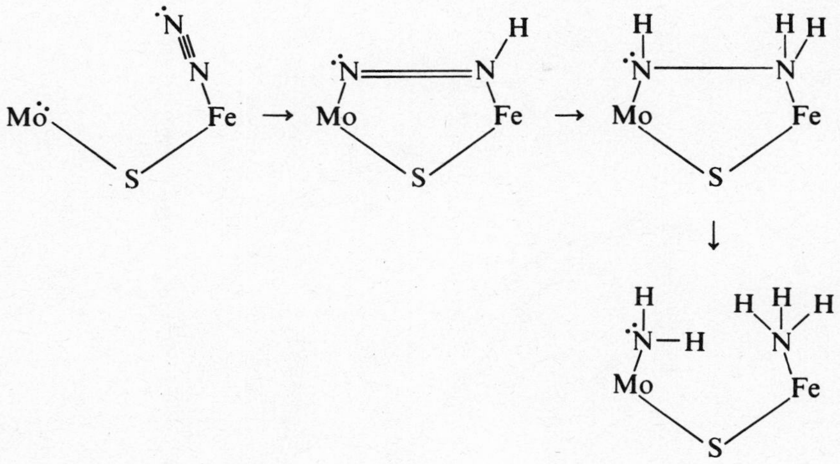

The proposed active site consists of an Mo atom and an Fe atom bridged by an S atom. In the initial step, N_2 forms a linear complex with the Fe atom. Transfer of a pair of electrons (or perhaps a hydride ion) from the Mo atom produces a binuclear diimide complex. Addition of two more electrons and protons yields a binuclear hydrazine complex, and addition of another pair of electrons and protons cleaves the N—N bond. Dissociation of NH_3 from the Fe atom leaves a site for coordination of another N_2 molecule, and hydrolysis of the Mo—NH_2 bond produces a second molecule of NH_3. Although the reactions of some synthetic model systems tend to support this mechanism, other mechanisms have been proposed,[2] and much more work will be necessary to definitely establish the mechanism.

Ferredoxins function as electron carriers in a wide variety of biological systems. They are proteins containing both iron and sulfur. The ferredoxin from *Micrococcus aerogenes* contains two cubic Fe_4S_4 groups situated in a protein chain with a very complicated conformation,[3] as shown in Fig. 18.5. The four

[1] Hardy, Burns, and Parshall, op. cit.

[2] A. B. Gilchrist, G. W. Rayner-Canham, and D. Sutton, *Nature*, **235**, 42 (1972); R. W. F. Hardy, R. C. Burns, and G. W. Parshall, in G. Eichhorn (ed.), "Inorganic Biochemistry," vol. 2, p. 745, Elsevier, Amsterdam, 1973.

[3] K. T. Yasunobu and M. Tanaka, in W. Lovenberg (ed.), "Iron-Sulfur Proteins," vol. 2, chap. 2, Academic, New York, 1973.

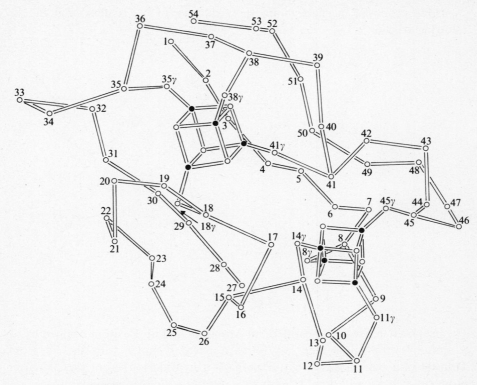

FIGURE 18.5

The skeleton of *M. aerogenes* ferredoxin. [*Reproduced with permission from K. T. Yasunobu and M. Tanaka, in W. Lovenberg (ed.), "Iron-Sulfur Proteins," vol. 2, chap. 2, Academic, New York, 1973.*]

tetrahedrally arranged iron atoms in each Fe_4S_4 cluster are connected to the protein by four more sulfur atoms which are in cysteine groups.[1]

The complex $Fe_4S_4(SCH_2Ph)_4{}^{2-}$, which has the structure shown in Fig. 18.6, is a ferredoxin analog.[2] This complex is synthesized by the reaction of iron(III) chloride, sodium methoxide, sodium hydrosulfide, and benzyl mercaptan in methanol solution. Many of the properties of the complex (magnetic susceptibility, [57]Fe Mössbauer spectrum, electronic spectrum, and redox properties) are analogous to those of ferredoxins which contain Fe_4S_4 clusters.

[1] L. C. Sieker, E. Adman, and L. H. Jensen, *Nature*, **235**, 40 (1972).
[2] T. Herskovitz, B. A. Averill, R. H. Holm, J. A. Ibers, W. D. Phillips, and J. F. Weiher, *Proc. Natl. Acad. Sci. U.S.*, **69**, 2437 (1972). Also see B. V. DePamphilis, B. A. Averill, T. Herskovitz, L. Que, Jr., and R. H. Holm, *J. Am. Chem. Soc.*, **96**, 4159 (1974), and previous papers.

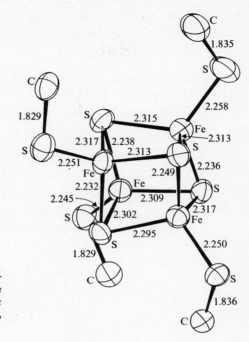

FIGURE 18.6
Structure of $Fe_4S_4(SCH_2Ph)_4{}^{2-}$. Hydrogen atoms and phenyl groups are omitted for clarity. [*From J. Herskovitz et al., Proc. Natl. Acad. Sci. U.S.*, **69**, 2437 (1972).*]

Although the compound might be formulated as a complex of two Fe(II) atoms and two Fe(III) atoms, all the physical methods, including X-ray photoelectron spectroscopy, indicate that the iron atoms are equivalent.[1] Hence a delocalized electronic description, in terms of identical iron atoms in the "+2.5" oxidation state, is probably appropriate.

PROBLEMS

18.1 Treatment of $Cr(CO)_6$ with $LiCH_3$, followed by $[(CH_3)_3O]BF_4$, yields the carbene

complex $(CO)_5CrC{\overset{\displaystyle CH_3}{\underset{\displaystyle OCH_3}{\big\langle}}}$. Propose a mechanism for this synthesis.

18.2 When $[(C_6H_5O)_3P]_4RhH$ is heated with a large excess of D_2 at 100°C, the ortho-hydrogen atoms undergo exchange to form $[(C_6H_3D_2O)_3P]_4RdD$. Explain.

18.3 Ethylene can be oxidized by an aqueous solution of $PdCl_4{}^{2-}$:

$$C_2H_4 + PdCl_4{}^{2-} + H_2O \rightarrow Pd + 2H^+ + 4Cl^- + CH_3CHO$$

[1] Ibid.

The rate law is

$$\text{Rate} = \frac{k(\text{PdCl}_4{}^{2-})(\text{C}_2\text{H}_4)}{(\text{H}^+)(\text{Cl}^-)^2}$$

Propose a plausible mechanism for the reaction.

18.4 The hydrosilylation reaction is

$$\text{RCH}{=}\text{CH}_2 + \text{HSiR}_3 \rightarrow \text{RCH}_2\text{CH}_2\text{SiR}_3$$

The reaction is catalyzed by $\text{Pd}(\text{PPh}_3)_4$. Suggest a mechanism.

18.5 Vitamin B_{12} coenzyme catalyzes the dehydration of diols:

$$\text{RCHOHCH}_2\text{OH} \rightarrow \text{RCH}_2\text{CHO} + \text{H}_2\text{O}$$

Suggest a mechanism. [Hint: Assume homolytic cleavage of the Co—C bond to form Co(II) and a radical which abstracts a hydrogen atom from the diol to form another radical which then undergoes rearrangement.] How could you test your mechanism, using tritium as a hydrogen tracer?

KINETICS AND MECHANISMS OF REACTIONS OF TRANSITION-METAL COMPLEXES

SUBSTITUTION REACTIONS OF OCTAHEDRAL COMPLEXES[1]

Lability

The rates of replacement of the ligands of hexacoordinate complexes by other ligands vary over an extremely wide range—from very high rates ($k \approx 10^9 \ M^{-1}$ s^{-1} at room temperature) to very low rates ($k \approx 10^{-9} \ M^{-1} \ s^{-1}$ at room temperature).[2] The complex $Co(NH_3)_6{}^{3+}$ is an example of a relatively inert, or slow-reacting, complex. Although the thermodynamic driving force of the following reaction is enormous,

$$Co(NH_3)_6{}^{3+} + H^+ + Cl^- \rightarrow Co(NH_3)_5Cl^{2+} + NH_4{}^+$$

the complex must be heated in 6 M hydrochloric acid for many hours to obtain any perceptible amount of the chloro complex. On the other hand, $Cu(NH_3)_4$-$(H_2O)_2{}^{2+}$ is a labile, or fast-reacting, complex. When a solution containing this

---

[1] F. Basolo and R. G. Pearson, "Mechanisms of Inorganic Reactions," 2d ed., Wiley, New York, 1967; R. G. Wilkins, "The Study of Kinetics and Mechanism of Reactions of Transition Metal Complexes," Allyn and Bacon, Boston, 1974.

[2] A rate constant of $10^9 \ M^{-1} \ s^{-1}$ corresponds to a half-life of 10^{-9} s with 1 M reagents, and $k = 10^{-9} \ M^{-1} \ s^{-1}$ corresponds to a half-life of about 30 years under the same conditions.

blue species is added to concentrated hydrochloric acid, the solution turns green essentially as rapidly as the solutions are mixed.

$$Cu(NH_3)_4(H_2O)_2{}^{2+} + 4H^+ + 4Cl^- \rightarrow CuCl_4{}^{2-} + 4NH_4{}^+ + 2H_2O$$

A chemist who wishes to synthesize an octahedral complex must have some idea of the lability of the complex in order to choose appropriate experimental conditions for the synthesis. When a particular geometric or optical isomer is to be prepared, it is necessary that the rate of isomerization or racemization be slow enough to permit the isolation of a pure isomer. Obviously any generalizations or rules that can be used to predict the relative reactivities of complexes are important.

The size of the central atom affects the ease with which ligands can be replaced. The smaller the central atom (other factors being equal), the more tightly the ligands are held and the more inert the complex is. Thus, the rate constants for the exchange of solvent and coordinated water molecules in the aqueous ions $Mg(H_2O)_6{}^{2+}$, $Ca(H_2O)_6{}^{2+}$, and $Sr(H_2O)_6{}^{2+}$ are $\sim 10^5$, $\sim 2 \times 10^8$, and $\sim 4 \times 10^8$ s^{-1}, respectively, at 25°C. The charge of the central atom is also important; the higher the charge, the more inert is the complex. This effect can be seen in the water exchange rate constants for the following aqueous ions:

$$\begin{cases} Na(H_2O)_6{}^+ \, (k \approx 8 \times 10^9 \text{ s}^{-1}; r_{Na^+} = 0.95 \text{ Å}) \\ Ca(H_2O)_6{}^{2+} \, (k \approx 2 \times 10^8 \text{ s}^{-1}; r_{Ca^{2+}} = 0.99 \text{ Å}) \end{cases}$$

and

$$\begin{cases} Mg(H_2O)_6{}^{2+} \, (k \approx 10^5 \text{ s}^{-1}; r_{Mg^{2+}} = 0.65 \text{ Å}) \\ Ga(H_2O)_6{}^{3+} \, (k = 10^3 \text{ s}^{-1}; r_{Ga^{3+}} = 0.62 \text{ Å}) \end{cases}$$

The net effect of changes in both charge and size can be seen in the marked decrease in reactivity in the series of complexes $AlF_6{}^{3-}$, $SiF_6{}^{2-}$, $PF_6{}^-$, and SF_6. The addition of base to a solution of $AlF_6{}^{3-}$ causes the instant precipitation of $Al(OH)_3$, whereas SF_6 undergoes no detectable reaction with hot concentrated base solutions over long periods of time.

The degree of lability or inertness of a transition-metal complex can be correlated with the d electron configuration of the metal ion. Let us consider a simple qualitative approach first used by Taube.[1] If a complex contains electrons in the antibonding $e_g{}^*$ orbitals, the ligands are expected to be relatively weakly bound and to be easily displaced. If a complex contains an empty t_{2g} orbital, the four lobes of that orbital correspond to directions from which an incoming ligand (which is to displace one of the bound ligands) can approach the complex with relatively little electrostatic repulsion. Therefore one concludes that

[1] H. Taube, *Chem. Rev.*, **50**, 69 (1952).

a complex with one or more e_g* electrons or with fewer than three d electrons should be relatively labile, and that a complex with any other electronic configuration should be relatively inert. In Table 19.1, examples of octahedral complexes with all possible electronic configurations are listed and categorized as labile or inert, depending on whether ligand substitution takes place in less than or more than 1 min at room temperature with 0.1 M reactants. All the data are consistent with Taube's simple rule.

Electrostatic crystal-field theory can be used to predict the relative labilities of octahedral transition-metal complexes if the geometric configurations of the activated complexes of the substitution reactions are known or assumed. The change in crystal-field stabilization energy on going from the reactant complex to the activated complex corresponds to the crystal-field contribution to the activation energy of the reaction. If the reaction is assumed to proceed by a mechanism in which the rate-determining step involves considerable dissociation of the original complex, a 5-coordinate square pyramid activated complex is a reasonable assumption. The crystal-field contributions to the activation energy, based on that assumption, are given in Table 19.2 for complexes with various electronic configurations. If the reaction is assumed to proceed by a mechanism in which the rate-determining step involves association of the incoming ligand with the original complex, a 7-coordinate pentagonal bipyramid activated complex is a reasonable assumption. The corresponding crystal-field contributions to the activation energy are given in Table 19.3.

Table 19.1 KINETIC CLASSIFICATION OF OCTAHEDRAL COMPLEXES

Labile	
d^0	$CaEDTA^{2-}$, $Sc(H_2O)_5OH^{2+}$, $TiCl_6^{2-}$
d^1	$Ti(H_2O)_6^{3+}$, $VO(H_2O)_5^{2+}$, $MoCl_5^{2-}$
d^2	$V(phen)_3^{3+}$, $ReOCl_5^{2-}$
d^4 (high-spin)	$Cr(H_2O)_6^{2+}$
d^5 (high-spin)	$Mn(H_2O)_6^{2+}$, $Fe(H_2O)_4Cl_2^+$
d^6 (high-spin)	$Fe(H_2O)_6^{2+}$
d^7	$Co(NH_3)_6^{2+}$
d^8	$Ni(en)_3^{2+}$
d^9	$Cu(NH_3)_4(H_2O)_2^{2+}$
d^{10}	$Ga(C_2O_4)_3^{3-}$

Inert	
d^3	$V(H_2O)_6^{2+}$, $Cr(en)_2Cl_2^+$
d^4 (low-spin)	$Cr(CN)_6^{4-}$, $Mn(CN)_6^{3-}$
d^5 (low-spin)	$Mn(CN)_6^{4-}$, $Fe(CN)_6^{3-}$
d^6 (low-spin)	$Fe(CN)_6^{4-}$, $Co(en)_2(H_2O)_2^{3+}$

Table 19.2 CRYSTAL-FIELD STABILIZATION ENERGIES AND CONTRIBUTIONS TO ACTIVATION ENERGIES (IN UNITS OF Δ_o) FOR DISSOCIATIVE MECHANISMS (OCTAHEDRON → SQUARE PYRAMID)[†]

Electronic configuration	Octahedron	Square pyramid	Contribution to E_a
d^0	0	0	0
d^1	0.400	0.457	−0.057
d^2	0.800	0.914	−0.114
d^3	1.200	1.000	0.200
d^4 (high-spin)	0.600	0.914	−0.314
d^4 (low-spin)	1.600	1.457	0.143
d^5 (high-spin)	0	0	0
d^5 (low-spin)	2.000	1.914	0.086
d^6 (high-spin)	0.400	0.457	−0.057
d^6 (low-spin)	2.400	2.000	0.400
d^7 (high-spin)	0.800	0.914	−0.114
d^7 (low-spin)	1.800	1.914	−0.114
d^8	1.200	1.000	0.200
d^9	0.600	0.914	−0.314
d^{10}	0	0	0

† F. Basolo and R. G. Pearson, "Mechanisms of Inorganic Reactions," 2d ed., Wiley, New York, 1967.

Table 19.3 CRYSTAL-FIELD STABILIZATION ENERGIES AND CONTRIBUTIONS TO ACTIVATION ENERGIES (IN UNITS OF Δ_o) FOR ASSOCIATIVE MECHANISMS (OCTAHEDRON → PENTAGONAL BIPYRAMID)[†]

Electronic configuration	Octahedron	Pentagonal bipyramid	Contribution to E_a
d^0	0	0	0
d^1	0.400	0.528	−0.128
d^2	0.800	1.056	−0.256
d^3	1.200	0.774	0.426
d^4 (high-spin)	0.600	0.493	0.107
d^4 (low-spin)	1.600	1.302	0.298
d^5 (high-spin)	0	0	0
d^5 (low-spin)	2.000	1.830	0.170
d^6 (high-spin)	0.400	0.528	−0.128
d^6 (low-spin)	2.400	1.548	0.852
d^7 (high-spin)	0.800	1.056	−0.256
d^7 (low-spin)	1.800	1.266	0.534
d^8	1.200	0.774	0.426
d^9	0.600	0.493	0.107
d^{10}	0	0	0

† F. Basolo and R. G. Pearson, "Mechanisms of Inorganic Reactions," 2d ed., Wiley, New York, 1967.

It is remarkable that, in spite of the different assumptions involved, the data in Tables 19.1 to 19.3 are to a large extent in qualitative agreement. Both sets of crystal-field data indicate that complexes with d^3, d^8, and low-spin d^6 configurations should be inert. The main disagreement between the crystal-field results and the qualitative predictions of Taube is in the case of d^8 complexes, which are predicted to be labile by the Taube rule. However, examination of quantitative rate data shows that, in a sense, both predictions are correct. Consider the rate constant data[1] for $+2$ ions given in Table 19.4. Although the rate constants for $Ni(H_2O)_6{}^{2+}$ are not as low as the corresponding constants for the d^3 $V(H_2O)_6{}^{2+}$ ion, they are definitely lower than those for all the other $+2$ ions listed.

Mechanisms

The data in Table 19.4 and a large number of similar kinetic data have the following general features:

1 For a given aqueous ion, the rate of replacement of a coordinated water molecule is not strongly affected by the nature of the entering ligand.
2 The water exchange reaction is always faster than substitution reactions involving other ligands.

These results suggest that the reactions proceed in two steps, the first of which is the formation of an aquo ion–ligand "outer-sphere" complex, and the

[1] R. G. Wilkins, *Acc. Chem. Res.*, **3**, 408 (1970).

Table 19.4 LOG k FOR SUBSTITUTION REACTIONS OF AQUEOUS $+2$ IONS AT 25°C†

Metal ion	Entering ligand			
	H₂O‡	NH₃§	HF§	phen§
V^{2+}	2.0	. . . .	. . . .	0.5
Cr^{2+}	8.5	. . . .	. . . .	~8.0
Mn^{2+}	7.5	. . . .	6.3	~5.4
Fe^{2+}	6.5	. . . .	6.0	5.9
Co^{2+}	6.0	5.1	5.7	5.3
Ni^{2+}	4.3	3.7	3.5	3.4
Cu^{2+}	8.5	. . . .	7.5	7.9
Zn^{2+}	7.5	. . . .	. . . .	6.8

† R. G. Wilkins, *Acc. Chem. Res.*, **3**, 408 (1970).
‡ Bulk water-coordinated water exchange rate constants (s^{-1}).
§ Second-order rate constants $(M^{-1}\ s^{-1})$.

second of which is the exchange of an "inner-sphere" coordinated water molecule with the outer-sphere coordinated ligand.

$$M(H_2O)_6{}^{n+} + L \underset{}{\overset{K_1}{\rightleftharpoons}} M(H_2O)_6 \cdot L^{n+} \tag{19.1}$$

$$M(H_2O)_6 \cdot L^{n+} \overset{k_2}{\longrightarrow} M(H_2O)_5 L^{n+} + H_2O \tag{19.2}$$

When the overall rate constants $(K_1 k_2)$ are divided by estimated outer-sphere complex formation constants (K_1), the resulting values of k_2 are very close to the corresponding water exchange rate constants. This result shows that the formation of the activated complex in the second, rate-determining step consists mainly of the breaking of the $M-OH_2$ bond. The entering ligand influences k_2 only very slightly.

In some substitution reactions, the entering ligand is not involved in the dissociation step, and a 5-coordinate intermediate is formed. The reactions of the $Co(CN)_5H_2O^{2-}$ ion with various ligands are examples of such reactions:[1]

$$Co(CN)_5H_2O^{2-} \underset{k_2}{\overset{k_1}{\rightleftharpoons}} Co(CN)_5{}^{2-} + H_2O$$

$$Co(CN)_5{}^{2-} + X^- \underset{k_4}{\overset{k_3}{\rightleftharpoons}} Co(CN)_5X^{3-}$$

Kinetic data for $X^- = Br^-, I^-, SCN^-$, and $N_3{}^-$ give the same value, 1.6×10^{-3} s^{-1}, for k_1. In fact, essentially the same value has been obtained for the rate constant of the water exchange reaction.

The base hydrolyses of pentaamminecobalt(III) complexes have been extensively studied.

$$Co(NH_3)_5X^{2+} + OH^- \rightarrow Co(NH_3)_5OH^{2+} + X^-$$

The reactions are first order in the complex and first order in hydroxide ion, and one might reasonably suspect a mechanism involving an ion pair, analogous to reactions 19.1 and 19.2, or a simple bimolecular displacement in which the hydroxide ion enters the coordination sphere as the X^- ion leaves. However, an extensive set of experimental data indicates that these base hydrolyses proceed by the following mechanism:[2, 3]

$$Co(NH_3)_5X^{2+} + OH^- \overset{\text{fast}}{\rightleftharpoons} Co(NH_3)_4(NH_2)X^+ + H_2O$$

$$Co(NH_3)_4(NH_2)X^+ \overset{\text{slow}}{\longrightarrow} Co(NH_3)_4(NH_2)^{2+} + X^-$$

$$Co(NH_3)_4(NH_2)^{2+} + H_2O \overset{\text{fast}}{\longrightarrow} Co(NH_3)_5OH^{2+}$$

[1] A. Haim and W. K. Wilmarth, *Inorg. Chem.*, **1**, 573 (1962); A. Haim, R. J. Grassie, and W. K. Wilmarth, *Adv. Chem. Ser.*, **49**, 31 (1965).
[2] Basolo and Pearson, op. cit.; Wilkins, op. cit.
[3] M. L. Tobe, *Acc. Chem. Res.*, **3**, 377 (1970).

In this mechanism the rate-determining and dissociative step takes place after the removal of a proton from one of the coordinated ammonia molecules. Strong evidence for the mechanism is obtained from studies[1] of the hydrolysis in the presence of various anions (Y^-). These studies show that $Co(NH_3)_5Y^{2+}$ ions are formed in addition to $Co(NH_3)_5OH^{2+}$ ions. The product ratio, $Co(NH_3)_5Y^{2+}/Co(NH_3)_5OH^{2+}$, shows little dependence on the leaving group X^-, is independent of the OH^- concentration, but depends on the Y^- concentration and varies from one Y^- to another. The results clearly indicate a common reactive intermediate in all the reactions.

The following arguments have been offered to rationalize the mechanism. The X^- ion can dissociate from the deprotonated complex more easily than from the original complex because of the lower positive charge of the deprotonated complex. The 5-coordinate intermediate is probably somewhat stabilized by the deprotonation because the NH_2^- ligand can denote some of its π lone-pair electron density to the electron-deficient cobalt atom.

SUBSTITUTION REACTIONS OF SQUARE PLANAR COMPLEXES[2]

It is believed that substitution reactions of square planar complexes almost always proceed by an associative mechanism involving an intermediate which contains both the leaving ligand and the entering ligand. It is assumed that the entering ligand approaches the complex from one side of the plane, over the ligand to be displaced. The leaving ligand presumably moves down as the entering ligand approaches, so that the intermediate has a trigonal bipyramidal configuration.

A 5-coordinate intermediate, or at least a 5-coordinate activated complex, is quite reasonable in view of the fact that there are many stable 5-coordinate d^8 complexes.

[1] D. A. Buckingham, I. I. Olsen, and A. M. Sargeson, *J. Am. Chem. Soc.*, **88**, 5443 (1966).
[2] Basolo and Pearson, op. cit.; Wilkins, op. cit.

For reactions of Pt(II) complexes in aqueous solution, the rate law often has the form

$$\text{Rate} = k_1(\text{complex}) + k_2(\text{complex})(Y^-)$$

where Y^- is the entering ligand. The second term corresponds to the bimolecular displacement of the leaving ligand by the entering Y^- ligand, and the first term corresponds to a two-step path in which first the leaving ligand is displaced by a water molecule in a rate-determining step and then the coordinated water molecule is relatively rapidly displaced by Y^-. Kinetic data on a wide variety of Pt(II) complexes show that the rate of a ligand substitution reaction depends on the natures of the entering ligand, the leaving ligand, and the ligand trans to the leaving ligand. The relation between the reaction rate and the ligand trans to the leaving ligand is called the trans effect, and we shall discuss this effect before considering the effects of the entering and leaving ligands.

The Kinetic Trans Effect

Ligands can be put in order according to their abilities to labilize (make susceptible to substitution) trans leaving groups. In the following list, the ligands are given in the order of increasing trans effect (i.e., increasing trans labilizing ability): $H_2O < OH^- < F^- \approx RNH_2 \approx py \approx NH_3 < Cl^- < Br^- < SCN^- \approx I^- \approx NO_2^- \approx C_6H_5^- < SC(NH_2)_2 \approx CH_3^- < NO \approx H^- \approx PR_3 < C_2H_4 \approx CN^- \approx CO$.

The effect is illustrated by the reaction of the $PtCl_4^{2-}$ ion with two molecules of ammonia. After one chloride ion has been displaced, there are two kinds of chloride ions remaining in the complex: the chloride ion which is trans to an ammonia molecule and the chloride ions which are trans to each other. Because chloride is more trans-directing than ammonia, the chloride ions which are trans to each other are more labile than the chloride ion trans to the ammonia molecule. Hence the second molecule of ammonia displaces one of the more labile chloride ions and yields cis-Pt(NH$_3$)$_2$Cl$_2$.

Trans-Pt(NH$_3$)$_2$Cl$_2$ can be prepared by the reaction of the Pt(NH$_3$)$_4^{2+}$ ion with two chloride ions. The first step yields Pt(NH$_3$)$_3$Cl$^+$, which contains two ammonia molecules trans to each other and an ammonia molecule trans to the chloride ion. The latter ammonia molecule is labilized by the chloride ion and is displaced next:

$$
\begin{array}{ccc}
\begin{array}{cc} H_3N & NH_3 \\ & Pt \\ H_3N & NH_3 \end{array}
& \xrightarrow{Cl^-} &
\begin{array}{cc} Cl & NH_3 \\ & Pt \\ H_3N & NH_3 \end{array}
\quad \xrightarrow{Cl^-} \quad
\begin{array}{cc} Cl & NH_3 \\ & Pt \\ H_3N & Cl \end{array}
\end{array}
$$

The trans effect can be used as the basis for the synthesis of many other isomeric Pt(II) complexes. For example, *cis*- and *trans*-PtNH$_3$NO$_2$Cl$_2^-$ are prepared from PtCl$_4^{2-}$ by the following routes.

$$
\begin{array}{cc} Cl & Cl \\ & Pt \\ Cl & Cl \end{array}
\xrightarrow{NH_3}
\begin{array}{cc} Cl & NH_3 \\ & Pt \\ Cl & Cl \end{array}
\xrightarrow{NO_2^-}
\begin{array}{cc} Cl & NH_3 \\ & Pt \\ Cl & NO_2 \end{array}
$$

$$
\begin{array}{cc} Cl & Cl \\ & Pt \\ Cl & Cl \end{array}
\xrightarrow{NO_2^-}
\begin{array}{cc} Cl & NO_2 \\ & Pt \\ Cl & Cl \end{array}
\xrightarrow{NH_3}
\begin{array}{cc} Cl & NO_2 \\ & Pt \\ H_3N & Cl \end{array}
$$

Notice that a different isomer is obtained by simply reversing the order of introduction of the groups.

We shall now show that it is possible to rationalize the trans effect on the basis of a trigonal bipyramidal intermediate. In this intermediate, the leaving ligand, the entering ligand, and the trans ligand occupy the three equatorial positions of the trigonal bipyramid. We have already pointed out in Chap. 17 that, in a trigonal bipyramidal d^8 complex, there is a greater interaction of the filled $d\pi$ orbitals with the equatorial ligands than with the axial ligands. Therefore $d\pi \to \pi^*$ back bonding and $d\pi$-$p\pi$ nonbonding repulsions affect the bonding of the equatorial ligands more than that of the axial ligands. It also appears that the equatorial σ atomic orbitals of the metal are somewhat more electronegative than the axial σ atomic orbitals, perhaps because of greater s character in the equatorial orbitals. Consequently strongly basic ligands with relatively electropositive donor atoms tend to favor the equatorial sites over the axial sites (other factors being equal).

In the following reactions,

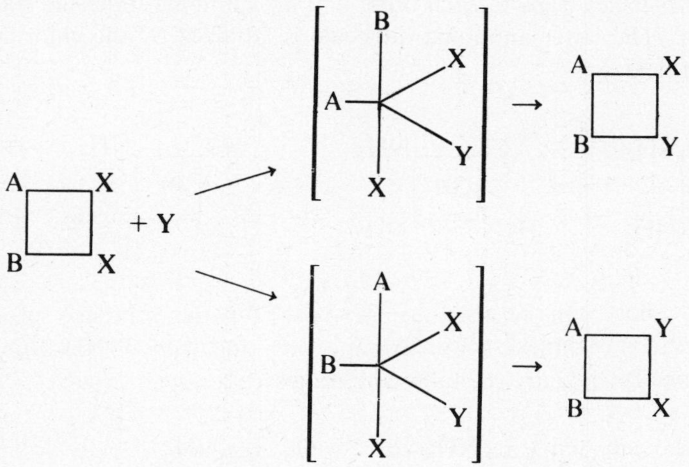

the geometry of the principal product is determined by the relative trans-directing abilities of A and B. The faster reaction corresponds to the more stable trigonal bipyramidal intermediate. Obviously the more strongly trans-directing ligand is the one which gives the more stable intermediate when it is in an equatorial, rather than an axial, position. Hence strong trans-directing ligands are those which engage in strong back bonding (CN^-, CO, olefins, etc.) or are very strong σ donors (CH_3^-, H^-, PR_3, etc.). Ligands which are strong π donors (H_2O, OH^-, NH_2^-, etc.) are weakly trans-directing.

Effect of Entering and Leaving Ligands[1]

If the entering ligand forms a stronger bond to the platinum atom than the leaving ligand, the rate-determining step is the formation of the bond between the platinum and the entering ligand. In cases like this the rate of reaction is a sensitive function of the nature of the entering ligand. For example, in reactions of the type

$$\text{trans-PtA}_2\text{Cl}_2 + \text{Y}^- \rightarrow \text{trans-PtA}_2\text{ClY} + \text{Cl}^-$$

the reaction rate increases in the following order of entering ligands:

$$OR^- < Cl^- \approx py \approx NO_2^- < N_3^- < Br^- < I^- < SO_3^{2-} < SCN^- < CN^-$$

[1] F. Basolo and R. G. Pearson, "Mechanisms of Inorganic Reactions," 2nd ed., Wiley, New York, 1967.

If the entering ligand forms a stronger bond than the leaving ligand, the rate is essentially independent of the nature of the leaving ligand, but if the leaving ligand forms a stronger bond than the entering ligand, the rate is a function of the nature of the leaving ligand. Both situations are found in reactions of thiourea with complexes of the type $Pt(dien)X^+$:

$$Pt(dien)X^+ + tu \rightarrow Pt(dien)tu^{2+} + X^-$$

The rate is essentially the same for the relatively weakly bound leaving ligands Cl^-, Br^-, and I^- but shows a definite trend with the more strongly bound leaving ligands: $Cl^- \approx Br^- \approx I^- \gg N_3^- \approx NO_2^- > SCN^- > CN^-$.

Pyridine is a relatively weakly bound ligand, and in reactions of the type

$$Pt(dien)X^+ + py \rightarrow Pt(dien)py^{2+} + X^-$$

the reaction rate shows a marked trend even among weakly bound leaving groups: $NO_3^- > H_2O > Cl^- > Br^- > I^- > N_3^- > SCN^- > NO_2^- > CN^-$.

The entering- and leaving-ligand series have a close similarity to the trans-effect series, but distinct differences can be seen. Some of these differences are probably due to the fact that the entering- and leaving-group series are affected by differences in the solvation of ligands, whereas the trans-effect series is independent of such solvation effects.

OXIDATION-REDUCTION REACTIONS[1, 2, 3]

Outer-Sphere Electron Transfer

Many complexes undergo oxidation-reduction reactions more rapidly than they undergo substitution reactions. When two such complexes engage in a redox reaction, it is clear that electron transfer must occur through the intact coordination shells of the metal ions. These reactions are called "outer-sphere" electron-transfer reactions. The following is an example of such a reaction.

$$Fe(CN)_6^{4-} + IrCl_6^{2-} \rightarrow Fe(CN)_6^{3-} + IrCl_6^{3-}$$

Even when one of the complexes is labile, the reaction will generally proceed by an outer-sphere mechanism if the inert complex does not possess a donor atom which can be used to form a bridge to the labile complex. For example, this is the situation in the reduction of $Co(NH_3)_6^{3+}$ by $Cr(H_2O)_6^{2+}$.

[1] Ibid.

[2] H. Taube, "Electron Transfer Reactions of Complex Ions in Solution," Academic, New York, 1970.

[3] A. G. Sykes, "Kinetics of Inorganic Reactions," Pergamon, Oxford, 1966.

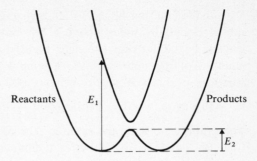

FIGURE 19.1
Potential energy of a symmetric electron exchange system along a particular configurational coordinate of the entire system.

The simplest type of outer-sphere redox reactions are those in which no net reaction occurs, i.e., electron exchange reactions such as the following.

$$Fe(H_2O)_6{}^{2+} + Fe(H_2O)_6{}^{3+} \rightarrow Fe(H_2O)_6{}^{3+} + Fe(H_2O)_6{}^{2+}$$

The rates of such reactions can be followed by the use of isotopic tracers. The transfer of an electron from one complex to another of the same formula and general structure can be looked upon as an electronic transition from one state of the entire system to another. The relative energies of the two states of the system depend on the geometries of the complexes. Figure 19.1 shows the energy levels as a function of a particular configurational coordinate of the system. This coordinate could be, for example, the average metal-ligand distance for one complex minus the average metal-ligand distance for the other complex, while maintaining a constant metal-metal distance. The most stable state for the reactants (i.e., the initial state) corresponds to the left-hand minimum, where the reduced species has a greater metal-ligand distance than the oxidized species. Electron transfer at this point would require absorption of the energy indicated by E_1 in Fig. 19.1. The two product species would have a high energy, inasmuch as the oxidized species would have a large metal-ligand distance and the reduced species would have a small metal-ligand distance. By first shortening the metal-ligand distance of the reactant-reduced species and lengthening the metal-ligand distance of the reactant-oxidized species, the electronic energy gap between the reactants and products can be reduced. The gap is minimized when the two complexes have the same dimensions. The barrier height at this point, indicated by E_2 in the figure, corresponds to the structural reorganizational contribution to the activation energy of the electron exchange. On this basis, one would expect that exchange between complexes of similar size would be faster than exchange between complexes of markedly different size. Relatively large changes in size can accompany the removal or addition of an $e_g{}^*$ electron, and electron exchange reactions which involve changes in the number of $e_g{}^*$ electrons are generally slower than those involving only changes in the number of t_{2g} electrons. This electronic effect can be seen in the rate-constant data of Table 19.5. The $Cr(H_2O)_6^{2+}$–$Cr(H_2O)_6^{3+}$ exchange and several of the Co(II)–Co(III) exchange reactions are very slow, whereas all the other exchange reactions are relatively fast. However the Co(II)–Co(III) exchange reactions require special comment.

The simple transfer of one electron from Co(II) to Co(III) leads to an excited state of each product ion and involves an asymmetric activated complex:

$$Co^{II}(t_{2g}{}^5 e_g{}^{*2}) + Co^{III}(t_{2g}{}^6) \rightarrow Co^{III}(t_{2g}{}^5 e_g{}^{*1}) + Co^{II}(t_{2g}{}^6 e_g{}^{*1})$$

Inasmuch as the symmetry of the reaction requires the steps leading to and from the activated complex to be symmetric, this process is unacceptable. It is believed that the reaction actually involves an excited state of one reactant. If the Co(III) is excited to the $t_{2g}{}^4 e_g{}^{*2}$ configuration, the following symmetric reaction can occur:

$$Co^{II}(t_{2g}{}^5 e_g{}^{*2}) + Co^{III}(t_{2g}{}^4 e_g{}^{*2}) \rightarrow Co^{III}(t_{2g}{}^4 e_g{}^{*2}) + Co^{II}(t_{2g}{}^5 e_g{}^{*2})$$

If the Co(II) is excited to the $t_{2g}{}^6 e_g{}^{*1}$ configuration, the exchange is also symmetric:

$$Co^{II}(t_{2g}{}^6 e_g{}^{*1}) + Co^{III}(t_{2g}{}^6) \rightarrow Co^{III}(t_{2g}{}^6) + Co^{II}(t_{2g}{}^6 e_g{}^{*1})$$

Table 19.5 RATE CONSTANTS OF ELECTRON EXCHANGE REACTIONS†

Reaction	$k, M^{-1}s^{-1}$ (25°C)
$Cr(H_2O)_6^{2+} + Cr(H_2O)_6^{3+}$	$\lesssim 2 \times 10^{-5}$
$Co(NH_3)_6^{2+} + Co(NH_3)_6^{3+}$	$< 10^{-9}$
$Co(en)_3^{2+} + Co(en)_3^{3+}$	1.4×10^{-4}
$CoEDTA^{2-} + CoEDTA^-$	1.4×10^{-4}
$Co(H_2O)_6^{2+} + Co(H_2O)_6^{3+}$	~ 5
$Co(o\text{-phen})_3^{2+} + Co(o\text{-phen})_3^{3+}$	1.1
$V(H_2O)_6^{2+} + V(H_2O)_6^{3+}$	1.0×10^{-2}
$Fe(H_2O)_6^{2+} + Fe(H_2O)_6^{3+}$	4
$Fe(CN)_6^{4-} + Fe(CN)_6^{3-}$	740
$Fe(o\text{-phen})_3^{2+} + Fe(o\text{-phen})_3^{3+}$	$> 10^5$
$Ru(NH_3)_6^{2+} + Ru(NH_3)_6^{3+}$	8×10^2
$Os(bipy)_3^{2+} + Os(bipy)_3^{3+}$	5×10^4
$IrCl_6^{3-} + IrCl_6^{2-}$	10^3

† F. Basolo and R. G. Pearson, "Mechanisms of Inorganic Reactions," 2d ed., Wiley, New York, 1967; R. G. Wilkins, "The Study of Kinetics and Mechanism of Reactions of Transition Metal Complexes," Allyn and Bacon, Boston, 1974; H. Taube, "Electron Transfer Reactions of Complex Ions in Solution," Academic, New York, 1970; Halpern, *Quart. Rev.*, **15**, 207 (1961).

In either case, the excitation energy contributes to the activation energy, and therefore most Co(II)–Co(III) exchange reactions are slow. However, in the case of $Co(H_2O)_6^{3+}$, the ligand-field strength is low and excitation to the high-spin state is fairly easy. Hence the $Co(H_2O)_6^{2+}$–$Co(H_2O)_6^{3+}$ exchange is rapid. In the case of $Co(o\text{-phen})_3^{2+}$, the ligand-field strength is very high and excitation to the low-spin state is easy. Hence the $Co(o\text{-phen})_3^{2+}$–$Co(o\text{-phen})_3^{3+}$ exchange is also fast.

Part of the enhanced rate of the $Co(o\text{-phen})_3^{2+}$–$Co(o\text{-phen})_3^{3+}$ exchange may be due to another effect, which also causes the $Fe(o\text{-phen})_3^{2+}$–$Fe(o\text{-phen})_3^{3+}$ exchange to be faster than other Fe(II)–Fe(III) exchanges. The later enhancement is probably due to the facts that (1) o-phenanthroline is an aromatic system which can be reduced relatively easily to a radical anion and (2) electron delocalization in the o-phen complexes allows the transferring electron to penetrate the coordination shells easily.

Outer-sphere redox reactions in which there is a net chemical change are usually faster than the corresponding electron exchange reactions. Some data are given in Table 19.6. Usually the greater the thermodynamic driving force

Table 19.6   A COMPARISON OF REDOX AND ELECTRON EXCHANGE RATE CONSTANTS (25°C)†

Reaction	$k, M^{-1}s^{-1}$
$Fe(CN)_6^{4-} + IrCl_6^{2-}$	3.8×10^5
$Fe(CN)_6^{4-} + Fe(CN)_6^{3-}$	7.4×10^2
$IrCl_6^{3-} + IrCl_6^{2-}$	10^3
$Cr(H_2O)_6^{2+} + Fe(H_2O)_6^{3+}$	$\sim 2 \times 10^3$
$Cr(H_2O)_6^{2+} + Cr(H_2O)_6^{3+}$	$\lesssim 2 \times 10^{-5}$
$Fe(H_2O)_6^{2+} + Fe(H_2O)_6^{3+}$	4
$Fe(o\text{-phen})_3^{2+} + Co(H_2O)_6^{3+}$	1.4×10^4
$Fe(o\text{-phen})_3^{2+} + Fe(o\text{-phen})_3^{3+}$	10^5
$Co(H_2O)_6^{2+} + Co(H_2O)_6^{3+}$	~ 5
$Fe(H_2O)_6^{2+} + Co(H_2O)_6^{3+}$	$10\ddagger$
$Fe(H_2O)_6^{2+} + Fe(H_2O)_6^{3+}$	$0.9\ddagger$
$Co(H_2O)_6^{2+} + Co(H_2O)_6^{3+}$	$\sim 1\ddagger$

† F. Basolo and R. G. Pearson, "Mechanisms of Inorganic Reactions," 2d ed., Wiley, New York, 1967; R. G. Wilkins, "The Study of Kinetics and Mechanism of Reactions of Transition Metal Complexes," Allyn and Bacon, Boston, 1974; A. G. Sykes, "Kinetics of Inorganic Reactions," Pergamon, New York, 1966.
‡ At 0°C.

for a redox reaction, the greater is the rate. A rough estimate of the rate constant k_{12} can be obtained from the following relation,[1]

$$k_{12} \approx (k_{11}k_{22}K_{12})^{1/2}$$

where k_{11} and k_{22} are the rate constants of the electron exchange reactions and K_{12} is the equilibrium constant.

Inner-sphere Redox Reactions[2]

In an "inner-sphere" redox reaction, the two metal ions are connected in the activated complex through a bridging ligand common to both coordination shells. The first definitive proof of such a mechanism was obtained by Taube and his coworkers,[3] who showed that aqueous chromium(II) is oxidized by a pentaamminecobalt(III) complex, $Co(NH_3)_5X^{2+}$, to give the species $Cr(H_2O)_5X^{2+}$. The result proves that, in the redox reaction, the X^- ion is transferred directly from Co(III) to Cr(II) and that in the activated complex the two metal atoms are bridged by the X^- ion, as shown in the following structure:

$$(NH_3)_5CoXCr(H_2O)_5{}^{4+}$$

This particular reaction system was chosen because the Co(III) complex retains its integrity in solution for long periods of time, both Cr(II) and Co(II) complexes are labile, and Cr(III) complexes are inert. Each of these chemical characteristics was essential to the success of the experiment.

It can be shown that the transfer of X^- from Co(III) to Cr(II) favors the transfer of the electron from Cr(II) to Co(III). The d electron structures of the two metals are shown in Fig. 19.2, stage A before electron transfer, stage B in the activated complex, and stage C immediately after electron transfer. In stage A, before X^- has shifted toward the chromium atom, the electron in the $d\sigma$ orbital has an energy below that of the $d\sigma$ orbitals of the cobalt atom. When X^- moves toward the chromium atom, the cobalt $d\sigma$-orbital energy levels split as their center of gravity is lowered and the chromium $d\sigma$-orbital energy levels move together as their center of gravity is raised. Electron transfer occurs when the energy of the lower cobalt $d\sigma$ orbital is approximately the same as that of the lower chromium $d\sigma$ orbital. Further shift of X^- traps the electron on the cobalt atom and completes the transfer of X^- to the chromium atom.

[1] R. A. Marcus, *Ann. Rev. Phys. Chem.*, **15**, 155 (1964).

[2] H. Taube, "Electron Transfer Reactions of Complex Ions in Solution," Academic, New York, 1970.

[3] H. Taube, H. Myers, and R. L. Rich, *J. Am. Chem. Soc.*, **75**, 4118 (1953); H. Taube and H. Myers, *J. Am. Chem. Soc.*, **76**, 2103 (1954).

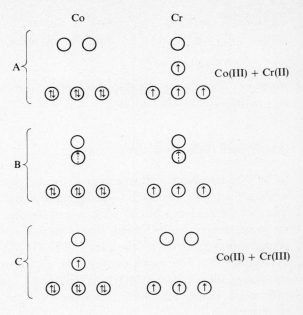

FIGURE 19.2
Changes in the *d* orbitals of Co and Cr as X⁻ moves from Co to Cr. Electron transfer occurs at stage B.

Electron Transfer Through Extended Bridges

The reduction of $Co(NH_3)_5py^{3+}$ by $Cr(H_2O)_6^{2+}$ ($k = 4.0 \times 10^{-3}\ M^{-1}\ s^{-1}$) yields $Cr(H_2O)_6^{3+}$ as the only Cr-containing product. The reaction undoubtedly proceeds by an outer-sphere mechanism. On the other hand, the complex

$$\left[(NH_3)_5Co-N\!\!\bigcirc\!\!-CONH_2 \right]^{3+}$$

reacts with $Cr(H_2O)_6^{2+}$ much more rapidly ($k = 17.4\ M^{-1}\ s^{-1}$) to give the following Cr(III) complex.

$$\left[HN\!\!\bigcirc\!\!-\underset{\underset{NH_2}{|}}{C}=OCr(H_2O)_5 \right]^{4+}$$

Separate experiments have verified that the latter species is the immediate, primary product; hence it is certain that the Cr(II) attacks the original complex at the remote keto group and that the electron is transferred to the Co(III) through the conjugated bonds of the isonicotinamide.[1]

Isied and Taube[2] prepared the complex

$$\left[(NH_3)_5CoO_2C-\bigcircN-Ru(NH_3)_4SO_4 \right]^{2+}$$

in the Co(III)—L—Ru(II) form and, by spectrophotometric measurements, determined the rate ($k \sim 1 \times 10^2$ s^{-1}) of the *intra*molecular electron transfer reaction:

$$Co(III)-L-Ru(II) \rightarrow Co(II)-L-Ru(III)$$

They found that the following complex, with a CH_2 group between the carboxylate group and the aromatic ring, underwent the analogous electron transfer much more slowly ($k \sim 2 \times 10^{-2}$ s^{-1}).

$$\left[(NH_3)_5CoO_2CCH_2-\bigcircN-Ru(NH_3)_4H_2O \right]^{4+}$$

The difference is undoubtedly mainly due to the blockage of the π bonding between the metal atoms. In fact, electron transfer in the latter complex may proceed by a different sort of route. Molecular models of the complex show that the carbonyl group is close to the π cloud of the ring, and thus the CH_2 linkage may be completely bypassed in the electron transfer.

Symmetric mixed-valence complexes are of considerable theoretical interest because of their close relation to electron transport phenomena. The complex

$$\left[(NH_3)_5Ru-N\bigcircN-Ru(NH_3)_5 \right]^{5+}$$ has an absorption band at 1560 nm

which is not seen in the spectra of the corresponding +4 and +6 complexes.[3] The band has been interpreted as an "intervalence transition" corresponding to E_1 in Fig. 19.1. This interpretation of the spectrum implies that the oxidation states of the metal atoms are +2 and +3, rather than +2.5 and +2.5. Indeed, the X-ray photoelectron spectrum of the complex shows peaks corresponding to

[1] H. Taube and E. S. Gould, *Acc. Chem. Res.*, **2**, 321 (1969).
[2] S. S. Isied and H. Taube, *J. Am. Chem. Soc.*, **95**, 8198 (1973).
[3] C. Creutz and H. Taube, *J. Am. Chem. Soc.*, **95**, 1086 (1973).

two types of Ru atoms, in agreement with this interpretation.[1] In contrast, various physical methods have shown that the iron atoms in the following mixed-valence biferrocenylene complex are essentially equivalent, and a $+2.5$, $+2.5$ oxidation-state description seems appropriate.[2]

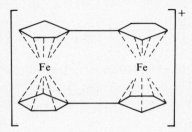

At this time there are no good rules for predicting the extent of electronic interaction between metal atoms in polynuclear complexes.

Unstable Intermediate Oxidation States[3]

Redox reactions involving changes in oxidation state of more than one unit often involve intermediates in unusual oxidation states. For example, the reaction of Tl^{3+} with Fe^{2+}

$$2Fe^{2+} + Tl^{3+} \rightarrow 2Fe^{3+} + Tl^+$$

is first order in each of the reactants during the first part of the reaction, suggesting the formation of either Tl(II) or Fe(IV) as an intermediate:

$$Fe^{2+} + Tl^{3+} \rightarrow Fe^{3+} + Tl(II)$$

or

$$Fe^{2+} + Tl^{3+} \rightarrow Fe(IV) + Tl^+$$

When the products Fe^{3+} and Tl^+ are added to the reaction mixture, Fe^{3+}, but not Tl^+, decreases the reaction rate. This result is in accord with the following mechanism:

$$Fe^{2+} + Tl^{3+} \rightleftharpoons Fe^{3+} + Tl(II)$$
$$Tl(II) + Fe^{2+} \rightarrow Tl^+ + Fe^{3+}$$

[1] P. H. Citrin, *J. Am. Chem. Soc.*, **95**, 6472 (1973).
[2] D. O. Cowan, C. LeVanda, J. Park, and F. Kaufman, *Acc. Chem. Res.*, **6**, 1 (1973).
[3] Sykes, op. cit.

The reaction of chromium(VI) with iodide and iron(II) in dilute acid solution has some remarkable features. The oxidation of iodide by chromium(VI) is normally slow, but in the presence of iron(II), which is rapidly oxidized by chromium(VI), iodine is rapidly formed. Iron(III) does not oxidize iodide rapidly in dilute acid; therefore one concludes that a reactive intermediate, presumably an intermediate oxidation state of chromium, is responsible for the oxidation of the iodide. When a large excess of iodide is present, two equivalents of iodide are oxidized per mole of Fe^{2+}:

$$Fe^{2+} + 2I^- + Cr(VI) \rightarrow Fe^{3+} + I_2 + Cr(III)$$

This result suggests that the first reactive intermediate formed is chromium(V),

$$Fe^{2+} + Cr(VI) \rightarrow Cr(V) + Fe^{3+}$$

and that the iodide reduces the Cr(V) to Cr(III):

$$Cr(V) + 2I^- \rightarrow Cr(III) + I_2$$

It is not known whether or not the latter reaction involves the intermediate formation of Cr(IV). This oxidation of iodide in the presence of iron(II) is an example of an *induced* reaction.

PROBLEMS

19.1 Arrange the following complexes in order of increasing lability. $Fe(H_2O)_6^{3+}$, $Cr(H_2O)_6^{3+}$, $Y(H_2O)_6^{3+}$, $Al(H_2O)_6^{3+}$, $Sr(H_2O)_6^{2+}$, $K(H_2O)_6^+$.

19.2 Explain why the following reaction is catalyzed by Co^{2+}:

$$Co(NH_3)_6^{3+} + 6CN^- \rightarrow Co(CN)_6^{3-} + 6NH_3$$

19.3 There are two isomers of $Pt(NH_3)_2Cl_2$, A and B. When A is treated with thiourea (tu), $Pt(tu)_4^{2+}$ is formed. When B is treated with thiourea, $Pt(NH_3)_2(tu)_2^{2+}$ is formed. Identify the isomers and explain the data.

19.4 Using the trans effect, predict the product of the reaction of 2 mol of ethylenediamine with 1 mol of $PtCl_6^{2-}$.

19.5 The reaction of $CrCl_3$ with liquid ammonia ordinarily gives principally $[Cr(NH_3)_5Cl]Cl_2$, but when a trace of KNH_2 is present, the main product is $[Cr(NH_3)_6]Cl_3$. Explain.

19.6 Suggest a method for preparing the complex $(NC)_5Fe-CN-Co(CN)_5^{6-}$.

19.7 The oxidation of Sn(II) by Fe^{3+} is first order in each reagent. How would you try to determine whether Sn(III) is involved in the mechanism?

19.8 Suggest methods for the preparation of the three isomers of $PtNH_3py(NO_2)Br$.

19.9 The rate of the aqueous vanadium(II)–vanadium(III) electron exchange reaction may be expressed Rate = $k(V^{II})(V^{III})$, where k shows the following dependence on hydrogen-ion concentration:

$$k = a + \frac{b}{(H^+)}$$

Explain this result in terms of two parallel reaction paths. (Hint: The "hydrolysis" of an aquo metal ion is really the ionization of a Brönsted acid.)

APPENDIX A
UNITS AND CONVERSION FACTORS

The tables in this section give values for some important physical constants (Table A.3) and conversion factors for units of energy (Table A.4). In recent years there has been a trend among scientists toward the use of the International System of Units[1] (SI units). These units have not been employed in this text, but these tables include values and conversion factors for both the older, traditional system of units and the SI units.

The International System is based on the basic units in Table A.1:

Table A.1

Physical quantity	Name	Symbol
Length	meter	m
Mass	kilogram	kg
Time	second	s
Electric current	ampere	A
Temperature	kelvin	K
Luminous intensity	candela	cd

[1] *Natl. Bur. Std. U.S., Spec. Publ.* 330, 1972.

The derived units in Table A.2 are also recommended:

Table A.2

Physical quantity	Name	Symbol
Force	newton	$N\ (kg\ m\ s^{-2})$
Energy	joule	$J\ (N\ m)$
Power	watt	$W\ (J\ s^{-1})$
Electric charge	coulomb	$C\ (A\ s)$
Electric potential	volt	$V\ (W\ A^{-1})$
Electric capacitance	farad	$F\ (A\ s\ V^{-1})$
Electric resistance	ohm	$\Omega\ (V\ A^{-1})$
Frequency	hertz	$Hz\ (s^{-1})$
Magnetic flux	weber	$Wb\ (V\ s)$
Magnetic flux density	tesla	$T\ (Wb\ m^{-2})$
Inductance	henry	$H\ (V\ s\ A^{-1})$

Table A.3 SOME USEFUL CONSTANTS

Constant and Symbol	Value, traditional units	Value, SI units
Gas constant, R	$1.9872\ cal\ deg^{-1}\ mol^{-1}$	$8.3143\ J\ K^{-1}\ mol^{-1}$
	$82.056\ ml\ atm\ deg^{-1}\ mol^{-1}$	
	$62,362\ ml\ torr\ deg^{-1}\ mol^{-1}$	
Ideal gas volume, std. cond.	$22,414\ ml\ atm\ mol^{-1}$	$2.241436 \times 10^{-2}\ m^3\ mol^{-1}$
Avogadro's number, N	$6.02209 \times 10^{23}\ mol^{-1}$	
Faraday constant, F	$96,487\ coulomb\ g\ equiv^{-1}$	$9.6487 \times 10^4\ C\ mol^{-1}$
	$23,061\ cal\ volt^{-1}\ equiv^{-1}$	
Boltzmann's constant, k	$1.3805 \times 10^{-16}\ erg\ deg^{-1}$	$1.3805 \times 10^{-23}\ J\ K^{-1}$
Ice point (0°C)	$273.150°K$	$273.150\ K$
Electron charge, e	$4.8030 \times 10^{-10}\ abs.\ esu$	$1.602 \times 10^{-19}\ C$
Electron mass, m	$9.1091 \times 10^{-28}\ g$	$9.1091 \times 10^{-31}\ kg$
	$5.486 \times 10^{-4}\ mu$	
	$0.5110\ MeV$	
Planck's constant, h	$6.6256 \times 10^{-27}\ erg\ sec$	$6.6256 \times 10^{-34}\ J\ s$
Velocity of light, c	$2.99795 \times 10^{10}\ cm\ sec^{-1}$	$2.99795 \times 10^8\ m\ s^{-1}$
Gravitational constant, g	$980.66\ cm\ sec^{-2}$	$9.8066\ m\ s^{-2}$
Bohr radius, a_0	$0.52918\ Å$	$0.52918 \times 10^{-10}\ m$
Proton mass, M_p	$1.6725 \times 10^{-24}\ g$	$1.6725 \times 10^{-27}\ kg$

Table A.4 UNITS OF MOLECULAR ENERGY

	erg molecule^{-1}	J molecule^{-1}	cal mol^{-1}	eV molecule^{-1}	wave number (cm^{-1})
1 erg molecule^{-1} =	1	10^{-7}	1.4394×10^{16}	6.2418×10^{11}	5.0345×10^{15}
1 J molecule^{-1} =	10^{7}	1	1.4394×10^{23}	6.2418×10^{18}	5.0345×10^{22}
1 cal mol^{-1} =	6.9473×10^{-17}	6.9473×10^{-24}	1	4.3363×10^{-5}	0.34976
1 eV molecule^{-1} =	1.6021×10^{-12}	1.6021×10^{-19}	23061	1	8065.7
1 wave number (cm^{-1}) =	1.9863×10^{-16}	1.9863×10^{-23}	2.8591	1.2398×10^{-4}	1

APPENDIX B
SYMMETRY AND GROUP THEORY

The elementary concepts of symmetry are intuitively understandable, and the various applications of symmetry in this book can be followed without any formal introduction to the subject. However, the inquiring reader may wonder about the source of group-theoretical symbols such as a_1, b_2, e, t_{2g}, etc., and may wish to learn some of the basic principles of symmetry and group theory. This appendix constitutes the barest introduction to these topics; if the material here serves to whet the reader's appetite for more extensive discussion, various texts can be consulted.[1]

SYMMETRY

A molecule is said to have "symmetry" if parts of it can be interchanged without producing a distinguishable change in the orientation of the molecule. More precisely, if a transformation of coordinates (a reflection, rotation, or a combination of these) produces an indistinguishable orientation of a molecule, the transformation is a "symmetry operation," and the molecule

---

[1] F. A. Cotton, "Chemical Applications of Group Theory," 2d ed., Wiley-Interscience, New York, 1971; L. H. Hall, "Group Theory and Symmetry in Chemistry," McGraw-Hill, New York, 1969; H. H. Jaffé and M. Orchin, "Symmetry in Chemistry," Academic, New York, 1965; J. P. Fackler, "Symmetry in Coordination Chemistry," Academic, New York, 1971; D. S. Schonland, "Molecular Symmetry," Van Nostrand, Princeton, N.J., 1965.

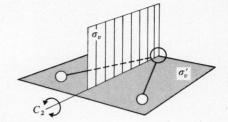

FIGURE B.1
The water molecule, showing the symmetry elements.

possesses a corresponding "symmetry element." A symmetry element is a point, line, or plane about which a symmetry operation is carried out. There are four symmetry elements and corresponding symmetry operations which can be applied to appropriate individual molecules:

1 *Plane of symmetry* (*mirror plane*), σ If reflection of all parts of a molecule through a plane yields an indistinguishable orientation of the structure, the plane is a plane of symmetry. For example, the water molecule (Fig. B.1) possesses two planes of symmetry: the plane in which the three atoms lie, and the plane perpendicular to that plane and bisecting the H—O—H angle.

2 *Center of symmetry* (*inversion center*), i If reflection through the center of a molecule yields an indistinguishable configuration, the center is a center of symmetry. Note that the square planar $PtCl_4^{2-}$ ion (Fig. B.2a) has a center of symmetry, whereas the tetrahedral SiF_4 molecule (Fig. B.2b) does not.

3 *n-Fold axis of symmetry* (*rotational axis*), C_n If rotation of a molecule by an angle of $2\pi/n$ about an axis gives an indistinguishable configuration, the axis is an axis of symmetry. Thus the water molecule possesses one twofold axis of symmetry (C_2), and the planar BF_3 molecule (Fig. B.3) possesses one threefold axis (C_3) and three twofold axes (C_2).

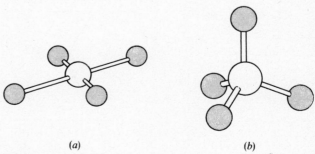

(a)   (b)

FIGURE B.2
The planar $PtCl_4^{2-}$ ion (a) and the tetrahedral SiF_4 molecules (b).

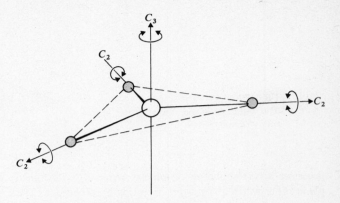

FIGURE B.3
The BF$_3$ molecule, showing the four symmetry axes.

4   n-Fold rotation-reflection axis (improper rotational axis), S_n If rotation by an angle $2\pi/n$ about an axis, followed by reflection through a plane perpendicular to the axis, yields an indistinguishable configuration, the axis is a rotation-reflection axis. For example, the staggered configuration of ethane has an S_6 axis coincident with the C_3 axis (Fig. B.4), and SiF$_4$ has three S_4 axes coincident with the three C_2 axes which bisect the F—Si—F angles (Fig. B.5). These symmetry elements are seen best with the aid of molecular models.

For the sake of completeness we include another symmetry operation, the "identity operation," denoted by the symbol E. This operation corresponds to doing nothing to a molecule, or to any sequence of operations which leaves the molecule in its original configuration.

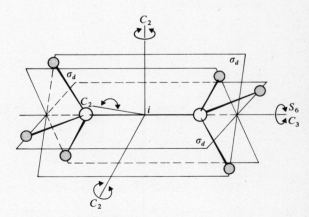

FIGURE B.4
The ethane molecule in a staggered configuration.

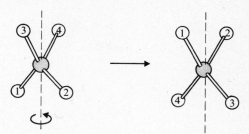

FIGURE B.5
The SiF_4 molecule, showing the changes in atomic positions accompanying an S_4 rotation-reflection.

Molecules can be classified into groups, called "point groups" or "symmetry groups," which correspond to certain combinations of symmetry elements. The point groups have labels, such as C_2, C_{2v}, D_{3h}, O_h, T_d, etc. For example, any molecule which possesses a C_2 axis, two planes of symmetry (σ_v and σ_v'), and no other symmetry elements (except that corresponding to the identity operation E) belongs to the C_{2v} point group. The species H_2O, SO_2F_2, and cis-SF_4Cl_2 belong to this point group. A list of the more important point groups and their symmetry elements is given in Table B.1.

GROUP THEORY

For each point group one can construct a corresponding "character table." At the top of a character table are listed horizontally the various classes of symmetry operations of the point group, including the identity operation. For example, the classes of symmetry operations

Table B.1 POINT GROUPS OF IMPORTANCE IN CHEMISTRY

Point group	Symmetry elements	Examples
C_s	Only a plane of symmetry, σ	$ONCl$, SO_4F
C_1	No symmetry	$SiFClBrI$, $SOFCl$
C_n	Only an n-fold rotational axis	H_2O_2, cis-$Co(en)_2Cl_2{}^+$
C_{nv}	n-Fold axis and n mirror planes passing through that axis	H_2S, NH_3, BrF_5
$C_{\infty v}$	Linear molecule without center of symmetry	CO, HCN
D_n	n-Fold axis and n C_2 axes perpendicular to that axis	$Co(en)_3{}^{3+}$
D_{nh}	Rotations as for D_n, one mirror plane perpendicular to C_n axis and n planes passing through C_n axis	BF_3, $PtCl_4{}^{2-}$
$D_{\infty h}$	Linear molecule with center of symmetry	H_2, C_3O_2
D_{nd}	Rotations as for D_n, n planes passing through C_n axis and bisecting C_2 axes	B_2Cl_4, staggered C_2H_6
T_d	Four C_3, three C_2, and three S_4 axes, and six mirror planes (tetrahedral symmetry)	SiH_4, $SO_4{}^{2-}$
O_h	Three C_4, four C_3, six C_2, three S_4, and four S_6 axes, 9 mirror planes, and a center of symmetry (octahedral symmetry)	SF_6
I_h	Six C_5, 10 C_3, and 15 C_2 axes, and 15 mirror planes (icosahedral symmetry)	$B_{12}H_{12}{}^{2-}$

Table B.2 C_{2v} **CHARACTER TABLE**

	E	C_2	$\sigma_v(xz)$	$\sigma_v'(yz)$
A_1	1	1	1	1
A_2	1	1	-1	-1
B_1	1	-1	1	-1
B_2	1	-1	-1	1

of the C_{2v} point group are E, C_2, $\sigma_v(xz)$, and $\sigma_v'(yz)$, as shown in the C_{2v} character table of Table B.2. Below the listing of the symmetry operations, there is a matrix of terms, such as 1, -1, 0, 2, etc. Each term is called a character, and the characters in each horizontal row constitute an "irreducible representation" of the point group. Each irreducible representation is labeled by a symbol, such as A_1, B_2, T_{2g}, etc.; these symbols are listed in a column on the left side of the table. The symbols are used for labeling the symmetry properties of atomic and molecular orbitals, vibrational modes, etc. In the case of an atomic orbital, each operation of the point group is carried out on the orbital in turn. A number is assigned to each operation. If the operation does not move or change the sign(s) of the orbital, the number is $+1$. If the operation causes all signs to change, the number is -1.

Table B.3   THE SYMMETRIES OF ATOMIC ORBITALS IN VARIOUS POINT GROUPS

Point group	s	p_z	p_x	p_y	d_{z^2}	$d_{x^2-y^2}$	d_{xy}	d_{yz}	d_{zx}
C_s	a'	a''	a'	a'	a'	a'	a'	a''	a''
C_2	a	a	b	b	a	a	a	b	b
C_3	a	a	e		a	e		e	
D_3	a_1	a_2	e		a_1	e		e	
C_{2v}	a_1	a_1	b_1	b_2	a_1	a_1	a_2	b_2	b_1
C_{3v}	a_1	a_1	e		a_1	e		e	
C_{4v}	a_1	a_1	e		a_1	b_1	b_2	e	
C_{5v}	a_1	a_1	e_1		a_1	e_2		e_1	
C_{6v}	a_1	a_1	e_1		a_1	e_2		e_1	
D_{2h}	a_g	b_{1u}	b_{3u}	b_{2u}	a_g	a_g	b_{1g}	b_{3g}	b_{2g}
D_{3h}	a_1'	a_2''	e'		a_1'	e'		e''	
D_{4h}	a_{1g}	a_{2u}	e_u		a_{1g}	b_{1g}	b_{2g}	e_g	
D_{5h}	a_1'	a_2''	e_1'		a_1'	e_2'		e_1''	
D_{6h}	a_{1g}	a_{2u}	e_{1u}		a_{1g}	e_{2g}		e_{1g}	
D_{2d}	a_1	b_2	e		a_1	b_1	b_2	e	
D_{3d}	a_{1g}	a_{2u}	e_u		a_{1g}	e_g		e_g	
D_{4d}	a_1	b_2	e_1		a_1	e_2		e_3	
D_{5d}	a_{1g}	a_{2u}	e_{2u}		a_{1g}	e_{2g}		e_{1g}	
T_d	a_1		t_2		e			t_2	
O_h	a_{1g}		t_{1u}		e_g			t_{2g}	
I_h	a_g		t_{1u}				h_g		
$C_{\infty v}$	a_1	a_1	e_1		a_1	e_2		e_1	
$D_{\infty h}$	$\sigma_g{}^+$	$\sigma_u{}^+$	π_u		$\sigma_g{}^+$	δ_g		π_g	

As an illustration of the procedure, we shall demonstrate how irreducible representations are assigned to the s and p atomic orbitals of the middle oxygen atom of the ozone molecule, which belongs to the C_{2v} point group. For ozone we shall follow the convention for planar C_{2v} molecules of assuming that the molecule lies in the yz plane. When the four operations are performed on the s orbital, the numbers 1, 1, 1, 1 are obtained; we see that the s orbital transforms according to the A_1 irreducible representation. Lowercase symbols are usually used for orbitals; hence the s orbital is said to have a_1 symmetry. When the C_{2v} operations are performed on the p_x orbital, the numbers 1, -1, 1, -1 are obtained; obviously this orbital has b_1 symmetry. Similarly, the p_y orbital yields 1, -1, -1, 1 corresponding to b_2 symmetry, and the p_z orbital yields 1, 1, 1, 1, corresponding to a_1 symmetry. The reader may, as an exercise, verify the symmetry assignments of the $p\pi$ group orbitals of ozone, $(1/\sqrt{2})(\phi_1 \pm \phi_3)$, discussed in Chap. 3.

By essentially the same process, the symmetries of orbitals in any other point group can be determined. The results for s, p, and d orbitals in the more important point groups are listed in Table B.3. From this table it can be seen, for example, that the d orbitals of the Co atom in $CoF_6{}^{3-}$ fall into two symmetry types, e_g and t_{2g}.

APPENDIX C
TERM SYMBOLS FOR FREE ATOMS AND IONS

The statement that a free atom or ion has a particular valence electron configuration, say np^2, is an incomplete description of the electronic state of the species. In order to completely describe the state, one must specify the values of the total *orbital* angular momentum quantum number L, the total *spin* angular momentum quantum number S, and the total angular momentum quantum number J.

The components, in a reference direction, of L and S can have the following quantized values:

$$M_L = L, L - 1, \ldots, -(L - 1), -L$$
$$M_S = S, S - 1, \ldots, -(S - 1), -S$$

Here M_L and M_S are the vector sums of the m_l and m_s values for the individual electrons:

$$M_L = m_{l_1} + m_{l_2} + m_{l_3} + \cdots$$
$$M_S = m_{s_1} + m_{s_2} + m_{s_3} + \cdots$$

The set of individual quantum states corresponding to a particular value of L and a particular value of S constitutes a "term" and is represented by a term symbol $^{2S+1}L_J$, where the value of L is indicated by one of the capital letters S, P, D, F, G, H, ... (corresponding to 0, 1, 2, 3, ...). The quantum number J is the vector sum of L and S and may have the values

$$J = L + S, L + S - 1, L + S - 2, \ldots, |L - S|$$

J can only have positive values or be zero. The components, in a reference direction, of J can have the values

$$M_J = J, J - 1, \ldots, -(J - 1), -J$$

For example, one of the terms of an atom with an np^2 configuration is the $^1D_2$ (pronounced singlet D two) term. This term can be represented by the following set of individual quantum states, in which the m_s values are indicated by arrows pointing up or down.

	$m_l = +1$	0	-1	M_L	M_S	M_J
$^1D_2$	⇅	○	○	2	0	2
	↑	↓	○	1	0	1
	○	⇅	○	0	0	0
	○	↑	↓	-1	0	-1
	○	○	⇅	-2	0	-2

The np^2 configuration also yields the terms $^3P_2$, $^3P_1$, and $^3P_0$, for which we can write the following individual quantum states:

	$m_l = +1$	0	-1	M_L	M_S	M_J
$^3P_2$	↑	↑	○	1	1	2
	↑	○	↑	0	1	1
	○	↑	↑	-1	1	0
	↓	○	↓	0	-1	-1
	○	↓	↓	-1	-1	-2
$^3P_1$	↓	↑	○	1	0	1
	↓	↓	○	1	-1	0
	○	↓	↑	-1	0	-1
$^3P_0$	↑	○	↓	0	0	0

The only other term of the np^2 configuration is the $^1S_0$ term:

	$m_l = +1$	0	-1	M_L	M_S	M_J
$^1S_0$	↓	○	↑	0	0	0

Note that we have now written all the possible combinations of m_l and m_s. It is important to recognize that most of the individual quantum state assignments are not unique. For example, the configurations ↓ ○ ↑ and ○ ⇅ ○ could just as well be interchanged, or suitable linear combinations assigned to particular terms.

It should be obvious that, because of limitations imposed by the Pauli principle, not all conceivable combinations of L and S are possible for a given polyelectronic system. A list of the allowed states for equivalent s, p, and d electrons is given in Table C.1.

Table C.1 ALLOWED STATES FOR EQUIVALENT s, p, AND
d **ELECTRONS**

Configuration	Terms
s	^2S
s^2	^1S
p or p^5	^2P
p^2 or p^4	^1S, ^1D, ^3P
p^3	^2P, ^2D, ^4S
p^6	^1S
d or d^9	^2D
d^2 or d^8	$^1(SDG)$, $^3(PF)$
d^3 or d^7	^2D, $^2(PDFGH)$, $^4(PF)$
d^4 or d^6	$^1(SDG)$, $^3(PF)$, $^1(SDFGI)$, $^3(PDFGH)$, ^5D
d^5	^2D, $^2(PDFGH)$, $^4(PF)$, $^4(SDFGI)$, $^4(DG)$, ^6S

Methods for determining the terms corresponding to particular configurations are described in various books.[1] In order to determine which term for a given configuration represents the ground state, we apply Hund's rules:

1 The most stable state is the one with the largest value of S.
2 Of a group of states with the same value of S, that with the largest value of L lies lowest.
3 Of a group of states with the same values of L and S, corresponding to a shell of electrons less than half-full, states with lower J lie lower. When the shell is more than half-full, the reverse applies.

In the X-ray photoelectron spectrum of argon (Fig. 1.10), the $2p$ peak is split into two components corresponding to the $^2P_{3/2}$ and $^2P_{1/2}$ ions which are formed. These components have relative intensities of $2:1$, corresponding to the ratio of the number of individual quantum states for the two states (i.e., the ratio of the $2J + 1$ values).

[1] H. B. Gray, "Electrons and Chemical Bonding," pp. 22–27, W. A. Benjamin, Inc., New York, 1964; C. J. Ballhausen, "Introduction to Ligand Field Theory," pp. 8–10, McGraw-Hill, New York, 1962; F. A. Cotton and G. Wilkinson, "Advanced Inorganic Chemistry," 3d ed., pp. 80–85, Wiley-Interscience, New York, 1972; J. E. Huheey, "Inorganic Chemistry," pp. 35–38, Harper & Row, New York, 1972; L. Pauling, "The Nature of the Chemical Bond," 3d ed., pp. 580–588, Cornell University Press, Ithaca, N.Y., 1960.

ANSWERS TO SELECTED PROBLEMS

CHAPTER 1

1.1   The probability of finding the electron at a radius r is proportional to the product of the electron density and the area of a sphere of radius r. Inasmuch as the latter area is zero when $r = 0$, the probability of finding the electron at a radius $r = 0$ is zero even though the electron density is greatest at $r = 0$.

1.2   The nodal surfaces of a $3p$ orbital are (1) a plane containing the nucleus and perpendicular to the axis of the orbital and (2) a spherical surface centered on the nucleus.

1.3 $I = 13.60(9)^2 = 1102$ eV.

1.4   Al, 1; S, 2; Sc^{3+}, 0; Cr^{3+}, 3; Ir^{3+}, 4; Dy^{3+}, 5.

1.5 Ru^{2+}.

1.12 Because $I(He) > I(Kr)$, the first reaction goes as written. Because $I(Cl) > I(Si)$, the second reaction goes as written. Because $A(Cl) > A(I)$, the third reaction goes in the reverse direction.

CHAPTER 2

2.1

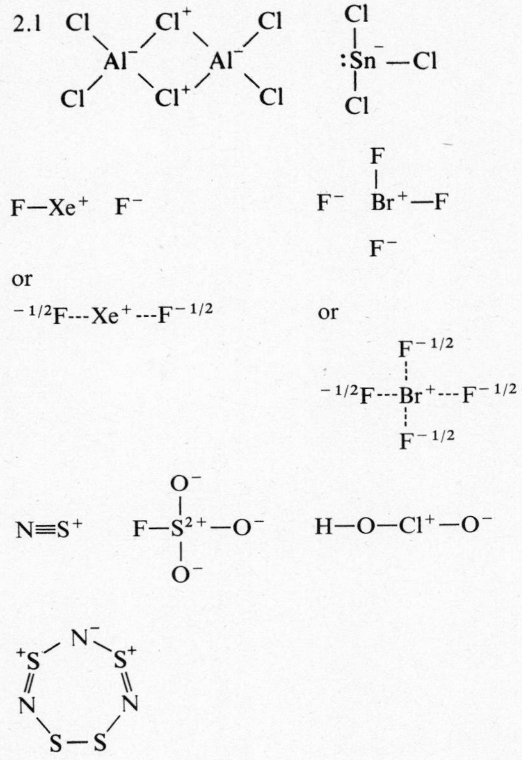

$F{-}Xe^+$ F^-

F^- $Br^+{-}F$

or

$^{-1/2}F{\cdots}Xe^+{\cdots}F^{-1/2}$

or

$^{-1/2}F{\cdots}Br^+{\cdots}F^{-1/2}$

$N{\equiv}S^+$ $F{-}S^{2+}{-}O^-$ $H{-}O{-}Cl^+{-}O^-$

2.2 Hyperconjugated resonance structures such as $O{=}S^{2+}F^-$ may be important in

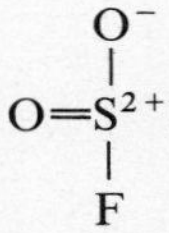

SO_2F_2, and therefore the S—O bond order may be comparable to that in SO_2, that is, 1.5. Lone-pair–lone-pair repulsions may somewhat weaken the bonds in SO_2.

2.5 The properties may be estimated from those for the following isoelectronic species: (a) NO_2, (b) N_2 and CO, (c) $H_3C{-}CH_3$, $H_2N{-}NH_2$, and $HO{-}OH$, (d) CO_2, (e) CH_3, (f) SiO_2, (g) SO_2.

2.7 Tin is tetracovalent (sp^3 hybridization) in tin metal and in SnX_4 molecules. The lower IS values for the latter molecules correspond to higher atomic changes and lower s-orbital occupations. In SnX_2 compounds, the tin atoms have lone-pair electrons with relatively high s-orbital character. Thus, although the tin atoms in these compounds have positive atomic charges, the IS values are even higher than they are in metallic tin.

2.8 The electronegativity of fluorine is greater than that of oxygen, and the Mn atoms in MnF_2 are more positive than those in MnO_2.

2.9 $q_N = -0.474$, $q_H = 0.158$, $q_C = 1.209$, $q_F = -0.302$.

2.12 SO_2, NF_3, SF_4, SF_2, N_2F_4, SiH_3Cl, O_3, BrF_5, S_2Cl_2.

2.14 $\Delta H_f^\circ(N_3H_5) = 58.5$ kcal mol^{-1}; $\Delta H_f^\circ[(GeH_3)_2Se] = 39$ kcal mol^{-1}.

CHAPTER 3

3.2 For linear H_3^+, one calculates the following energy levels (in order of increasing energy): $\alpha + \sqrt{2}\beta$, α, $\alpha - \sqrt{2}\beta$. The two electrons occupy the lowest level, and thus the total energy is $2\alpha + 2\sqrt{2}\beta$. For triangular H_3^+, the energy levels are $\alpha + 2\beta$, $\alpha - \beta$, and $\alpha - \beta$. With two electrons occupying the lowest level, the total energy is $2\alpha + 4\beta$. Thus we calculate that the triangular form is more stable by $4\beta - 2\sqrt{2}\beta$.

3.4 $\alpha_N = 14.4$ eV; $\alpha_O = 18.5$ eV; $p = 4.85$; $t = 0.058$; $S = 0.178$; $\beta = 5.1$ eV. The $\pi \to \pi^*$ transition energy is estimated as $\Delta\alpha/2 + \frac{1}{2}\sqrt{(\Delta\alpha)^2 + 8\beta^2}$, or 9.5 eV.

3.6 The bond orders and numbers of unpaired electrons are NeO^+, 1.5, 1; O_2^+, 2.5, 1; CN^+, 2, 0; BN, 2, 0; SiF^+, 3, 1; NO^-, 2, 2; PCl, 2, 2; I_2^+, 1.5, 1; NeH^+, 1, 0.

3.8 The P_4^{2+} ion might be expected to have a square structure, with a total of five bonds (one delocalized P—P π bond and four P—P σ bonds).

CHAPTER 4

4.1 $H_2 \overset{k}{\rightleftharpoons} 2H$

$H + p\text{-}H_2 \overset{k}{\longrightarrow} o\text{-}H_2 + H$

4.2 The rate of the reaction

$$F + H_2 \to HF + H$$

is comparable to that of the reaction

$$H + F_2 \to HF + F$$

(both reactions are exothermic), and therefore the H-atom concentration is comparable to the F-atom concentration. For this reason one must include the reactions $H + F \to HF$ and $H + H \to H_2$ in the mechanism.

4.4 $2NO \to N_2 + O_2$

$H_2 + N_2 \to N_2H_2$

$SO_2 + F_2 \to SO_2F_2$

4.6 Both nuclear-fission and branching-chain reactions are exothermic. The neutrons formed in nuclear fission are analogous to the chain-carrying radical species. More neutrons are produced in fission than are consumed, just as is the case with radicals in a branching-chain reaction. The escape of neutrons from the fissioning material is analogous to the chain-terminating steps at the reaction vessel wall. The "critical mass" of nuclear fission is analogous to the first explosion limit.

CHAPTER 5

5.1   a $CaH_2 + CO_3^{2-} + 2H_2O \rightarrow CaCO_3 + 2H_2 + 2OH^-$

      c $3LiAlH_4 + 4(C_2H_5)_2O \cdot BF_3 \rightarrow 2B_2H_6 + 3LiAlF_4 + 4(C_2H_5)_2O$

      e $AsH_3 + GeH_3^- \rightarrow AsH_2^- + GeH_4$

5.3 $NaNH_2 + GeH_4 \xrightarrow{\text{liq. } NH_3} GeH_3^- + Na^+ + NH_3$

      $GeH_3^- + CH_3I \rightarrow CH_3GeH_3 + I^-$

      $CH_3GeH_3 + HBr \rightarrow CH_3GeH_2Br + H_2$

5.5   If the exchange reaction does not proceed by a mechanism involving the rate-determining step

$$PH_3 + OD^- \xrightarrow{k_1} PH_2^- + HOD$$

then the rate of the actual rate-determining step must be greater than that of the above reaction. Hence $k_1 \leq 0.4 \ M^{-1} \ s^{-1}$ and $pK_{PH_3} \geq 27$.

5.7 $NH_3 < CH_3GeH_3 < AsH_3 < C_2H_5OH < H_4SiO_4 < HSeO_4^- < HSO_4^- < H_3O^+$
      $< HSO_3F < HClO_4$.

5.9   Slowest: $NH_3 + OH^- \rightarrow H_2O + NH_2^-$

      Fastest: $HSO_4^- + OH^- \rightarrow H_2O + SO_4^{2-}$

CHAPTER 6

6.1   The π bonding between the boron atom and the fluorine atoms of BF_3 is stronger than the analogous π bonding in BCl_3. (The $p\pi$-$p\pi$ overlap for two first-row atoms is greater than that for a first-row atom and a second-row atom.) Hence it is more difficult to convert the boron atom of BF_3 to sp^3 σ bonding by coordination of a Lewis base because of the greater loss of π bonding energy.

6.2   b $Li_3N + 3NH_4^+ \rightarrow 3Li^+ + 4NH_3$

      d $H_3BO_3 + H_2SO_4 \rightarrow 3H_3O^+ + B(HSO_4)_4^- + 2HSO_4^-$

6.3 $Hphth^- + OH^- \xrightarrow{H_2O} phth^{2-} + H_2O$

      $Hphth^- + H^+ \xrightarrow{HOAc} H_2phth$

6.5   The OH^- is strongly hydrogen-bonded to the dissolved water molecules, thus enhancing the solubility of the KOH.

6.7   b and c.

6.9   In the absence of steric effects, $(CH_3)_3N$ would be expected to be the strongest base. However, in $(CH_3)_3BN(CH_3)_3$, the repulsions between methyl groups on adjacent atoms are so great as to make the energy of interaction of $B(CH_3)_3$ with $(CH_3)_3N$ effectively weaker than that of $B(CH_3)_3$ with $(CH_3)_2NH$.

CHAPTER 7

7.1   a $2KI + IO_3^- + 6H^+ + 6Cl^- \rightarrow 3ICl_2^- + 6H_2O + 2K^+$

      c $3Cl_2O + 6OH^- \rightarrow 4Cl^- + 2ClO_3^- + 3H_2O$

      e $4MnO_4^- + 3H_2PO_2^- + 7H^+ \rightarrow 4MnO_2 + 3H_3PO_4 + 2H_2O$

      g $S + SO_3^{2-} \rightarrow S_2O_3^{2-}$

7.3 $E° = -1.73$ V.

7.5 Prepare a sample of H_2SO_5 in which the —OOH oxygens are labeled randomly with a certain concentration of ^{18}O. Mix this with an equal amount of ordinary H_2SO_5 and carry out the decomposition. If mass-spectrometric examination of the O_2 shows that one-half of the molecules have the ^{18}O content of the labeled —OOH groups and that one-half of the molecules are isotopically normal, the first mechanism is indicated. If practically all the O_2 molecules contain at least one ^{16}O atom and if the concentration of ^{18}O is one-half of that in the labeled —OOH groups, then the second mechanism is indicated.

7.7 1 (*c*); 2 (*b*); 3 (*a*); 4 (*d*).

CHAPTER 8

8.1 $HI \rightarrow \frac{1}{2}H_2 + \frac{1}{2}I_2$

8.3 Anode: $e^-_{am} \rightarrow e^-$
Cathode: $e^- \rightarrow e^-_{am}$

8.5 *a* $CH_3GeH_3 + e^-_{am} \rightarrow \frac{1}{2}H_2 + CH_3GeH_2^-$
b $I_2 + 2e^-_{am} \rightarrow 2I^-$
c $(C_2H_5)_2S + 2e^-_{am} + NH_3 \rightarrow C_2H_5S^- + C_2H_6 + NH_2^-$

CHAPTER 9

9.1

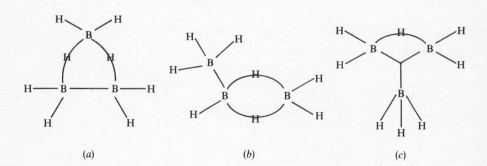

(*a*)   (*b*)   (*c*)

Structure (*a*) is actually observed. In each of the other structures, one boron atom is rather loosely connected to the others by a single bond.

9.3 $b = h = n$, and $q = -2$. Hence $\alpha = n + 3$ and $\beta = n - 2$. If we assume that there are n B—H bonds, then the number of B—B bonds is 3 and the number of B—B—B bonds is $n - 2$.

9.5

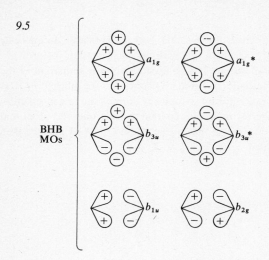

BHB MOs

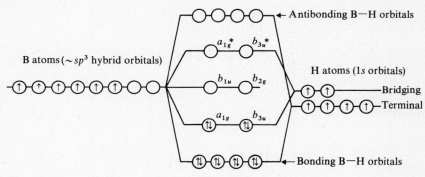

CHAPTER 10

10.1
	$\Delta H°$, kcal
$NH_4{}^+(g) + Cl^-(g) \rightarrow NH_4Cl(s)$	-158.7
$NH_4Cl(s) \rightarrow \frac{1}{2}N_2(g) + 2H_2(g) + \frac{1}{2}Cl_2(g)$	75.2
$\frac{1}{2}Cl_2(g) \rightarrow Cl(g)$	28.6
$Cl(g) + e^-(g) \rightarrow Cl^-(g)$	-83.5
$N(g) + 3H(g) \rightarrow NH_3(g)$	-279
$\frac{1}{2}N_2(g) \rightarrow N(g)$	113
$2H_2(g) \rightarrow 4H(g)$	208.4
$H(g) \rightarrow H^+(g) + e^-(g)$	313.6
$NH_4{}^+(g) \rightarrow NH_3(g) + H^+(g)$	218

10.4   Possible coordination numbers of M and X are 3 and 1, 6 and 2, 9 and 3, etc., respectively.

10.6   The minimum value of r_M/r_X is 0.5275.

10.8   The greater lattice energy of $Mg^{2+}O^{2-}$, compared with that of Mg^+O^-, more than compensates for the higher ionization potential and lower electron affinity. Mg^+O^- would be expected to be paramagnetic, whereas MgO is diamagnetic.

10.10 The S^{2-} and Se^{2-} ions are so much larger than the metal ions that anion-anion contact occurs in the crystals, and the metal ions "rattle" in the anion lattices. For S^{2-}, $r = \frac{1}{2}\sqrt{2}\,(2.60) = 1.84$ Å.

CHAPTER 11

11.1   Increase in the pressure of Zr favors the type I structure (d^3s^1 configuration) over the type II structure ($d^2s^1p^1$ configuration). Increase in the pressure of Fe favors the type III structure ($d^5s^1p^2$ configuration) over the type I structure (d^7s^1 configuration).

11.3   The electrical conductivity of a solid-solution alloy passes through a minimum on varying the composition from one pure metal to the other. This result is probably caused by the relative disorder in the lattice in the alloy.

11.5   Ag, III; Ag_3Al, I; Ag_9Al_4, γ-brass structure; Ag_5Al_3, II; Al, III.

11.7   (a) yes; (b) no, Re melts too high; (c) no, Pt (a d-electron-rich metal) and Ta (a d-electron-poor metal) would react to form a very stable alloy.

11.9   Eu and Yb have cores like those of Gd and Lu, respectively. Hence Eu and Yb crystals involve two bonding electrons per atom, whereas the other lanthanide metals involve three bonding electrons per atom. Thus the "abnormal" densities of Eu and Yb are related to the extraordinary stability of the half- and completely filled $4f$ electronic shell.

CHAPTER 12

12.1   11,300 Å.

12.3   (a) p; (b) p; (c) n; (d) p; (e) n.

12.5   (c) $\ll$ (b) $<$ (a) $<$ (d).

CHAPTER 14

14.2   The 2-coordinate silver atoms are Ag(I), and the 4-coordinate silver atoms are Ag(III). These types of coordination are characteristic of d^{10} and d^8 ions, respectively.

14.4   (a) A total of three isomers, including a pair of optical isomers. (b) A total of five isomers, including a pair of optical isomers. (c) Three isomers.

14.6 $Co(en)_3^{3+}$, trigonal distortion; $trans$-$Co(NH_3)_4Cl_2^+$, tetragonal distortion.

14.8   The Pd(II) and Pt(II) atoms are big enough to accommodate four ligands (even bulky ligands) in a square planar configuration. The smaller Ni(II) atom must coordinate tetrahedrally when the ligands are very bulky.

CHAPTER 15

15.2   In spite of the negative charge of Cl^-, Δ_o for Cl^- is less than Δ_o for NH_3 because of the repulsive interaction between the $p\pi$ electrons of Cl^- and the $d\pi$ electrons of Cr^{3+} (Fig. 15.11). Δ_o for CN^- is very high because of back bonding i.e., the shift of electron density from the $d\pi$ orbitals of Cr^{3+} to the antibonding π^* orbitals of CN^-. The latter interaction stabilizes the $d\pi$ orbitals and increases Δ_o (Fig. 15.12).

15.4   Fe—C bond order 1.5; C—N bond order 2.5. Formal charges of Fe, C and N: -1, 0, and -0.5, respectively.

CHAPTER 16

16.2   The $Mn(H_2O)_6^{2+}$ ion is a weak-field, high-spin d^5 complex with the spectroscopic symbol $^6A_1$ (Table 16.1). There is only one combination of quantum numbers corresponding to five unpaired d electrons. Inasmuch as transitions involving a change in multiplicity are forbidden, there are no allowed transitions. The pale color of $Mn(H_2O)_6^{2+}$ is due to weak forbidden transitions.

16.4 CrF_6^{3-}, green; $Cu(NH_3)_4^{2+}$, blue; FeO_4^{2-}, blue-red (purple or magenta); $Co(en)_3^{2+}$, orange.

16.5 CoF_6^{3-}: 16,400 cm^{-1} ($^5T_{2g} \rightarrow {}^5E_g$). $NiCl_4^{2-}$: 2400 cm^{-1} [$^3T_1(F) \rightarrow {}^3T_2$]; 5400 cm^{-1} [$^3T_1(F) \rightarrow {}^3A_2$]; 17,300 cm^{-1} [$^3T_1(F) \rightarrow {}^3T_1(P)$]. $Fe(H_2O)_6^{2+}$: 8200 cm^{-1} ($^5T_{2g} \rightarrow {}^5E_g$).

16.7 $CoCl_4^{2-}$ would be expected to have a greater Δ and hence a lower μ. (See Eq. 16.1 and Table 16.7.)

16.9   The NiF_6^{2-} ion in K_2NiF_6 is a low-spin d^6 octahedral complex of Ni(IV). The compound $Ni(NH_3)_2Cl_2$ is probably not a 4-coordinate Ni(II) complex but rather an octahedrally coordinated Ni(II) compound, with bridging Cl^- ions. $Ni(PEt_3)_2Cl_2$ is a square planar complex, and $Ni(Ph_3AsO)_2Cl_2$ is tetrahedral.

16.11   Flattening of the tetrahedron splits the t_2 energy levels into an upper level of energy 2δ and two lower levels each of energy $-\delta$. In a flattened d^8 complex, the ligand-field stabilization energy is $3\delta - 2\delta = \delta$; in a flattened d^9 complex, the stabilization energy is $4\delta - 2\delta = 2\delta$. Elongation of the tetrahedron splits the t_2 levels into two upper levels each of energy δ and a lower level of energy -2δ. In an elongated d^8 complex, the stabilization energy is $4\delta - 2\delta = 2\delta$; in a flattened d^9 complex, the stabilization energy is $4\delta - 3\delta = \delta$.

16.14   The ligand-field stabilization energy of a d^7 tetrahedral complex is greater than that of a d^8 tetrahedral complex, and the LFSE of a d^8 square planar complex is greater than that of a d^7 square planar complex.

16.16 MnO is antiferromagnetic. The crystal contains two interpenetrating sets of Mn^{2+} ions. In one set, all the magnetic moments are aligned in one direction; in the other set, the magnetic moments are aligned in the opposite direction. Spin polarization of the oxide ions is believed to aid the magnetic coupling between the two sets of Mn^{2+} ions.

CHAPTER 17

17.2

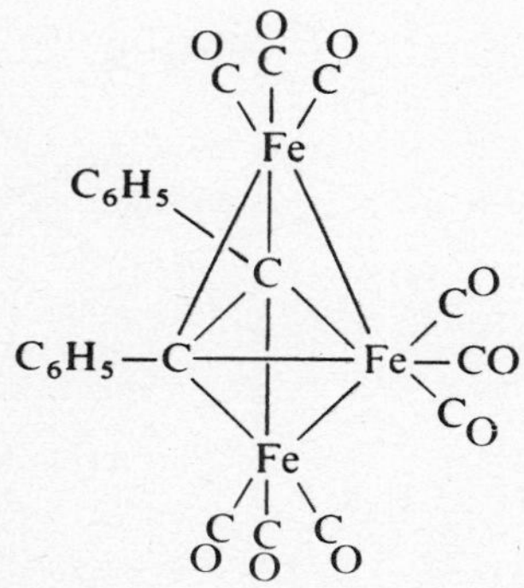

17.4 Either a trigonal bipyramidal $Os_5(CO)_{16}$ or a square pyramidal $Os_5(CO)_{17}$ would fit the rules.

17.6 Diamagnetic $NiCl_4^{2-}$ would be square planar; such a structure is unstable with respect to the tetrahedral form, probably because of repulsions between the chlorines. $Ti(CO)_4^{4+}$ is unstable because there are no $d\pi$ electrons to engage in back bonding. $Cd(CO)_3$ would be expected to be unstable because the d electrons are held so tightly (high nuclear charge of Cd) that they cannot engage in back bonding to the CO groups. AuF_5^{2-} is unstable with respect to AuF_4^- because fluoride ions cannot engage in back bonding and cannot remove the negative charge which builds up on the Au atom when five ligands are coordinated to it.

CHAPTER 18

18.1 $(OC)_5Cr=C=O + LiCH_3 \rightarrow (OC)_5Cr=C-O^-$
          $| \quad Li^+$
          CH_3

          $\downarrow (CH_3)_3O^+$

          $(OC)_5Cr=C-OCH_3$
          $|$
          CH_3

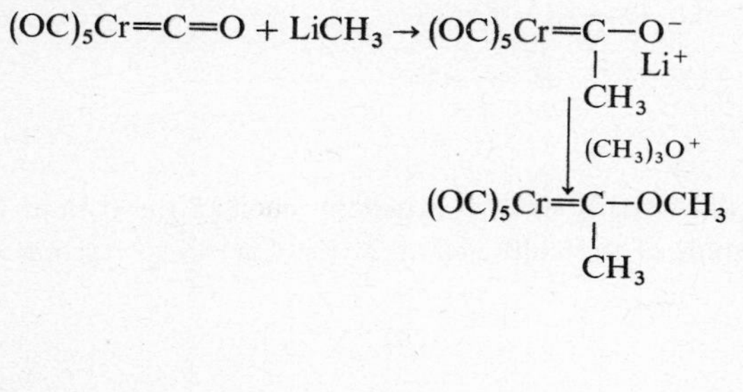

18.3 $PdCl_4^{2-} + C_2H_4 + H_2O \rightleftharpoons PdCl_2(C_2H_4)OH^- + 2Cl^- + H^+$

$$\begin{bmatrix} \begin{array}{cc} \overset{\displaystyle H}{\underset{\displaystyle O}{}} & CH_2 \\ & \| \\ \diagup\;Pd\diagup & CH_2 \\ Cl\quad\; & Cl \end{array} \end{bmatrix}^{-} \xrightarrow{\text{slow}} Pd + 2Cl^- + CH_3CHO + H^+$$

The slow step may involve the formation of an unstable intermediate of the type $Cl_2(H_2O)Pd{-}CH_2{-}CH_2OH^-$.

18.5 The following is just one possible mechanism:

CH₂—R bonded to Co (on surface) ⟶ ·CH₂—R + Co(II) (on surface)

$$RCHOHCH_2OH + \cdot CH_2R \rightarrow \underset{\displaystyle OH}{\overset{\displaystyle H\;\;H}{RC{-}\underset{\displaystyle \cdot}{C}{-}OH}} + CH_3R$$

$$\underset{\displaystyle OH}{\overset{\displaystyle H\;\;H}{RC{-}\underset{\displaystyle \cdot}{C}{-}OH}} \rightarrow \overset{\displaystyle H\;\;H}{RC{=}C{-}OH} + \cdot OH$$

$$\cdot OH + CH_3R \rightarrow H_2O + \cdot CH_2R$$

·CH₂R + Co(II) (on surface) ⟶ CH₂R bonded to Co (on surface)

$$\begin{array}{c} \overset{\displaystyle H\;\;H}{RC{=}C{-}O} \\ | \\ H{-}O\cdots H \\ | \\ H \end{array} \rightarrow \begin{array}{c} \overset{\displaystyle H\;\;H}{RC{-}C{=}O} \\ H\quad O{-}H \\ | \\ H \end{array}$$

If the CH_2 group of the diol were labeled with tritium, most of the tritium would end up in the water, and some of it would end up in the $Co{-}CH_2{-}$ group of the coenzyme.

CHAPTER 19

19.1 $Cr(H_2O)_6^{3+} < Al(H_2O)_6^{3+} < Fe(H_2O)_6^{3+} < Y(H_2O)_6^{3+} < Sr(H_2O)_6^{2+} < K(H_2O)_6^{+}$.

19.3 A is the cis complex; B is the trans complex. In the cis complex, the ammonia molecules are labilized by the *trans*-chloride ions; replacement of ammonia molecule by thiourea causes the *trans*-chloride to be labilized. Hence all positions become occupied by thiourea. In the trans complex, the chloride ions labilize each other; only these ligands are placed by thiourea molecules.

19.5 The NH_2^- ions catalyze the displacement of the last coordinated chloride ion by ammonia.

$$Cr(NH_3)_5Cl^{2+} + NH_2^- \rightleftharpoons Cr(NH_3)_4(NH_2)Cl^+ + NH_3$$

$$Cr(NH_3)_4(NH_2)Cl^+ \rightarrow Cr(NH_3)_4(NH_2)^{2+} + Cl^-$$

$$Cr(NH_3)_4(NH_2)^{2+} + 2NH_3 \rightarrow Cr(NH_3)_6^{3+} + NH_2^-$$

19.7 If the reaction rate is reduced by the addition of Fe^{2+} to the system, the formation of Sn(III) in a reversible primary step would be strongly indicated:

$$Sn(II) + Fe^{3+} \rightleftharpoons Sn(III) + Fe^{2+}$$

$$Sn(III) + Fe^{3+} \rightarrow Sn(IV) + Fe^{2+}$$

19.9 The data suggest the following reaction paths:

$$V(H_2O)_6^{2+} + V(H_2O)_6^{3+} \xrightarrow{k_1} V(H_2O)_6^{3+} + V(H_2O)_6^{2+}$$

$$V(H_2O)_6^{2+} + V(H_2O)_5OH^{2+} \xrightarrow{k_2} V(H_2O)_5OH^{2+} + V(H_2O)_6^{2+}$$

If we let K equal the acid ionization constant of $V(H_2O)_6^{3+}$, then $k = k_1 + k_2 K/(H^+)$.

INDEX